Non-coding RNAs for Crop Improvement

Kavita Goswami • Temesgen Assefa Gelaw
Neeti Sanan-Mishra
Editors

Non-coding RNAs for Crop Improvement

Concepts and Applications

Springer

Editors
Kavita Goswami
Plant RNAi Biology
International Centre for Genetic Engineering and Biotechnology
New Delhi, India

Temesgen Assefa Gelaw
Department of Biotechnology
College of Agriculture and Natural Resources Science, Debre Birhan University
Debre Birhan, Ethiopia

Neeti Sanan-Mishra
Plant RNAi Biology
International Centre for Genetic Engineering and Biotechnology
New Delhi, India

ISBN 978-981-97-7212-4 ISBN 978-981-97-7213-1 (eBook)
https://doi.org/10.1007/978-981-97-7213-1

This Springer imprint is published by the registered company Springer Nature Singapore Pte Ltd.
The registered company address is: 152 Beach Road, #21-01/04 Gateway East, Singapore 189721, Singapore

We dedicate this book to Dr. Sunil K. Mukherjee (1950–2023) as a token of our appreciation and respect for him and his achievements

Preface

The study of non-coding RNAs (ncRNAs) is a promising new development in the rapidly developing field of agricultural research. The key to unlocking new opportunities for sustainable agriculture and global food security lies in our ability to understand the complex functions. These mysterious molecules will play a crucial role as we navigate the challenges of crop improvement.

This book, *Non-coding RNAs for Crop Improvement: Concepts and Applications*, offers an extensive overview of the complex field of non-coding RNAs and its potential to revolutionize agricultural biotechnology. Its pages provide knowledge, findings, and techniques that open new possibilities for using ncRNAs to improve crop resilience, productivity, and nutritional value.

In our capacity as the editors of this groundbreaking work, we have made an effort to assemble a group of chapters written by eminent specialists and investigators at the forefront of agricultural ncRNA research. This book offers a comprehensive review of the most recent state-of-the-art research, approaches, and applications in the subject, thanks to the contributors' varied perspectives and collective expertise.

The subjects covered in the chapters of this volume are wide-ranging, from the basic principles of ncRNA synthesis and function to their useful applications in stress tolerance enhancement, crop breeding, and nutritional fortification. Understanding the intricacies of ncRNA-mediated regulatory networks in crops requires a thorough comprehension of the important discoveries, empirical evidence, and practical considerations presented in each chapter.

Furthermore, by combining knowledge from genetics, molecular biology, bioinformatics, and agronomy, this book crosses academic boundaries. Our goal is to stimulate synergistic advances in the field of ncRNA research toward breakthrough discoveries and significant applications in crop improvement through multidisciplinary discourse and collaboration.

Our collective goal is to position this book as a catalyst for revolutionary change in agricultural biotechnology, enabling scholars, decision-makers, and interested parties to unlock the full potential of ncRNAs in addressing the major issues confronting the world's agricultural sector. We would like to extend our sincerest gratitude to each and every contributor who so generously shared their knowledge and perspectives in making this book a reality. Their commitment to advancing agricultural research and their enthusiasm for doing so are prime examples of the creativity and teamwork that propel progress in our sector.

It is an honor for us as editors to introduce this collection to the scientific community and beyond. We invite readers to join us on a voyage of inspiration, discovery, and adventure as we explore the complex world of non-coding RNAs for crop improvement.

New Delhi, India Kavita Goswami
Debre Birhan, Ethiopia Temesgen Assefa Gelaw
New Delhi, India Neeti Sanan-Mishra

Acknowledgments

We extend our heartfelt gratitude to all those who contributed to the creation of this book, and we appreciate their efforts immensely. First and foremost, we extend our sincerest gratitude to the authors whose dedication and expertise have brought forth the content within this book, enriching its pages with insight, creativity, and knowledge. Your commitment to excellence has made this project truly exceptional.

We are deeply thankful to the Springer Nature editorial team for their unwavering support and tireless efforts throughout the publishing process. Their meticulous attention to detail and commitment to excellence have ensured the clarity and coherence of this book.

Special thanks go to the designers and production staff whose creative talents have transformed words into a visually captivating work of art. Their contributions have significantly enhanced its aesthetic appeal and overall impact.

We would also like to acknowledge the reviewers and advisors whose insightful feedback and guidance have played a significant role in shaping the content and scholarly rigor of this book.

We sincerely thank the International Centre for Genetic Engineering and Biotechnology (ICGEB), New Delhi, for providing us with all the necessary facilities.

Finally, we would like to express our gratitude to our readers for their interest and support. It is your engagement with the material that ultimately gives purpose to our endeavors.

This book serves as a testament to the collaborative spirit and collective dedication of all those involved. It is with profound appreciation that we recognize every individual who contributed to its realization.

Contents

Editors and Contributors

About the Editors

Kavita Goswami is a highly qualified professional with a Ph.D. from SHUATS, Prayagraj, and an expertise in bioinformatics. Throughout her academic career, she has consistently demonstrated extraordinary academic achievements. She holds a B.Sc. and a M.Sc. degree in bioinformatics from CCS University, Meerut. Dr. Kavita has actively participated in training programs and symposiums, demonstrating her commitment to staying abreast of the latest developments in her field. She has also worked toward certifications, demonstrating her dedication to continuous learning. She holds a PG diploma in IPR and a six-month certificate in Computer Concepts from IGNOU and a certificate course in Advance Excel from Iraedu. Dr. Kavita has expertise in a variety of bioinformatics tools and programming languages, such as HTML, R, and Perl. Her expertise with next-generation sequence analysis and programs such as ARMOUR database emphasizes her technical abilities.

She has made substantial contributions to scientific literature, with numerous publications in national and international journals in the field of microRNA (miRNA) regulation in plants, particularly focusing on stress responses and development. Through comparative analyses, database creation, and exploration of miRNA–target interactions, she has elucidated the roles of miRNAs in mediating plant responses to various stressors such as salt, heat, and viral infections. These studies provide valuable insights into the molecular mechanisms underlying stress adaptation in plants, paving the way for future advancements in crop resilience and agricultural sustainability. Dr. Kavita has also presented her research findings at various conferences and workshops, showcasing her ability to disseminate knowledge and engage with the scientific community.

Temesgen Assefa Gelaw is a faculty (Assistant Professor), Department of Biotechnology, College of Agriculture and Natural Resource Sciences, Debre Birhan University, Ethiopia. He has more than nine years of experience in industries and higher education. He has completed his B.Sc. in applied biology from Wolkite University, Ethiopia, and his M.Sc. in biotechnology from Adama Science and Technology University, Ethiopia, with very great distinctions and awards. Throughout his academic career, he has demonstrated astonishing academic achievements. He was awarded the Arturo Falaschi Fellowship from International Centre for Genetic Engineering and Biotechnology for his Ph.D. in plant biotechnology from Jawaharlal Nehru University, India. In his doctoral thesis, Dr. Temesgen delved into the functional roles of miRNAs in drought stress tolerance in maize and rice. This work was recognized through best poster presentation awards at National Conference of Plant Physiology, India, as well as by Springer Nature. He has published more than 15 manuscripts in national and international journals. His main research areas include stress physiology, molecular biology, plant tissue culture, and nanotechnology. He also serves as a reviewer for national and international journals.

Neeti Sanan-Mishra is a prominent scientist in the field of molecular biology, currently serving as Group Leader at the International Centre for Genetic Engineering and Biotechnology (ICGEB) in New Delhi, India. She is working on understanding the role of RNA interference (RNAi) mechanisms in plant responses to abiotic stress factors and viral infections, with a focus on the miRNAs. She has a strong academic background, having completed her graduation in Botany from the University of Delhi and postgraduation in Life Science from Jawaharlal Nehru University. She obtained her doctoral degree in Molecular Plant Physiology from Jawaharlal Nehru University, where her research focused on unraveling the molecular and electrical signaling mechanisms of light.

At ICGEB, Dr. Neeti leads a dynamic team engaged in groundbreaking research projects on the use of RNAi technology to address pressing climate change-associated challenges in agriculture. Her research group is focusing on discovering the functions of miRNAs in plant development and response to salinity, high temperatures, drought, and viral infections. This can been seen in the long list of publications in leading international journals. The development of publicly available ARMOUR database, for studying miRNA–mRNA interactions in rice, is another noteworthy accomplishment of her group. Dr. Neeti is particularly known for her expertise in leveraging cutting-edge biotechnology tools such as synthetic biology and gene editing to manipulate miRNAs for trait improvement and crop protection.

In addition to her research endeavors, Dr. Neeti is actively involved in scientific outreach and education. She mentors doctoral students and postdoctoral researchers, nurturing the next generation of scientific talent and fostering a collaborative and supportive research environment. Dr. Neeti's contributions have been recognized through awards and grants, underscoring her impact on advancing scientific knowledge.

Contributors

Mohd Firdaus Abdullah Malaysian Cocoa Board, Wisma SEDCO, Kota Kinabalu, Sabah, Malaysia

Titilayo Ajose Fruits Research Programme, National Horticultural Research Institute, Jericho Reservation Area, Idi-Ishin, Ibadan, Oyo State, Nigeria

Abreham Chebte Alemu Department of Biotechnology, College of Agriculture and Natural Resource Sciences, Debre Birhan University, Debre Birhan, Ethiopia

Festetics Doctoral School, Hungarian University of Agriculture and Life Sciences, Georgikon Campus, Keszthely, Hungary

Olawale Arogundade Department of Crop Protection and Environmental Biology, University of Ibadan, Ibadan, Oyo State, Nigeria

Fruits Research Programme, National Horticultural Research Institute, Jericho Reservation Area, Idi-Ishin, Ibadan, Oyo State, Nigeria

Sekinat O. Azeez Department of Botany, Obafemi Awolowo University, Ile-Ife, Osun State, Nigeria

Abisola Banji Department of Crop Protection and Environmental Biology, University of Ibadan, Ibadan, Oyo State, Nigeria

Ankur R. Bhardwaj Department of Botany, University of Delhi, New Delhi, India

Ragini Bhardwaj ICAR-National Bureau of Plant Genetic Resources, Pusa, New Delhi, India

Department of Bioscience and Biotechnology, Banasthali Vidyapith University, Tonk, Rajasthan, India

Sambhavana Chauhan Department of Genetics, University of Delhi, South Campus, New Delhi, India

Kumar Durgesh Division of Genetics, ICAR-Indian Agriculture Research Institute, New Delhi, India

Shachi Gahoi Department of Computational Biology and Bioinformatics, Jacob Institute of Biotechnology and Bioengineering, Sam Higginbottom University of Agriculture, Technology and Sciences, Prayagraj, Uttar Pradesh, India

Budhayash Gautam Department of Computational Biology and Bioinformatics, Jacob Institute of Biotechnology & Bioengineering, Sam Higginbotton University of Agriculture, Technology & Sciences, Prayagraj, Uttar Pradesh, India

Temesgen Assefa Gelaw Department of Biotechnology, College of Agriculture and Natural Resources Science, Debre Birhan University, Debre Birhan, Ethiopia

Kavita Goswami Plant RNAi Biology, International Centre for Genetic Engineering and Biotechnology, New Delhi, Delhi, India

Sharad Kumar Jaiswal Department of Computational Biology and Bioinformatics, Jacob Institute of Biotechnology & Bioengineering, Sam Higginbotton University of Agriculture, Technology & Sciences, Prayagraj, Uttar Pradesh, India

Shailendra K. Jha Division of Genetics, ICAR-Indian Agriculture Research Institute, New Delhi, India

M. G. K. Jones State Agricultural Biotechnology Centre, Murdoch University, Perth, WA, Australia

Ambreen Khan Plant RNAi Biology Group, International Centre for Genetic Engineering and Biotechnology, New Delhi, India

Sudhir Kumar Plant RNAi Biology Group, International Centre for Genetic Engineering and Biotechnology, New Delhi, India

Anita Kumari Department of Botany, University of Delhi, New Delhi, India

Sweta Kumari Division of Biochemistry, ICAR-Indian Agriculture Research Institute, New Delhi, India

Jafar K. Lone ICAR-National Bureau of Plant Genetic Resources, Pusa, New Delhi, India

Department of Biochemistry, School of Bioengineering and Biosciences, Lovely Professional University, Phagwara, Punjab, India

Joshua O. Matthew Fruits Research Programme, National Horticultural Research Institute, Jericho Reservation Area, Idi-Ishin, Ibadan, Oyo State, Nigeria

M. S. Nimmy ICAR-National Institute of Plant Biotechnology, New Delhi, India

Ritu Pandey Department of Botany, Hansraj College, University of Delhi, New Delhi, India

Adeeb Rahman Plant RNAi Biology, International Centre for Genetic Engineering and Biotechnology, New Delhi, India

Neeti Sanan-Mishra Plant RNAi Biology, International Centre for Genetic Engineering and Biotechnology, New Delhi, India

Lekshmy Sathee Division of Plant Physiology, ICAR-Indian Agriculture Research Institute, New Delhi, India

Vaishali Saxena Plant RNAi Biology Group, International Centre for Genetic Engineering and Biotechnology, New Delhi, India

Hemakshi Sharma Department of Botany, University of Rajasthan, Jaipur, Rajasthan, India

Archana Singh Division of Biochemistry, ICAR-Indian Agriculture Research Institute, New Delhi, India

Chandra Pal Singh Department of Botany, University of Rajasthan, Jaipur, Rajasthan, India

Shivani Singh Department of Biological Sciences, Faculty of Science, Sam Higginbotton University of Agriculture, Technology & Sciences, Prayagraj, Uttar Pradesh, India

Aarti Soni Department of Botany, University of Delhi, New Delhi, India

Sudhir K. Sopory International Centre for Genetic Engineering and Biotechnology, New Delhi, India

Rakesh Srivastava Department of Botany, University of Delhi, New Delhi, India

C. L. Tan Malaysian Cocoa Board, Wisma SEDCO, Kota Kinabalu, Sabah, Malaysia

Ruby Tiwari Department of Genetics, University of Delhi, South Campus, New Delhi, India

Anita Tripathi NerdGene Healthcare Pvt. Ltd, Pune, Maharashtra, India

T. Vinutha Division of Biochemistry, ICAR-Indian Agriculture Research Institute, New Delhi, India

Rekha Yadav Department of Botany, University of Rajasthan, Jaipur, Rajasthan, India

Exploring the RNAi Vistas: From Discovery to Emerging Frontiers

1

Anita Kumari, Aarti Soni, Rakesh Srivastava, Ritu Pandey, and Ankur R. Bhardwaj

Abstract

RNA interference (RNAi) stands as a universally preserved molecular process pivotal in orchestrating posttranscriptional gene suppression throughout a broad spectrum of organisms. This chapter commences with the seminal revelation of RNAi mechanism, tracing its inception from petunias to its ubiquitous occurrence across species. We further dive into the different categories of regulatory noncoding small RNAs, such as small interfering RNAs (siRNAs), microRNAs (miRNAs), and Piwi-interacting RNAs (piRNAs), which have broadened the horizon of RNAi-mediated gene control beyond traditional comprehension. We also explore the significance of RNAi in plant functional genomics and its applications in enhancing crop yield. Finally, we discuss the incorporation of RNAi with other biotechnological methods like gene editing, which has the capacity to influence the trajectory of agriculture and foster sustainable food production in future.

Keywords

RNA interference · ncRNA · PTGS · miRNAs · siRNAs

A. Kumari · A. Soni · R. Srivastava
Department of Botany, University of Delhi, Delhi, India

R. Pandey
Department of Botany, Hansraj College, University of Delhi, Delhi, India

A. R. Bhardwaj (✉)
Department of Botany, Ramjas College, University of Delhi, Delhi, India
e-mail: ankurbhardwaj@ramjas.du.ac.in

K. Goswami et al. (eds.), *Non-coding RNAs for Crop Improvement*,
https://doi.org/10.1007/978-981-97-7213-1_1

1.1 Introduction

The discovery and subsequent understanding of RNAi have transformed our knowledge of gene regulation in the field of molecular biology and have offered highly effective tools for gene manipulation. Since its first discovery in the late 1990s, RNAi has become a major tool in biological research, providing several possibilities for therapeutic intervention, functional genomics, and gene function investigation. When RNAi was first found in petunia, researchers became fascinated by this natural process (Napoli et al. 1990; Van der Krol et al. 1990). The RNAi mechanism is present across a wide range of organisms and regulates gene expression by generating short double-stranded RNA (dsRNA) molecules. Specialized ribonucleases (RNases) are synthesized to selectively degrade target RNA, including viral RNA, messenger RNA (mRNA), and transposon RNA, thereby inhibiting their translation or replication. These target dsRNA molecules are synthesized from initial single-stranded RNA molecules by RNA-dependent RNA polymerases (RdRP). This system acts as a natural defense against molecular invaders, which can severely damage a host genome.

RNAi encompasses the functions of siRNAs, miRNAs, and piRNAs, which induce gene silencing by cleaving mRNAs and or inhibiting protein synthesis. siRNAs are double-stranded RNA molecules typically 19–21 base pairs in length, originating from long dsRNA molecules produced either externally or internally within the organism, whereas miRNAs are single-stranded RNAs, typically 21–24 nucleotides long. The miRNAs are generated from endogenous primary miRNAs (pri-miRNAs) transcribed by RNA polymerase II from miRNA genes, representing a diverse family of noncoding RNAs. In contrast, piRNAs are single-stranded RNAs, generally around 23–36 nucleotides in length (Zhu and Palli 2020).

In this chapter, we present a review of the significant discoveries in the field of RNAi. We provide an overview of the biogenesis and mechanisms of RNAi pathways. Additionally, we highlight the applications of RNAi in crop improvement as well as the challenges and limitations associated with its implementation.

1.2 Posttranscriptional Gene Silencing (PTGS) in Plants

In 1990, Napoli and colleagues introduced the chimeric gene "chalcone synthase," leading to the reversible co-suppression of homologous genes in trans. Chalcone synthase (CHS) is an essential enzyme in flavonoid synthesis pathway. To enhance pigmentation in petunia petals, they constructed a chimeric CHS gene and introduced multiple copies into three petunia genotypes: a pale pink inbred (R18), a commercial hybrid variety (Pink Cascade), and a deep violet inbred (V26). Surprisingly, instead of darker color, many of the transformed flowers were purely white, and some exhibited novel variegated patterns with more white areas than the pigmented parents. RNase protection assays on purely white flowers revealed a 50-fold reduction in mRNA levels. Further analysis revealed that white flowers underwent somatic reversion to phenotypically parental-pigmented flowers. The production of purely white flowers occurred due to co-suppression of endogenous CHS genes and the introduced chimeric *CHS* gene. Although the exact mechanism underlying reversible co-suppression of homologous genes in trans was not fully elucidated, Napoli et al. (1990) suggested that DNA methylation might be responsible. Another report by Van Der Krol et al. (1990) in the same plant species also showed a substantial decrease in expression of DFR or CHS gene. Hamilton and Baulcombe (1999), while working with five tomatoes and three tobacco lines, showed that the antisense RNA molecule involved in PTGS was of an unvarying length of ~25 nucleotides, and that its accumulation necessitated the replication of RNA viruses or transgene sense transcription because no 25-nt RNA molecule was accumulated in the absence of PTGS. Because the 25-nt antisense RNA molecule has antisense polarity, this study further showed that it was not a byproduct of target RNA breakdown. Consequently, the antisense RNA of 25 nucleotides, which is probably generated from a template of RNA, maybe the specificity determinant of PTGS.

1.2.1 Quelling and RNAi

In the early 1990s, several reports described the filamentous fungus *Neurospora crassa* exhibiting transgene-induced gene silencing, termed "quelling," and initially suggested that methylation might be involved. Treatment with 5-azacytidine, an inhibitor of cytosine methylation, appeared to enhance the reactivation frequency in certain silenced transformants, as transgenic regions were often methylated (Pandit and Russo 1992; Romano and Macino 1992). However, Cogoni and colleagues provided insights into the molecular mechanisms involved in quelling in 1996. Their results demonstrated that quelling occurs at the posttranscriptional stage and incorporates a dominant, trans-acting molecule rather than methylation. They selected the visual marker gene, albino-1 (al-1), involved in carotenoid biosynthesis, for their study. Silencing of this gene resulted in an albino phenotype. In 24% of transformants employing an al-1+ host, the transforming DNA, including a full-length al-1 gene and promoter region, produced an albino phenotype, suggesting that both the transgene and the native al-1 gene were silenced. Deletion variants of the al-1 gene revealed that a transgene requires at least ~132 bp of sequence similarity to the transcribed region of the native gene to induce quelling. Interestingly, a mutant deficient in cytosine methylation (dim-2) exhibited normal frequencies and levels of silencing, indicating that cytosine methylation is not responsible for quelling. Silencing was found to be a dominant trait in heterokaryotic strains containing both non-transgenic and transgenic nuclei, suggesting that quelling is mediated by a diffusible, trans-acting molecule. Sense RNA from transgenes was present in quelled strains but absent in their revertants, indicating that transcription of the transgene(s) is necessary to silence both endogenous and ectopic gene copies (Cogoni et al. 1996).

1.2.2 Virus-Induced Gene Silencing

Virus-induced gene silencing (VIGS) is a type of post-transcriptional gene silencing based on RNA-mediated antiviral defense mechanisms in plants (Zulfiqar et al. 2023). In 1995, Kumagai and colleagues demonstrated cytoplasmic inhibition of phytoene desaturase (PDS) expression, a thylakoid membrane-bound enzyme in *Nicotiana benthamiana*, by viral delivery of antisense RNA. Plants transfected with antisense PDS sequences exhibited an albino phenotype and substantial phytoene accumulation in their leaves. Another study by Teresa et al. (1998) unveiled the silencing of the phytoene desaturase (PDS) gene in *Nicotiana benthamiana* infected with potato virus X (PVX) vectors having a green fluorescent protein (GFP) transgene and PDS inserts in plants infected with PVX-GFP. They confirmed that VIGS is cytoplasmic and posttranscriptional because it targets the exons of viral and PDS RNAs instead of their introns. Anandalakshmi et al. (1998) demonstrated in the same year that a viral sequence that was first recognized as a facilitator of synergistic viral disease inhibits posttranscriptional gene silencing produced by both viruses and transgenes.

1.2.3 Andrew Fire and Craig Mello's Contribution: Recognition of RNAi

In 1998, Fire and Mello published an article titled "Potent and specific genetic interference by double-stranded RNA in *Caenorhabditis elegans*," where they discovered that small dsRNA molecules can interfere with the function of endogenous genes and silence their expression. They were awarded the Nobel Prize in Physiology or Medicine in 2006 for this groundbreaking discovery. Working with *C. elegans*, they selected the gene "unc-22," encoding an abundant but nonessential myofilament protein. A reduction in unc-22 activity caused an increasingly severe twitching phenotype, and a complete loss-of-function led to additional muscle structural defects and impaired motility. The researchers showed that a sense-antisense mixture of dsRNA was highly effective in interfering with endogenous gene activity, being at least two orders times more effective than either single strand alone. Their work established a new conceptual framework for RNA-mediated gene silencing by emphasizing the role of dsRNA. However, questions remained about the mechanism by which dsRNA could degrade endogenous mRNA (Fire et al. 1998). The major milestones in the field of RNAi are depicted in Fig. 1.1.

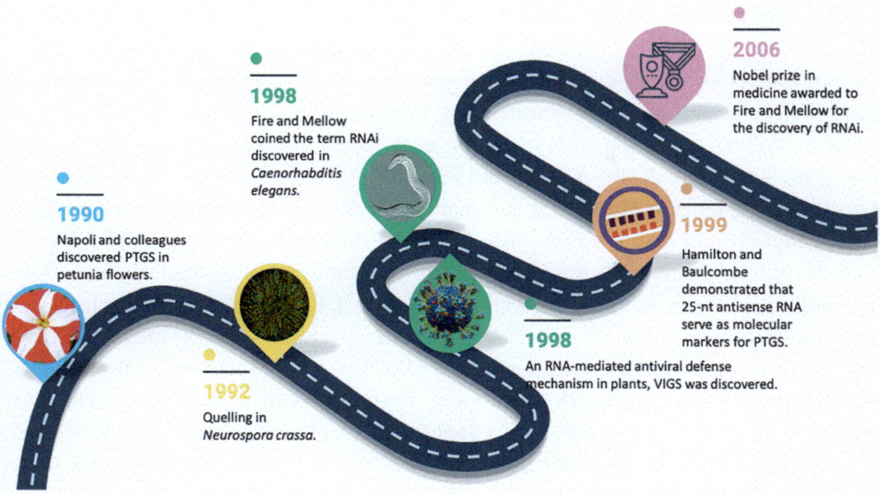

Fig. 1.1 Timescale depicting the early observations and key milestones in the discovery of RNAi

1.3 Biogenesis of Small RNAs

The biogenesis of small RNAs (sRNAs) typically entails the enzymatic cleavage of longer RNA molecules, resulting in 20–24 nucleotide (nt) sRNA duplexes mediated by Dicer-like (DCL) endoribonucleases. These precursor molecules appear either as hairpin-like RNA structures formed by the self-complementary folding of RNAs or as double-stranded RNAs (dsRNAs) synthesized by RNA-dependent RNA polymerases (RDRs). Typically, one strand from the DCL-processed short dsRNA molecules is selectively loaded onto Argonaute (AGO) proteins. This selective loading mechanism targets endogenous or exogenous RNAs through sequence complementarity, thereby enabling the silencing of genes or transposons. The cellular sources of miRNA and siRNA exhibit some variation: miRNAs originate from the genome, while siRNAs can either be endogenous or result from external factors such as viral infection. Another significant distinction lies in the dsRNA precursors: siRNA duplexes exhibit precise base-pairing, whereas miRNA helices include mismatches and longer terminal loops. Despite their different origins, these processing pathways converge when either type of RNA is incorporated into the RNA-induced silencing complex (RISC).

1.3.1 MicroRNA Biogenesis

Plant microRNAs are generated by endogenous *MIRNA* (*MIR*) genes. Most of these genes are located in intergenic, noncoding loci. However, in a few reports, intronic regions of protein-coding genes have also been found to encode miRNAs (Rajagopalan et al. 2006; Yan et al. 2012; Yang et al. 2012). The *MIR* gene transcription is carried out by DNA-dependent RNA polymerase II (Pol II), leading to the generation of primary miRNAs. These pri-miRNAs are single-stranded poly-adenylated RNA transcripts that possess self-complementary characteristics, allowing them to fold into hairpin-like structures. DCL1, through recognition and cleavage of these folded pri-miRNAs, generates mature miRNA duplexes. Thus, each duplex comprises a miRNA strand and a miRNA* strand, with the miRNA* strand typically undergoing degradation. The retained miRNA strand associates with an AGO protein, forming RISC. Plant miRNAs predominantly exhibit a length of 21 nucleotides, although there are reports of 20 or 22 nucleotide-length miRNAs (Wu et al. 2010; Axtell and Meyers 2018).

The transcription process of *MIR* genes mirrors that of protein-coding genes, necessitating the recruitment of RNA polymerase II (Pol II) and co-transcriptional processes such as 5′ capping and 3′ polyadenylation of primary transcripts. Studies have revealed numerous transcription factors that control the transcription of miRNAs. For example, *Arabidopsis* SQUAMOSA PROMOTER BINDING PROTEIN-LIKE (SPL) family of transcription factors plays a pivotal role in the regulation of miR172 (Wu et al. 2009). In *Arabidopsis*, various proteins are linked to the RNA Pol. II machinery have been identified in the regulation of miRNA transcription such as SMALL1 (SMA1, a DEAD-box domain spliceosomal protein), PROTEIN PHOSPHATASE4 (PP4) complex, and EMBRYONIC FLOWER2 (EMF2) complex. These molecules have been shown to interact with *MIR* gene promoters to control their transcription (Xu et al. 2016; Li et al. 2018; Wang et al. 2019). Additionally, NOT2a and NOT2b (NEGATIVE ON TATA LESS2) proteins, SAC3A, and THP1 (part of THREE PRIME REPAIR EXONUCLEASE-2, TREX-2) complex work with Pol II to facilitate *MIR* transcription (Wang et al. 2013; Zhang et al. 2020).

Plant primary miRNAs exhibit imperfect fold-back structures and undergo cleavage by DCL1 in the nucleus, resulting in the generation of miRNA–miRNA∗ duplexes (Kurihara and Watanabe 2004; Park et al. 2005). This process differs from the animal scheme in which the pri-miRNAs are first processed in the nucleus by Drosha and later in the cytoplasm by Dicer (Treiber et al. 2019). The specific secondary structure of pri-miRNAs plays a significant part in governing the sites of cleavage within the pri-miRNAs (Bologna et al. 2013; Zhu et al. 2013). The final two cuts result in the release of a miRNA–miRNA∗ duplex (Bologna et al. 2013, Zhu et al. 2013). D-bodies (nuclear dicing bodies) are known to perform key steps in the biogenesis of miRNAs. These D-bodies are formed by association of HYL1 (HYPONASTIC LEAVES1), SE (SERRATE) and DCL1 (Han et al. 2004; Vazquez et al. 2004; Lobbes et al. 2006; Yang et al. 2006a, b; Fang and Spector 2007). Both HYL1 and SE contribute to the precise processing of primary miRNAs by DCL1

(Kurihara et al. 2006; Dong et al. 2008). Mutations in SE and HYL1, which lead to loss of function, have shown a drastic decline in the abundance of the miRNA population. Also, these mutants exhibit severe defects related to normal growth and development. Many characteristics of defective developmental growth are found in both mutants (SE and HYL1), indicating their involvement in the same scheme (Yang et al. 2006a, b). Other factors that play a crucial role in miRNA biogenesis are CHR2 (CHROMATIN REMODELING FACTOR2) and HST (HASTY) (Fang et al. 2015a, b; Cambiagno et al. 2021). It has been seen that many proteins that remain linked to the transcription machinery are also interacting with D-bodies, which in turn indicate that pri-miRNAs undergo co-transcriptional processing. Moreover, R loops that are DNA: RNA hybrids promote the co-transcriptional processing of pri-miRNAs (Gonzalo et al. 2022). The regulation of D-bodies component accumulation is intricately controlled. The regulation of D-bodies component accumulation is intricately controlled. The nuclear protein XAP5 CIRCADIAN TIMEKEEPER (XCT) plays a part in endorsing DCL1 transcription and facilitates the processing of primary microRNAs (Fang et al. 2015a, b; Li et al. 2018). Furthermore, DCL1 transcripts are directly targeted by miR162, a miRNA produced as a result of DCL1 activity, indicating a negative feedback regulation mechanism (Xie et al. 2003). During photomorphogenesis, destabilization of DCL1 has been seen, which is mediated by PHYTOCHROME-INTERACTING FACTOR4 (PIF4, a helix-loop-helix transcription factor), which consequently leads to inhibition of pri-miRNA processing (Sun et al. 2018). The processing of pri-miRNA is also ceased in RNA DEBRANCHING ENZYME1 (DBR1) mutant. DBR1 is a protein that takes part in the degradation of intronic lariat RNAs. This type of inhibition is most probably due to the D-bodies sequestration by more lariat RNAs, presenting a unique method by which the activity of DCL1 is regulated (Li et al. 2016).

Some *MIR* genes from plants are located within introns, while others have introns within the *MIR* genes themselves (Lu et al. 2008; Szarzynska et al. 2009; Yan et al. 2012). Several proteins have dual functions in both splicing and primary microRNA processing, and some splicing factors co-localize or interact with D-bodies, influencing pri-miRNA processing. Examples of such factors include RS40, RS41 (ARGININE/SERINE-RICH SPLICING FACTOR), SMA1, STA1 (STABILIZED1), GRP7 (GLYCINE-RICH RBP7) (Ben Chaabane et al. 2013; Koester et al. 2014; Chen et al. 2015; Li et al. 2018). Additionally, intronic miRNA processing is regulated by proteins namely LUC7rl (LETHAL UNLESS CBC 7 RL), PRP40b, PRP40a, and PRP39b (Knop et al. 2017). Primary microRNAs also undergo 3′ tailing processes, such as uridylation and cytidylation, which aid in facilitating pri-miRNA processing (Song et al. 2019). The HESO1 (HEN1 SUPPRESSOR1) plays a crucial role in mediating uridylation as well as cytidylation of pri-miRNA. Additionally, NTP6 and NTP7 (which are nucleotidyl transferases) take part in pri-miRNA cytidylation (Song et al. 2019). RNA methylation, specifically N6-methyladenosine (m6A), has recently gained recognition as a significant method of gene regulation (Frye et al. 2018). In *Arabidopsis*, MTA (adenosine methylase) mRNA is responsible for depositing m6A on pri-miRNAs, thereby facilitating pri-miRNA processing and influencing D-body assembly (Bhat et al. 2020). Following primary microRNA processing, HUA ENHANCER 1 (HEN1, a 2′-O-methyltransferase) takes over from SERRATE (SE) and interacts with DCL1 and HYPONASTIC LEAVES1 (HYL1). HEN1 catalyzes 2′-O-methylation at the 3′ ends of miRNA duplexes, contributing to the stabilization of miRNAs (Yu et al. 2005; Baranauskė et al. 2015). Loss-of-function mutations in hen1 result in reduced abundance of nearly all miRNAs, and the remaining miRNAs exhibit heterogeneity in size, revealing phenomena like 3′ to 5′ truncation and 3′ uridylation of miRNAs (Li et al. 2005; Yu et al. 2005; Yang et al. 2006a, b). In wild-type and hen1 mutant lines, SDN1, SDN2, and SDN3 (SMALL RNA–DEGRADING NUCLEASE genes) degrade miRNA as well as siRNA (Yu et al. 2017). Collectively, the present studies underscore that the function of miRNAs as well as their turnover is regulated by HEN1. HEN1 performs methylation of miRNAs, thus preventing uridylation, thereby preventing miRNA degradation and preserving their regulatory roles.

1.3.2 siRNA Biogenesis

1.3.2.1 Heterochromatic-siRNA (Hc-siRNA) Biogenesis

The hc-siRNAs (heterochromatic small interfering RNAs) are derived from transposons located in the heterochromatic regions of the genome. These are usually 24 nucleotides long and play a crucial role in RNA-directed DNA methylation (RdDM). In plants, particularly in *Arabidopsis*, hc-siRNAs are the major constituents among the small RNA population, accounting for more than 90% (Zhang et al. 2007; Mosher et al. 2008). The biogenesis of hc-siRNAs begins with transcription by Pol IV (Herr et al. 2005; Onodera et al. 2005) and subsequently, a second RNA strand is synthesized by RDR2 (Xie et al. 2004). The resulting dsRNAs are then processed by DCL3 into hc-siRNA duplexes, followed by methylation. Finally, a single strand is loaded onto Argonaute (AGO4/AGO6/AGO9) to form RISCs (Zilberman et al. 2003; Li et al. 2005; Havecker et al. 2010). Pol IV has the capacity to start the process of transcription at different positions featuring the pyrimidine-purine sequences, generating multiple short transcripts (typically 26–45 nt) from the same single locus (Blevins et al. 2015; Zhai et al. 2015). The Pol IV is recruited for transcription by SAWADEE HOMEODOMAIN HOMOLOG1 (SHH1)/DNA-BINDING TRANSCRIPTION FACTOR1 (DTF1) and CLASSY1 (SNF2 domain–containing chromatin-remodeling factor) (Law et al. 2011, 2013). It has been found through biochemical analysis that Pol IV and RDR2 interact physically (Haag et al. 2012; Fukudome et al. 2021). Structural studies have identified tight coupling of the functions of Pol IV and RDR2. In this coupling, transcripts of Pol IV are directly channeled to RDR2, and RDR2 subsequently synthesizes Pol IV–RDR2 (P4R2) RNAs (Huang et al. 2021). DCL3 dices the double-stranded P4R2 RNAs to yield one hc-siRNA per precursor (Blevins et al. 2015; Zhai et al. 2015). In vitro experiments have demonstrated that it is possible to produce hc-siRNA by combining the basic requirements of a DNA template, ribonucleotide triphosphates, Pol IV, DCL3, and RDR2 (Singh et al. 2019). Studies have reported a plethora of siRNAs (24 nt long) from ovules and seeds named "sirens." These sirens are produced through a small number of loci, which are usually near the protein-coding genes and migrate from the seed coat to the filial tissue where they bring about the in-trans DNA methylation of protein-coding genes. The biogenesis of sirens is dependent on Pol IV and RDR2 and follows the same scheme as of hc-siRNAs. The presence of sirens has been documented in *Arabidopsis*, *Brassica rapa*, and rice, and further investigation is needed to explore their existence in other plants (Rodrigues et al. 2013; Grover et al. 2020; Burgess et al. 2022; Li et al. 2022; Zhou et al. 2022).

1.3.2.2 Secondary Small Interfering RNA Biogenesis (tasiRNA, phasiRNA, easiRNAs)

The biogenesis of secondary small interfering RNAs is triggered by other miRNAs or siRNAs. PhasiRNAs are those secondary siRNAs that are produced from long RNA precursors. They exhibit head-to-tail arrangement known as phasing. The biogenesis of phasiRNAs involves the cleavage of precursors by RISC at specific sites. The resulting fragments are subsequently converted into a double strand that is cleaved by DCL protein in a sequential manner. PHAS precursors, mRNAs, or long noncoding RNAs, are products of RNA Pol. II. Protein-coding genes giving rise to phasiRNAs come from various families. There are two main classes of phasiRNAs generated from noncoding loci namely tasiRNA (trans-acting siRNAs) and reproductive phasiRNAs (Fei et al. 2013; Liu et al. 2020). In *Arabidopsis*, a total of eight tasiRNA-generating loci (TAS1-TAS4) have been identified. The biogenesis of tasiRNAs involves miRNA-mediated cleavage of a TAS precursor followed by conversion to a double-stranded RNA by RDR6. DCL4, assisted by DRB4 (DOUBLE-STRANDED RNA–BINDING PROTEIN 4), then iteratively cleaves the double-stranded RNA to generate 21 nt duplexes of phasiRNAs. These duplexes are sorted and subsequently loaded onto AGO proteins.

There are two pathways for tasiRNA biogenesis, differing in the miRNA triggers and AGO proteins involved. The more common one-hit pathway uses a 22 nt miRNA as a trigger with a 5′ U, directing miRNA loading to AGO1, while the less common two-hit pathway involves AGO7-miR390 cleaving a precursor at two sites (Fei et al. 2013; Liu et al. 2020). Reproductive phasiRNAs, initially characterized in grass species, are classified into two sizes: 21-nt and 24-nt phasiRNAs. Their biogenesis follows the one-hit model where miR2118 triggers the generation of 21-nt phasiRNAs and miR2275 triggers the generation of 24-nt phasiRNAs. In maize, transcription factors viz. OCL4 (OUTER CELL LAYER4), MALE STERILE23, MALE STERILE32, and bHLH122 are implicated in regulating the biogenesis of 21-and 24-nucleotide phasiRNAs, respectively. Secondary siRNAs can also be generated without phasing, exemplified by a class known as epigenetically activated siRNAs (easiRNAs). easiRNAs are typically 21 or 22 nt in length and are abundantly found in the pollen vegetative cells (Slotkin et al. 2009). Studies have shown that easiRNAs biogenesis is triggered by DCL1-dependent miRNAs from activated transposons. Loss-of-function mutants of DCL1, DCL4, and RDR6 exhibit a drastic decrease in 21 nt easiRNAs (Creasey et al. 2014). Other factors that play a significant role in biogenesis include DCL2, AGO1/2/6, and Pol IV (McCue et al. 2012; Nuthikattu et al. 2013; McCue et al. 2015). The easiRNA biogenesis is suggested to initiate during or after meiosis. However, the easiRNA biogenesis pathway is majorly not well understood, and majority of the studies are done on *Arabidopsis* only.

1.3.2.3 Naturally Antisense Transcript siRNAs (natsiRNA) Biogenesis

Endogenous self-complementary transcripts can form double-stranded RNA through processing by a DCL protein to generate naturally antisense transcript small interfering RNAs (natsiRNAs) and trigger gene silencing (Reis and Poirier 2021). While the functional roles of individual trans-natsiRNAs derived from non-overlapping loci remain largely uncharacterized in plants, the discussion here centers on cis-natsiRNAs. Cis-natsiRNAs originate from transcripts transcribed from the same genomic locus but in opposite directions, and their biogenesis pathways appear to be diverse. For reported cis-natsiRNAs, one pathway relies on RNA Pol IV, RDR2, and DCL3, like the biogenesis of hc-siRNAs contributing to gene silencing (Swiezewski et al. 2007). Other pathways involve DCL1, DCL2, and/or DCL3, along with other proteins (Borsani et al. 2005; Zhang et al. 2012). Under salt stress in *Arabidopsis*, a 24-nt siRNA is produced from an antisense transcript, SRO5, leading to the downregulation of the P5CDH gene (encoding a -pyrroline-5-carboxylate dehydrogenase). P5CDH transcripts, in turn, generate DCL1-dependent 21-nt phasiRNAs (Zhang et al. 2012). In response to *Pseudomonas syringae* infection, a 22-nt natsiRNA arises from overlapping ATGB2 and PPRL transcripts. This natsiRNA silences PPRL transcripts, contributing to disease resistance (Katiyar-Agarwal et al. 2006). In *Arabidopsis* pollen, overlapping transcripts of KOKOPELLI (KPL) and ARIADNE14 (ARI14) genes generate a sperm-specific natsiRNA that silences ARI14, facilitating successful fertilization (Ron et al. 2010). natsiRNAs have been implicated in mRNA cleavage and are suggested to mediate DNA and/or histone methylation (Swiezewski et al. 2007; Wan et al. 2016). The biogenesis mechanisms and molecular activities of cis-natsiRNAs are diverse and complex, and the categorization of natsiRNAs as a uniform sRNA type has been questioned, especially considering the limited number of cis-natsiRNA pairs producing detectable siRNAs and their relatively low-density in mapped regions (Zhang et al. 2012).

1.4 siRNAs Amplification

One remarkable aspect of RNA interference is its high efficacy. A robust and efficient RNAi response can be achieved with just a few dsRNA transcripts per cell (Mello and Conte Jr 2004). In *C. elegans*, the primary dsRNA trigger can initiate the production of secondary siRNAs when target mRNA is present (Meister and Tuschl 2004). These secondary siRNAs can significantly amplify and sustain the RNAi response. In plants and nematodes, siRNA amplification can extend throughout the entire organism, resulting in systemic silencing. Many eukaryotes capable of RNAi possess recognizable RdRP-encoding genes in their genomes, although insects and vertebrates are notable exceptions. One functional outcome of siRNA amplification by RdRPs is transitive RNAi, which involves generating siRNAs that correspond to regions of the mRNA transcript not initially targeted by the original dsRNA trigger. This can lead to the silencing of multiple transcripts, especially those with highly conserved sequences. The absence of transitive RNAi in vertebrates and insects enhances RNAi specificity, allowing for the targeting of individual alternatively spliced mRNA isoforms from the same locus.

Comprehensive documentation of secondary siRNAs has primarily been achieved in worms (Pak and Fire 2007; Sijen et al. 2007), revealing some unexpected findings. Firstly, most of these secondary siRNAs align with the antisense strand of the mRNA targeted by the primary dsRNA trigger. This is puzzling if secondary siRNAs were generated through a dsRNA intermediate processed by Dicer, as thermodynamic siRNA asymmetry would produce a mix of sense and antisense guide strands. Secondly, secondary siRNAs possess di-or triphosphate groups at their 5′ ends, suggesting they are likely primary, unprimed products of RdRP and not processed by Dicer. Consequently, it appears that not all siRNAs originate directly from double-stranded precursors. It is plausible that secondary siRNAs follow a different pathway, bypassing the RISC assembly process typical of primary siRNAs and, instead, may directly load into Ago proteins.

1.5 RNAi-Mediated Gene Silencing for Resilient Crops

RNAi plays a crucial role in the sphere of agricultural biotechnology. It is currently being exploited in diverse fields ranging from nutritional content enhancement to mitigating the undesirable effects of various biotic and abiotic stresses. The process involves identifying target genes, creating the necessary constructs and developing transgenic plants that are subsequently examined and evaluated for the engineered traits. The effectiveness of RNAi relies on critical factors like the selection of promoters driving gene expression, strategic use of markers to monitor transgene integration, and the transformation technique. We sum up certain studies that utilized RNAi for crop improvement in Table 1.1.

Table 1.1 Studies utilizing RNAi-mediated gene silencing for crop improvement

Organism	Trait	Gene targeted	Reference
Oryza sativa	Biomass and yield	OsDWARF4	Feldmann (2006)
	Yield	OsGA20ox2	Qiao et al. (2007)
	Glutelin (low)	GluA, GluB	Kusaba et al. (2003)
	Grain size, quality and yield	OsSPL16, OsSPL14	Wang et al. (2012)
	Floral development	OsGAMYB	Tsuji et al. (2006)
	Growth and development	OsARF18	Huang et al. (2006)
	Cd tolerance	OsPCS1	Li et al. (2007)
	Panicle architecture and branching	LAX1, LAX2, RCN2, OsRA2, OsSPL	Yang et al. (2019)
	Male sterility	OsGEN-L	Moritoh et al. (2005)
	Resistance to *Nilaparvata lugens* Resistance to *Helicoverpa armigera*	Nlsid-1, Nlaub AChE	Zha et al. (2011)
	Resistance to *Magnaporthe grisea*	OsFAD7, OsFAD8, OsNRAMP6, OsSSI2	Yara et al. (2007), Jiang et al. (2009), Campo et al. (2013)
	Drought stress response	OsRACK1	Li et al. (2009)
	ABA response	OsARF	Liu et al. (2008)
Solanum lycopersicum	Carotenoid and flavonoid	SlDET1	Davuluri et al. (2005)
	Lycopene and carotene	SlNCED1	Sun et al. (2012)
	Prolonged shelf-life	ACC oxidase, ACS synthase	Gupta et al. (2013)
	Seedless fruits	SlARF7, SlPad-, SlAUCSIA	De Jong et al. (2009), Molesini et al. (2009), Matsuo et al. (2020)
	Resistance to *fusarium oxysporum*	Pbs2, Hog1, Fmk1	Pareek and Rajam (2017)
	Tomato leaf curl New Delhi virus resistance	AV1 and AV2	Van et al. (2013)
	Drought tolerance	SlDREB2	Tao et al. (2022)
Glycine max	Isoflavone enhancement	GmFSH	Jiang et al. (2014)
	Resistance to *Sclerotinia sclerotiorum*	Lignin production genes	Peltier et al. (2009)
	Meloidogyne incognita resistance	TP, MSP	Ibrahim et al. (2011)
	Drought response	GmNFYA3	Ni et al. (2013)

(continued)

Table 1.1 (continued)

Organism	Trait	Gene targeted	Reference
Zea mays	Starch accumulation in leaves	AtGWD	Weise et al. (2012)
	Resistance to *aspergillus flavus*	Polyketide synthase	Thakare et al. (2017)
	Resistance to *Diabrotica virgifera*	DvSnf7	Khajuria et al. (2018)
	ABA response	PLD	Wei et al. (2009)
	Salt and drought response	NFY-AT	Luan et al. (2015)

1.6 RNAi and CRISPR-Cas9

Disrupting normal gene expression and studying the resulting phenotype has been the key to understanding its primary function. The RNAi technique has predominated for nearly two decades as an efficient tool to decipher gene functionality. One of the major advantages of this technology is that the silencing machinery is a natural pathway regulating cellular gene expression in eukaryotic cells, therefore it does not require any prior genetic modification. Nevertheless, RNAi frequently results in hypomorphic phenotypes, which may not precisely replicate the complete loss of functionality observed as in the case of genetic mutations. Additionally, siRNAs also induce the silencing of non-target transcripts with minimal sequence complementarity (Carthew and Sontheimer 2009). A single siRNA can potentially repress hundreds of sequence-dependent transcripts whose phenotype may supersede that of the intended gene (Franceschini et al. 2014). Moreover, the artificial introduction of siRNAs can impair endogenous miRNA functions causing alterations in off-target gene expression levels (Khan et al. 2009). RNAi also triggers the methylation of homologous DNA, including the transcribed transgene itself, which may extend to the neighboring sequences like the transgene promoter, potentially leading to a reduction in RNAi efficiency over successive generations (Casacuberta et al. 2015). These practical challenges aroused the need for the creation of novel reverse genetics technologies.

The discovery of zinc finger nucleases (ZFNs), transcription activator-like effector nucleases (TALENs), and, more recently, the CRISPR/Cas toolbox as a reverse genetics approach has ushered in a new era of functional studies (Gaj et al. 2013). The CRISPR/Cas system generates small targeted insertions or deletions on the DNA, which causes a frameshift in the coding sequence leading to loss-of-function mutation. Additionally, modified CRISPR approaches, CRISPRi (interference), and CRISPRa (activation) lead to the transcriptional repression and overexpression, respectively, of the target gene (Cheng et al. 2013; Gilbert et al. 2013). In contrast with RNAi, CRISPR almost always creates a null mutant (Housden and Perrimon 2016). Therefore, it is anticipated that CRISPR phenotypes will be more consistent and generate robust screening results. However, while

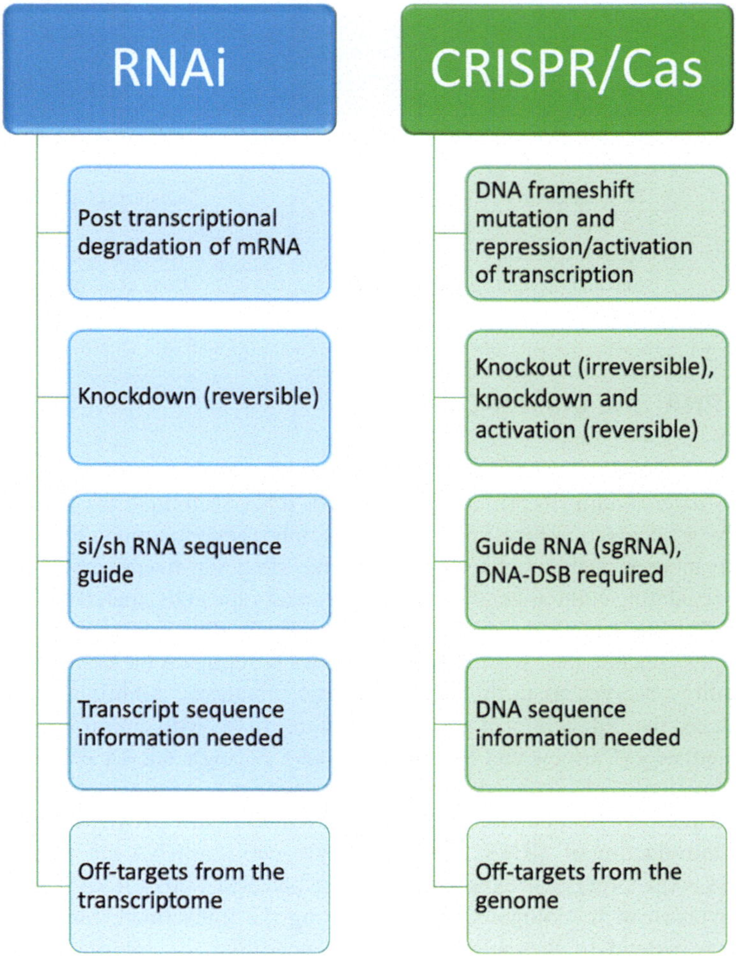

Fig. 1.2 Block diagram representing the fundamental differences between RNAi and CRISPR/Cas technologies

complete knockouts may be desirable, partial gene knockdowns assist researchers in understanding the function of lethal genes. A basic comparison between the two techniques has been depicted in Fig. 1.2.

Given the swift advancement of CRISPR technology, a pertinent inquiry comes to mind—whether the use of RNAi-based approach is nearing an end? Assessing the pros of RNAi highlights that it is rapid and saves both the time and money of the investigators. Moreover, the transcripts targeted by RNAi are present in the cytoplasm and hence more accessible as compared to CRISPR-targeted genomic DNA in the nucleus whose accessibility can also be hindered by chromatin states (Wu et al. 2014). Further, gene silencing through RNAi can be employed to virtually target any protein-coding gene while the application of CRISPR/Cas depends upon the presence of a protospacer adjacent motif (PAM) at the target site. However, variations in the PAM can be tolerated by certain Cas9 variants (Kleinstiver et al. 2015). Taken together both RNAi and CRISPR/Cas serve as potent instruments for genetic manipulation. Although CRISPR/Cas technique stands out as a more versatile and efficient method, RNAi continues to hold a significant position in the context of various experimental platforms. We believe that a coexistence of both methods is needed for better scientific advancements.

1.7 Regulatory and Ethical Considerations in RNAi Research

RNAi has been effectively employed to silence undesirable genes contributing to enhanced crop resilience against stresses. Prior to reaching the market shelves, these products must undergo certain regulatory evaluations established by individual countries in accordance with their governing laws (Table 1.2). The regulatory

Table 1.2 Regulatory framework for genetically modified crops in various countries

Country	Regulatory authority	Biosafety legislation	Regulatory standards for release
USA	USDA-APHIS, FDA, and EPA	Coordinated Framework for the Regulation of Biotechnology (1986)	Risk assessment
European Union	EFSA	Biosafety Directives and Regulation (2003)	Risk assessment, risk management, coexistence, monitoring, labeling, and detection methods
India	GEAC, IBC and RCGM	Environment Protection Act (1986)	Contained research, confined field trials, food safety assessment, and environmental risk assessment
China	OAGEBA and CFDA	Regulation on Protection of New Varieties of Plants (1997)	Risk assessment, risk management, coexistence, monitoring, labeling, and detection methods
UK	DEFRA	Biosafety Directives and Regulations (1990)	Risk assessment, risk management, coexistence, monitoring, labeling, and detection methods
Japan	Ministry of environment	Cartagena Act (2003)	Risk assessment, risk management, labeling, and monitoring
Australia	OGTR	Food Standards Australia New Zealand Act (1991) and Gene Technology Act (2000)	Risk assessment, risk management, and monitoring

Note: *USDA-APHIS* US Department of Agriculture-Animal and Plant Health Inspection Service, *FDA* The Food and Drug Administration, *EPA* The Environmental Protection Agency, *EFSA* European Food Safety Authority, *GEAC* Genetic Engineering Approval Committee, *IBC* Institutional Biosafety Committee, *RCGM* Review Committee on Genetic Manipulation, *OAGEBA* Office of Agricultural Genetic Engineering Biosafety Administration, *CFDA* China Food and Drug Administration, *DEFRA* Department for Environment Food and Rural Affairs, and *OGTR* Office of the Gene Technology Regulator assessments necessitate experimental findings that support the biosafety of the output for the environment as well as for humans and animals. Certain risks associated with the RNAi strategy exist. Due to the conserved nature of RNAi, both intended target genes and unintended off-targets may undergo silencing in the genetically modified (GM) plant as well as the organisms that consume or infect the plant (Knip et al. 2014). Considering that silencing off-targets presents a new risk, management strategies need to be implemented. Performing genome-wide bioinformatic analyses to examine sequence homology between the expressed sRNAs and all mRNAs can aid in identifying off-targets both within the modified plant and non-target organisms. However, its applicability in risk evaluation is restricted because of incomplete crop genome sequencing and genomic variations among cultivars.

1.8 Challenges and Limitations

RNAi is a potent tool for gene regulation that has been extensively studied and utilized in plants. However, like any technology, RNAi in plants has its own set of challenges and limitations. One of the primary concerns is the likelihood of unintended effects, where off-target genes are silenced due to sequence similarity with the target gene. This can lead to unpredictable and unintended changes in the plant's phenotype (Gebremichael et al. 2021). Indeed, meticulous dsRNA design will limit the probability of non-target effects owing to sequence homology during the initial phases of the development of the invasive RNAi mechanism. Generating a distinct siRNA/dsRNA that has low DNA similarity with other genetic loci significantly reduces the likelihood of off-target effects (Jackson et al. 2003; Lundgren and Lundgren and Duan 2013; Gebremichael et al. 2021). The choice of a suitable target gene has a major impact on how RNAi effects turn out. The optimal RNAi gene target must be highly expressed. To make the silence effect more noticeable, there should not be any functional redundancy. Many plant genes have redundant functions, meaning that silencing one gene may not lead to the desired phenotype if other related genes can compensate for its loss. This can make it difficult to achieve the desired phenotype through RNAi (Lawrence and Pikaard 2003; Mamta and Rajam 2017). The effectiveness of RNAi may fluctuate based on factors such as the targeted gene, the delivery method of the RNAi construct, and the plant species. Achieving consistent and reliable silencing across different genes and plant species can be challenging. While various efficient delivery methods such as Agrobacterium-mediated transformation, particle bombardment, nanoparticles, micro-injection, spraying, and viral vectors have been used, each method has its own advantages, which depend on plant species or tissues selected (Lundgren and Lundgren and Duan 2013; Mamta and Rajam 2017; Sharma et al. 2023). RNAi-mediated gene silencing is often transient in plants, meaning that the effect may diminish over time as the RNAi molecules are degraded. This limits the long-term stability of gene silencing and may require repeated applications or other strategies to maintain the desired phenotype (Zhu and Palli 2020). Reaching efficient RNAi poses challenges when the trigger, typically dsRNA or siRNA, undergoes rapid degradation within an organism. Overcoming this hurdle often necessitates the application of a high dosage of dsRNA or siRNA (Zhu and Palli 2020).

When using exogenous RNAi techniques, dsRNA introduction can occasionally cause unanticipated epigenetic changes that lead to plants with changed epigenetics (Gebremichael et al. 2021). For example, Dubrovina et al. 2019 used transgenic *Arabidopsis thaliana* with a GFP/NPTII cassette, utilizing in vitro generated dsRNA targeting GFP and NPTII genes (Dubrovina et al. 2019). It revealed DNA methylation in the relevant coding area in addition to downregulation of the GFP and NPTII mRNAs (Dubrovina et al. 2019). This suggests a more thorough procedure and highlights the necessity of closely evaluating any possible epigenetic changes when using exogenous RNA interference. Even in the absence of transgenes, plants treated with foreign dsRNAs may experience epigenetic alterations.

1.9 Conclusion and Future Perspectives

Since its discovery in the late 1990s, RNAi has revolutionized the course of molecular biology studies. From its humble beginnings in petunias to its Nobel-prize-winning recognition by Fire and Mello, this technology has paved the way for significant advancements in gene regulation, functional genomics, and therapeutic interventions. The process involves mRNA degradation to trigger gene silencing, which has been methodically utilized either via *Agrobacterium*-mediated transformation or alternative techniques to modulate gene expression. Diverse categories of small regulatory RNAs, including miRNAs, siRNAs, tasiRNAs, phasiRNAs, easiRNAs and natsiRNAs, collaborate to coordinate this posttranscriptional and epigenetic regulatory process. The utilization of RNAi for crop enhancement is diverse, encompassing various aspects like the production of seedless fruits, prolonged shelf-life, alteration of plant architecture, resistance to abiotic and biotic stresses, and improved nutritional value. Furthermore, the importance of RNAi in understanding gene functions is highlighted by its potential to unveil complex regulatory networks governing intricate traits. Nevertheless, the adoption of RNAi methods requires careful attention to biosafety considerations. RNAi will, however, continue to be a fundamental tool in the functional genomics arsenal coupled with genome editing technologies like CRISPR-Cas9. Additionally, prospective efforts should focus on advancing our comprehension of RNAi mechanisms, delivery techniques, and challenges like off-target effects.

Looking ahead, it is imperative for researchers to concentrate on tackling the obstacles and venturing into novel avenues of RNAi investigation. Research endeavors should prioritize enhancing the sRNA designing principles to further ensure the precise targeting of intended genes while avoiding non-target ones. Sophisticated bioinformatic tools and algorithms should be devised to increase the accuracy of off-target prediction. Furthermore, it is also crucial to conduct thorough investigations into the stability and longevity of traits introduced by RNAi. This will aid in both the regulatory assessments as well as widespread commercial adoption. Additionally, the epigenetic alterations triggered by RNAi should also be explored.

Gaining insights into how RNAi-induced silencing affects the DNA methylation patterns, chromatin and histone modifications will help understand the regulatory mechanisms influenced by RNAi.

References

Anandalakshmi R, Pruss GJ, Ge X, Marathe R, Mallory AC, Smith TH, Vance VB (1998) A viral suppressor of gene silencing in plants. Proc Natl Acad Sci USA 27 95(22):13079–13084

Axtell MJ, Meyers BC (2018) Revisiting criteria for plant microRNA annotation in the era of big data. Plant Cell 30(2):272–284

Baranauskė S, Mickutė M, Plotnikova A, Finke A, Venclovas Č, Klimašauskas S, Vilkaitis G (2015) Functional mapping of the plant small RNA methyltransferase: HEN1 physically interacts with HYL1 and DICER-LIKE 1 proteins. Nucleic Acids Res 43(5):2802–2812

Ben Chaabane S, Liu R, Chinnusamy V, Kwon Y, Park JH, Kim SY, Zhu JK, Yang SW, Lee BH (2013) STA1, an *Arabidopsis* pre-mRNA processing factor 6 homolog, is a new player involved in miRNA biogenesis. Nucleic Acids Res 41(3):1984–1997

Bhat SS, Bielewicz D, Gulanicz T, Bodi Z, Yu X et al (2020) mRNA adenosine methylase (MTA) deposits m6A on pri-miRNAs to modulate miRNA biogenesis in *Arabidopsis* thaliana. PNAS 117(35):21785–21795

Blevins T, Podicheti R, Mishra V, Marasco M, Wang J, Rusch D, Tang H, Pikaard CS (2015) Identification of pol IV and RDR2-dependent precursors of 24 nt siRNAs guiding de novo DNA methylation in *Arabidopsis*. elife 4:e09591

Bologna NG, Schapire AL, Zhai J, Chorostecki U, Boisbouvier J, Meyers BC, Palatnik JF (2013) Multiple RNA recognition patterns during microRNA biogenesis in plants. Genome Res 23(10):1675–1689

Borsani O, Zhu J, Verslues PE, Sunkar R, Zhu J-K (2005) Endogenous siRNAs derived from a pair of natural cis-antisense transcripts regulate salt tolerance in *Arabidopsis*. Cell 123(7):1279–1291

Burgess D, Chow HT, Grover JW, Freeling M, Mosher RA (2022) Ovule siRNAs methylate protein coding genes in trans. Plant Cell 34(10):3647–3664

Cambiagno DA, Giudicatti AJ, Arce AL, Gagliardi D, Li L, Yuan W, Lundberg DS, Weigel D, Manavella PA (2021) HASTY modulates miRNA biogenesis by linking pri-miRNA transcription and processing. Mol Plant 14(3):426–439

Campo S, Peris-Peris C, Siré C, Moreno AB, Donaire L, Zytnicki M, Notredame C, Llave C, San Segundo B (2013) Identification of a novel micro RNA (mi RNA) from rice that targets an alternatively spliced transcript of the N ramp6 (N atural resistance-associated macrophage protein 6) gene involved in pathogen resistance. New Phytol 199(1):212–227

Carthew RW, Sontheimer EJ (2009) Origins and mechanisms of miRNAs and siRNAs. Cell 136:642–655

Casacuberta JM, Devos Y, Du Jardin P, Ramon M, Vaucheret H, Nogué F (2015) Biotechnological uses of RNAi in plants: risk assessment considerations. Trends Biotechnol 33(3):145–147

Chen T, Cui P, Xiong L (2015) The RNA-binding protein HOS5 and serine/arginine-rich proteins RS40 and RS41 participate in miRNA biogenesis in *Arabidopsis*. Nucleic Acids Res 43(17):8283–8298

Cheng AW, Wang H, Yang H, Shi L, Katz Y, Theunissen TW, Rangarajan S, Shivalila CS, Dadon DB, Jaenisch R (2013) Multiplexed activation of endogenous genes by CRISPR-on, an RNA-guided transcriptional activator system. Cell Res 23:1163–1171

Cogoni C, Irelan JT, Schumacher M, Schmidhauser TJ, Selker EU, Macino G (1996) Transgene silencing of the al-1 gene in vegetative cells of Neurospora is mediated by a cytoplasmic effector and does not depend on DNA-DNA interactions or DNA methylation. EMBO J 15(12):3153–3163

Creasey KM, Zhai J, Borges F, Van Ex F, Regulski M et al (2014) miRNAs trigger widespread epigenetically activated siRNAs from transposons in *Arabidopsis*. Nature 508(7496):411–415

Davuluri GR, Van Tuinen A, Fraser PD, Manfredonia A, Newman R, Burgess et al (2005) Fruit-specific RNAi-mediated suppression of DET1 enhances carotenoid and flavonoid content in tomatoes. Nat Biotechnol 23:890–895

De Jong M, Wolters-Arts M, Feron R, Mariani C, Vriezen WH (2009) The Solanum lycopersicum auxin response factor 7 (SlARF7) regulates auxin signaling during tomato fruit set and development. Plant J 57:160–170

Dong Z, Han M-H, Fedoroff N (2008) The RNA-binding proteins HYL1 and SE promote accurate in vitro processing of pri-miRNA by DCL1. PNAS 105(29):9970–9975

Dubrovina AS, Aleynova OA, Kalachev AV, Suprun AR, Ogneva ZV, Kiselev KV (2019) Induction of transgene suppression in plants via external application of synthetic dsRNA. Int J Mol Sci 20:1585

Fang Y, Spector DL (2007) Identification of nuclear dicing bodies containing proteins for microRNA biogenesis in living *Arabidopsis* plants. Curr Biol 17(9):818–823

Fang X, Cui Y, Li Y, Qi Y (2015a) Transcription and processing of primary microRNAs are coupled by Elongator complex in *Arabidopsis*. Nat Plants 1:15075

Fang X, Shi Y, Lu X, Chen Z, Qi Y (2015b) CMA33/XCT regulates small RNA production through modulating the transcription of Dicer-like genes in *Arabidopsis*. Mol Plant 8(8):1227–1236

Fei Q, Xia R, Meyers BC (2013) Phased, secondary, small interfering RNAs in posttranscriptional regulatory networks. Plant Cell 25(7):2400–2415

Feldmann KA (2006) Steroid regulation improves crop yield. Nat Biotechnol 24:46–47

Fire A, Xu S, Montgomery MK, Kostas SA, Driver SE, Mello CC (1998) Potent and specific genetic interference by double-stranded RNA in Caenorhabditis elegans. Nature 391(6669):806–811

Franceschini A, Meier R, Casanova A, Kreibich S, Daga N, Andritschke D, Dilling S, Ramo P, Emmenlauer M, Kaufmann A et al (2014) Specific inhibition of diverse pathogens in human cells by synthetic microRNA-like oligonucleotides inferred from RNAi screens. Proc Natl Acad Sci USA 111:4548–4553

Frye M, Harada BT, Behm M, He C (2018) RNA modifications modulate gene expression during development. Science 361(6409):1346–1349

Fukudome A, Singh J, Mishra V, Reddem E, Martinez-Marquez F et al (2021) Structure and RNA template requirements of *Arabidopsis* RNA-DEPENDENT RNA POLYMERASE 2. PNAS 118(51):e2115899118

Gaj T, Gersbach CA, Barbas CF (2013) ZFN, TALEN, and CRISPR/Cas-based methods for genome engineering. Trends Biotechnol 31:397–405

Gebremichael DE, Haile ZM, Negrini F, Sabbadini S, Capriotti L, Mezzetti B, Baraldi E (2021) RNA interference strategies for future management of plant pathogenic fungi: prospects and challenges. Plan Theory 10:650

Gilbert LA, Larson MH, Morsut L, Liu Z, Brar GA, Torres SE, Stern-Ginossar N, Brandman O, Whitehead EH, Doudna JA et al (2013) CRISPR-mediated modular RNA-guided regulation of transcription in eukaryotes. Cell 154:442–451

Gonzalo L, Tossolini I, Gulanicz T, Cambiagno DA, Kasprowicz-Maluski A, Smolinski DJ, Mammarella MF, Ariel FD, Marquardt S, Szweykowska-Kulinska Z, Jarmolowski A (2022) R-loops at microRNA encoding loci promote co-transcriptional processing of pri-miRNAs in plants. Nat Plants 8(4):402–418

Han M-H, Goud S, Song L, Fedoroff N (2004) The *Arabidopsis* double-stranded RNA-binding protein HYL1 plays a role in microRNA-mediated gene regulation. PNAS 101(4):1093–1098

Grover JW, Burgess D, Kendall T, Baten A, Pokhrel S, King GJ, Meyers BC, Freeling M, Mosher RA (2020) Abundant expression of maternal siRNAs is a conserved feature of seed development. Proc Natl Acad Sci 117(26):15305–15315

Gupta A, Pal RK, Rajam MV (2013) Delayed ripening and improved fruit processing quality in tomato by RNAi-mediated silencing of three homologs of 1-aminopropane-1-carboxylate synthase gene. J Plant Physiol 170:987–995

Haag JR, Ream TS, Marasco M, Nicora CD, Norbeck AD, Pasa-Tolic L, Pikaard CS (2012) In vitro transcription activities of Pol IV, Pol V, and RDR2 reveal coupling of Pol IV and RDR2 for dsRNA synthesis in plant RNA silencing. Mol Cell 48(5):811–818

Hamilton AJ, Baulcombe DC (1999) A species of small antisense RNA in posttranscriptional gene silencing in plants. Science 286(5441):950–952

Havecker ER, Wallbridge LM, Hardcastle TJ, Bush MS, Kelly KA, Dunn RM, Schwach F, Doonan JH, Baulcombe DC (2010) The *Arabidopsis* RNA-directed DNA methylation argonautes functionally diverge based on their expression and interaction with target loci. Plant Cell 22(2):321–334

Herr AJ, Jensen MB, Dalmay T, Baulcombe DC (2005) RNA polymerase IV directs silencing of endogenous DNA. Science 308(5718):118–120

Housden BE, Perrimon N (2016) Comparing CRISPR and RNAi-based screening technologies. Nat Biotechnol 34(6):621–623

Huang J, Li Z, Zhao D (2006) Deregulation of the OsmiR160 target gene OsARF18 causes growth and developmental defects with an alteration of auxin signaling in rice. Sci Rep 6:1–14

Huang K, Wu XX, Fang CL, Xu ZG, Zhang HW, Gao J, Zhou CM, You LL, Gu ZX, Mu WH, Feng Y (2021) Pol IV and RDR2: a two-RNA-polymerase machine that produces double-stranded RNA. Science 374(6575):1579–1586

Ibrahim HM, Alkharouf NW, Meyer SL, Aly MA, Abd El Kader Y, Hussein EH, Matthews BF (2011) Post-transcriptional gene silencing of root-knot nematode in transformed soybean roots. Exp Parasitol 127(1):90–99

Jackson AL, Bartz SR, Schelter J, Kobayashi SV, Burchard J, Mao M, Li B, Cavet G, Linsley PS (2003) Expression profiling reveals off-target gene regulation by RNAi. Nat Biotechnol 21:635–637

Jiang CJ, Shimono M, Maeda S, Inoue H, Mori M, Hasegawa M, Sugano S, Takatsuji H (2009) Suppression of the rice fatty-acid desaturase gene OsSSI2 enhances resistance to blast and leaf blight diseases in rice. Mol Plant-Microbe Interact 22(7):820–829

Jiang Y, Hu Y, Wang B, Wu T (2014) Bivalent RNA interference to increase isoflavone biosynthesis in soybean (Glycine max). Braz Arch Biol Technol 57:163–170

Katiyar-Agarwal S, Morgan R, Dahlbeck D, Borsani O, Villegas A Jr, Zhu JK, Staskawicz BJ, Jin H (2006) A pathogen-inducible endogenous siRNA in plant immunity. Proc Natl Acad Sci 103(47):18002–18007

Khajuria C, Ivashuta S, Wiggins E, Flagel L, Moar W, Pleau M, Miller K, Zhang Y, Ramaseshadri P, Jiang C, Hodge T (2018) Development and characterization of the first dsRNA-resistant insect population from western corn rootworm, *Diabrotica virgifera* virgifera LeConte. PloS one 13(5):e0197059

Khan AA, Betel D, Miller ML, Sander C, Leslie CS, Marks DS (2009) Transfection of small RNAs globally perturbs gene regulation by endogenous microRNAs. Nat Biotechnol 27:549–555

Kleinstiver BP, Prew MS, Tsai SQ, Nguyen NT, Topkar VV, Zheng Z, Joung JK (2015) Broadening the targeting range of Staphylococcus aureus CRISPR-Cas9 by modifying PAM recognition. Nat Biotechnol 33(12):1293–1298

Knip M, Constantin ME, Thordal-Christensen H (2014) Trans-kingdom cross-talk: small RNAs on the move. PLoS Genet 10(9):e1004602

Knop K, Stepien A, Barciszewska-Pacak M, Taube M, Bielewicz D, Michalak M, Borst JW, Jarmolowski A, Szweykowska-Kulinska Z (2017) Active 5′ splice sites regulate the biogenesis efficiency of *Arabidopsis* microRNAs derived from intron-containing genes. Nucleic Acids Res 45(5):2757–2775

Koester T, Meyer K, Weinholdt C, Smith LM, Lummer M, Speth C, Grosse I, Weigel D, Staiger D (2014) Regulation of pri-miRNA processing by the hnRNP-like protein at GRP7 in *Arabidopsis*. Nucleic Acids Res 42(15):9925–9936

Kurihara Y, Watanabe Y (2004) *Arabidopsis* micro-RNA biogenesis through Dicer-like 1 protein functions. PNAS 101(34):12753–12758

Kurihara Y, Takashi Y, Watanabe Y (2006) The interaction between DCL1 and HYL1 is important for efficient and precise processing of pri-miRNA in plant microRNA biogenesis. RNA 12(2):206–212

Kusaba M, Miyahara K, Iida S, Fukuoka H, Takano T, Sassa H, Nishimura M, Nishio T (2003) Low glutelin content1: a dominant mutation that suppresses the glutelin multigene family via RNA silencing in rice. Plant Cell 15(6):1455–1467

Law JA, Vashisht AA, Wohlschlegel JA, Jacobsen SE (2011) SHH1, a homeodomain protein required for DNA methylation, as well as RDR2, RDM4, and chromatin remodeling factors, associate with RNA polymerase IV. PLoS Genet 7(7):e1002195

Law JA, Du J, Hale CJ, Feng S, Krajewski K, Palanca AMS, Strahl BD, Patel DJ, Jacobsen SE (2013) Polymerase IV occupancy at RNA-directed DNA methylation sites requires SHH1. Nature 498(7454):385–389

Lawrence RJ, Pikaard CS (2003) Transgene-induced RNA interference: a strategy for overcoming gene redundancy in polyploids to generate loss-of-function mutations. Plant J 36:114–121

Li J, Yang Z, Yu B, Liu J, Chen X (2005) Methylation protects miRNAs and siRNAs from a 3-end uridylation activity in *Arabidopsis*. Curr Biol 15(16):1501–1507

Li JC, Guo JB, Xu WZ, Ma M (2007) RNA interference-mediated silencing of phytochelatin synthase gene reduce cadmium accumulation in rice seeds. J Integr Plant Biol 49:1032–1037

Li DH, Hui LIU, Yang YL, Zhen PP, Liang JS (2009) Down-regulated expression of RACK1 gene by RNA interference enhances drought tolerance in rice. Rice Sci 16:14–20

Li Z, Wang S, Cheng J, Su C, Zhong S, Liu Q, Fang Y, Yu Y, Lv H, Zheng Y, Zheng B (2016) Intron lariat RNA inhibits microRNA biogenesis by sequestering the dicing complex in *Arabidopsis*. PLoS Genet 12(11):e1006422

Li S, Xu R, Li A, Liu K, Gu L, Li M, Zhang H, Zhang Y, Zhuang S, Wang Q, Gao G (2018) SMA1, a homolog of the splicing factor Prp28, has a multifaceted role in miRNA biogenesis in *Arabidopsis*. Nucleic Acids Res 46(17):9148–9159

Li C, Gent JI, Xu H, Fu H, Russell SD, Sundaresan V (2022) Resetting of the 24-nt siRNA landscape in rice zygotes. Genome Res 32(2):309–323

Liu HH, Tian X, Li YJ, Wu CA, Zheng CC (2008) Microarray-based analysis of stress-regulated microRNAs in *Arabidopsis* thaliana. RNA 14:836–843

Liu Y, Teng C, Xia R, Meyers BC (2020) PhasiRNAs in plants: their biogenesis, genic sources, and roles in stress responses, development, and reproduction. Plant Cell 32(10):3059–3080

Lobbes D, Rallapalli G, Schmidt DD, Martin C, Clarke J (2006) SERRATE: a new player on the plant microRNA scene. EMBO Rep 7(10):1052–1058

Lu C, Jeong DH, Kulkarni K, Pillay M, Nobuta K, German R, Thatcher SR, Maher C, Zhang L, Ware D, Liu B (2008) Genome-wide analysis for discovery of rice microRNAs reveals natural antisense microRNAs (nat-miRNAs). Proc Natl Acad Sci 105(12):4951–4956

Luan M, Xu M, Lu Y, Zhang L, Fan Y, Wang L (2015) Expression of zma-miR169 miRNAs and their target ZmNF-YA genes in response to abiotic stress in maize leaves. Gene 555:178–185

Lundgren JG, Duan JJ (2013) RNAi-based insecticidal crops: potential effects on nontarget species. Bioscience 63:657–665

Mamta B, Rajam MV (2017) RNAi technology: a new platform for crop pest control. Physiol Mol Biol Plants 23:487–501

Matsuo S, Miyatake K, Endo M, Urashimo S, Kawanishi T, Negoro S, Shimakoshi S, Fukuoka H (2020) Loss of function of the Pad-1 aminotransferase gene, which is involved in auxin homeostasis, induces parthenocarpy in Solanaceae plants. Proc Natl Acad Sci 117(23):12784–12790

McCue AD, Nuthikattu S, Reeder SH, Slotkin RK (2012) Gene expression and stress response mediated by the epigenetic regulation of a transposable element small RNA. PLoS Genet 8(2):e1002474

McCue AD, Panda K, Nuthikattu S, Choudury SG, Thomas EN, Slotkin RK (2015) ARGONAUTE 6 bridges transposable element mRNA-derived siRNAs to the establishment of DNA methylation. EMBO J 34(1):20–35

Meister G, Tuschl T (2004) Mechanisms of gene silencing by double-stranded RNA. Nature 431(7006):343–349

Mello CC, Conte D Jr (2004) Revealing the world of RNA interference. Nature 431(7006):338

Molesini B, Pandolfini T, Rotino GL, Dani V, Spena A (2009) Aucsia gene silencing causes parthenocarpic fruit development in tomato. J Plant Physiol 149:534–548

Moritoh S, Miki D, Akiyama M, Kawahara M, Izawa T, Maki H, Shimamoto K (2005) RNAi-mediated silencing of OsGEN-L (OsGEN-like), a new member of the RAD2/XPG nuclease family, causes male sterility by defect of microspore development in rice. Plant Cell Physiol 46(5):699–715

Mosher RA, Schwach F, Studholme D, Baulcombe DC (2008) PolIVb influences RNA-directed DNA methylation independently of its role in siRNA biogenesis. PNAS 105(8):3145–3150

Napoli C, Lemieux C, Jorgensen R (1990) Introduction of a chimeric chalcone synthase gene into petunia results in reversible co-suppression of homologous genes in trans. Plant Cell 2(4):279–289

Ni Z, Hu Z, Jiang Q, Zhang H (2013) GmNFYA3, a target gene of miR169, is a positive regulator of plant tolerance to drought stress. Plant Mol Biol 82:113–129

Nuthikattu S, McCue AD, Panda K, Fultz D, DeFraia C, Thomas EN, Slotkin RK (2013) The initiation of epigenetic silencing of active transposable elements is triggered by RDR6 and 21-22 nucleotide small interfering RNAs. Plant Physiol 162(1):116–131

Onodera Y, Haag JR, Ream T, Nunes PC, Pontes O, Pikaard CS (2005) Plant nuclear RNA polymerase IV mediates siRNA and DNA methylation-dependent heterochromatin formation. Cell 120(5):613–622

Pak J, Fire A (2007) Distinct populations of primary and secondary effectors during RNAi in C. Elegans. Science 315(5809):241–244

Pandit NN, Russo VEA (1992) Reversible inactivation of a foreign gene, hph, during the asexual cycle in Neurospora crassa transformants. MGG Mol General Genetics 234(3):412–422

Pareek M, Rajam MV (2017) RNAi-mediated silencing of MAP kinase signalling genes (Fmk1, Hog1, and Pbs2) in fusarium oxysporum reduces pathogenesis on tomato plants. Fungal Biol 121:775–784

Park MY, Wu G, Gonzalez-Sulser A, Vaucheret H, Poethig RS (2005) Nuclear processing and export of microRNAs in *Arabidopsis*. PNAS 102(10):3691–3696

Peltier AJ, Hatfield RD, Grau CR (2009) Soybean stem lignin concentration relates to resistance to Sclerotinia sclerotiorum. Plant Dis 93:149–154

Qiao F, Yang Q, Wang CL, Fan YL, Wu XF, Zhao KJ (2007) Modification of plant height via RNAi suppression of OsGA20ox2 gene in rice. Euphytica 158:35–45

Rajagopalan R, Vaucheret H, Trejo J, Bartel DP (2006) A diverse and evolutionarily fluid set of microRNAs in *Arabidopsis* thaliana. Genes Dev 20(24):3407–3425

Reis RS, Poirier Y (2021) Making sense of the natural antisense transcript puzzle. Trends Plant Sci 26(11):1104–1115

Rodrigues JA, Ruan R, Nishimura T, Sharma MK, Sharma R, Ronald PC, Fischer RL, Zilberman D (2013) Imprinted expression of genes and small RNA is associated with localized hypomethylation of the maternal genome in rice endosperm. Proc Natl Acad Sci 110(19):7934–7939

Romano N, Macino G (1992) Quelling: transient inactivation of gene expression in Neurospora crassa by transformation with homologous sequences. Mol Microbiol 6(22):3343–3353

Ron M, Alandete Saez M, Eshed Williams L, Fletcher JC, McCormick S (2010) Proper regulation of a sperm-specific cis-nat-siRNA is essential for double fertilization in *Arabidopsis*. Genes Dev 24(10):1010–1021

Sharma S, Kumar A, Kumari N, Walia A (2023) RNAi as a tool to enhance crop yield and biotic stress management in the plants. Plant Cell Tissue Organ Cult 152:437–454

Sijen T, Steiner FA, Thijssen KL, Plasterk RH (2007) RETRACTED: secondary siRNAs result from unprimed RNA synthesis and form a distinct class. Science 315(5809):244–247

Singh J, Mishra V, Wang F, Huang H-Y, Pikaard CS (2019) Reaction mechanisms of pol IV, RDR2, and DCL3 drive RNA channeling in the siRNA-directed DNA methylation pathway. Mol Cell 75(3):576–89.e5

Slotkin RK, Vaughn M, Borges F, Tanurdžić M, Becker JD, Feijó JA, Martienssen RA (2009) Epigenetic reprogramming and small RNA silencing of transposable elements in pollen. Cell 136(3):461–472

Song J, Wang X, Song B, Gao L, Mo X, Yue L, Yang H, Lu J, Ren G, Mo B, Chen X (2019) Prevalent cytidylation and uridylation of precursor miRNAs in *Arabidopsis*. Nat Plants 5(12):1260–1272

Sun G, Stewart CN Jr, Xiao P, Zhang B (2012) MicroRNA expression analysis in the cellulosic biofuel crop switchgrass (Panicum virgatum) under abiotic stress. PLoS One 7:e32017

Sun Z, Li M, Zhou Y, Guo T, Liu Y, Zhang H, Fang Y (2018) Coordinated regulation of *Arabidopsis* microRNA biogenesis and red light signaling through Dicer-like 1 and phytochrome-interacting factor 4. PLoS Genet 14(3):e1007247

Swiezewski S, Crevillen P, Liu F, Ecker JR, Jerzmanowski A, Dean C (2007) Small RNA-mediated chromatin silencing directed to the 3 region of the *Arabidopsis* gene encoding the developmental regulator, FLC. PNAS 104:3633–3638

Szarzynska B, Sobkowiak L, Pant BD, Balazadeh S, Scheible WR, Mueller-Roeber B, Jarmolowski A, Szweykowska-Kulinska Z (2009) Gene structures and processing of *Arabidopsis* thaliana HYL1-dependent pri-miRNAs. Nucleic Acids Res 37(9):3083–3093

Tao L, Yu G, Chen H, Wang B, Jiang L, Han X, Lin G, Cheng XG (2022) SlDREB2 gene specifically recognizing to the universal DRE elements is a transcriptional activator improving drought tolerance in tomato. Sci Hortic 295:110887

Teresa RM, Voinnet O, Baulcombe DC (1998) Initiation and maintenance of virus-induced gene silencing. Plant Cell 10(6):937–946

Thakare D, Zhang J, Wing RA, Cotty PJ, Schmidt MA (2017) Aflatoxin-free transgenic maize using host-induced gene silencing. Sci Adv 3:e1602382

Treiber T, Treiber N, Meister G (2019) Regulation of microRNA biogenesis and its crosstalk with other cellular pathways. Nat Rev Mol Cell Biol 20(1):5–20

Tsuji H, Aya K, Ueguchi-Tanaka M, Shimada Y, Nakazono M, Watanabe R, Nishizawa NK, Gomi K, Shimada A, Kitano H, Ashikari M (2006) GAMYB controls different sets of genes and is differentially regulated by microRNA in aleurone cells and anthers. Plant J 47(3):427–444

Van Der Krol AR, Mur LA, Beld M, Mol JNM, Stuitje AR (1990) Flavonoid genes in petunia: addition of a limited number of gene copies may lead to a suppression of gene expression. Plant Cell 2(4):291–299

Van VT, Choudhury NR, Mukherjee SK (2013) Transgenic tomato plants expressing artificial microRNAs for silencing the pre-coat and coat proteins of a begomovirus, tomato leaf curl New Delhi virus, show tolerance to virus infection. Virus Res 172:35–45

Vazquez F, Gasciolli V, Crete P, Vaucheret H (2004) The nuclear dsRNA binding protein HYL1 is required for microRNA accumulation and plant development, but not posttranscriptional transgene silencing. Curr Biol 14(4):346–351

Wan Q, Guan X, Yang N, Wu H, Pan M, Liu B, Fang L, Yang S, Hu Y, Ye W, Zhang H (2016) Small interfering RNA s from bidirectional transcripts of Gh MML 3_A12 regulate cotton fiber development. New Phytol 210(4):1298–1310

Wang S, Wu K, Yuan Q, Liu X, Liu Z, Lin X, Zeng R, Zhu H, Dong G, Qian Q, Zhang G (2012) Control of grain size, shape and quality by OsSPL16 in rice. Nat Genet 44(8):950–954

Wang L, Song X, Gu L, Li X, Cao S, Chu C, Cui X, Chen X, Cao X (2013) NOT2 proteins promote polymerase II–dependent transcription and interact with multiple microRNA biogenesis factors in *Arabidopsis*. Plant Cell 25(2):715–727

Wang S, Quan L, Li S, You C, Zhang Y, Gao L, Zeng L, Liu L, Qi Y, Mo B, Chen X (2019) The PROTEIN PHOSPHATASE4 complex promotes transcription and processing of primary microRNAs in *Arabidopsis*. Plant Cell 31(2):486–501

Wei L, Zhang D, Xiang F, Zhang Z (2009) Differentially expressed miRNAs potentially involved in the regulation of defense mechanism to drought stress in maize seedlings. Int J Plant Sci 170:979–989

Weise SE, Aung K, Jarou ZJ, Mehrshahi P, Li Z, Hardy AC, Carr DJ, Sharkey TD (2012) Engineering starch accumulation by manipulation of phosphate metabolism of starch. Plant Biotechnol J 10(5):545–554

Wu G, Park MY, Conway SR, Wang JW, Weigel D, Poethig RS (2009) The sequential action of miR156 and miR172 regulates developmental timing in *Arabidopsis*. Cell 138(4):750–759

Wu L, Zhou H, Zhang Q, Zhang J, Ni F, Liu C, Qi Y (2010) DNA methylation mediated by a microRNA pathway. Mol Cell 38(3):465–475

Xie Z, Kasschau KD, Carrington JC (2003) Negative feedback regulation of Dicer-Like1 in *Arabidopsis* by microRNA-guided mRNA degradation. Curr Biol 13(9):784–789

Xie Z, Johansen LK, Gustafson AM, Kasschau KD, Lellis AD, Zilberman D, Jacobsen SE, Carrington JC (2004) Genetic and functional diversification of small RNA pathways in plants. PLoS Biol 2(5):e104

Xu M, Hu T, Smith MR, Poethig RS (2016) Epigenetic regulation of vegetative phase change in *Arabidopsis*. Plant Cell 28(1):28–41

Yan K, Liu P, Wu CA, Yang GD, Xu R, Guo QH, Huang JG, Zheng CC (2012) Stress-induced alternative splicing provides a mechanism for the regulation of microRNA processing in *Arabidopsis* thaliana. Mol Cell 48(4):521–531

Yang Z, Ebright YW, Yu B, Chen X (2006a) HEN1 recognizes 21–24 nt small RNA duplexes and deposits a methyl group onto the 2 OH of the 3 terminal nucleotide. Nucleic Acids Res 34(2):667–675

Yang L, Liu Z, Lu F, Dong A, Huang H (2006b) SERRATE is a novel nuclear regulator in primary microRNA processing in *Arabidopsis*. Plant J 47(6):841–850

Yang GD, Yan K, Wu BJ, Wang YH, Gao YX, Zheng CC (2012) Genomewide analysis of intronic microRNAs in rice and *Arabidopsis*. J Genet 91(3):313–324

Yang X, Wang J, Dai Z, Zhao X, Miao X, Shi Z (2019) miR156f integrates panicle architecture through genetic modulation of branch number and pedicel length pathways. Rice 12:1–11

Yara A, Yaeno T, Hasegawa M, Seto H, Montillet JL, Kusumi K, Seo S, Iba K (2007) Disease resistance against Magnaporthe grisea is enhanced in transgenic rice with suppression of ω-3 fatty acid desaturases. Plant Cell Physiol 48(9):1263–1274

Yu B, Yang Z, Li J, Minakhina S, Yang M, Padgett RW, Steward R, Chen X (2005) Methylation as a crucial step in plant microRNA biogenesis. Science 307(5711):932–935

Yu Y, Ji L, Le BH, Zhai J, Chen J, Luscher E, Gao L, Liu C, Cao X, Mo B, Ma J (2017) ARGONAUTE10 promotes the degradation of miR165/6 through the SDN1 and SDN2 exonucleases in *Arabidopsis*. PLoS Biol 15(2):e2001272

Zha W, Peng X, Chen R, Du B, Zhu L, He G (2011) Knockdown of midgut genes by dsRNA-transgenic plant-mediated RNA interference in the hemipteran insect *Nilaparvata lugens*. PLoS One 6:e20504

Zhai J, Bischof S, Wang H, Feng S, Lee TF, Teng C, Chen X, Park SY, Liu L, Gallego-Bartolome J, Liu W (2015) A one precursor one siRNA model for pol IV-dependent siRNA biogenesis. Cell 163(2):445–455

Zhang X, Henderson IR, Lu C, Green PJ, Jacobsen SE (2007) Role of RNA polymerase IV in plant small RNA metabolism. PNAS 104(11):4536–4541

Zhang X, Xia J, Lii YE, Barrera-Figueroa BE, Zhou X, Gao S, Lu L, Niu D, Chen Z, Leung C, Wong T (2012) Genome-wide analysis of plant nat-siRNAs reveals insights into their distribution, biogenesis and function. Gen Biol 13:1–16

Zhang B, You C, Zhang Y, Zeng L, Hu J, Zhao M, Chen X (2020) Linking key steps of microRNA biogenesis by TREX-2 and the nuclear pore complex in *Arabidopsis*. Nat Plants 6(8):957–969

Zhou M, Coruh C, Xu G, Martins LM, Bourbousse C, Lambolez A, Law JA (2022) The CLASSY family controls tissue-specific DNA methylation patterns in *Arabidopsis*. Nat Commun 13(1):244

Zhu KY, Palli SR (2020) Mechanisms, applications, and challenges of insect RNA interference. Annu Rev Entomol 65:293–311

Zhu H, Zhou Y, Castillo-González C, Lu A, Ge C, Zhao YT, Duan L, Li Z, Axtell MJ, Wang XJ, Zhang X (2013) Bidirectional processing of pri-miRNAs with branched terminal loops by *Arabidopsis* Dicer-like1. Nat Struct Mol Biol 20(9):1106–1115

Zilberman D, Cao X, Jacobsen SE (2003) ARGONAUTE4 control of locus-specific siRNA accumulation and DNA and histone methylation. Science 299(5607):716–719

Zulfiqar S, Farooq MA, Zhao T, Wang P, Tabusam J, Wang Y, Xuan S, Zhao J, Chen X, Shen S, Gu A (2023) Virus-induced gene silencing (VIGS): a powerful tool for crop improvement and its advancement towards epigenetics. Int J Mol Sci 24(6):5608

Understanding Plant ncRNAs and Transcriptional Gene Regulation in Plant

2

Sharad Kumar Jaiswal, Shivani Singh,
and Budhayash Gautam

Abstract

The intricate world of biological processes, where gene expression is regulated like a finely tuned symphony, depends heavily on noncoding RNAs (ncRNAs). In addition to their well-known roles as messenger, ribosomal, and transfer RNAs, the science of plant biology has recognized the important regulatory role played by diverse ncRNAs. These ncRNAs act as riboregulators, contributing to the control of plant growth, development, and stress response by specifically targeting nucleic acids through homologous sequence interactions. Regulatory ncRNAs vary in length, encompassing short and long noncoding RNAs (lncRNAs), and they modulate a range of biological processes. Over time, ncRNAs have evolved into various forms based on their biogenesis, including microRNAs (miRNAs), small interfering RNAs (siRNAs), miRNA variations (isomiRs), long noncoding RNAs, circular RNAs (cirRNAs), and ncRNA derivatives. This article explores the biogenesis, different types of ncRNAs, databases, and prediction tool in plant growth. It also investigates how regulatory ncRNAs can be utilized to enhance crops with essential agricultural traits. Molecular biology has undergone a revolutionary transformation because of recent developments in high-throughput transcriptome sequencing and expression analysis, made possible by next-generation sequencing (NGS) technology. Despite the complexity arising from coordinated cross-talk and commonalities in ncRNA

S. K. Jaiswal · B. Gautam (✉)
Department of Computational Biology and Bioinformatics, Jacob Institute of Biotechnology & Bioengineering, Sam Higginbotton University of Agriculture, Technology & Sciences, Prayagraj, Uttar Pradesh, India
e-mail: budhayash.gautam@shiats.edu.in

S. Singh
Department of Biological Sciences, Faculty of Science, Sam Higginbotton University of Agriculture, Technology & Sciences, Prayagraj, Uttar Pradesh, India

K. Goswami et al. (eds.), *Non-coding RNAs for Crop Improvement*,
https://doi.org/10.1007/978-981-97-7213-1_2

synthesis, this article summarizes our present understanding of plant miRNAs and lncRNAs, emphasizing their genesis, synthesis, methods of action, and critical role in plant responses. It is known that several ncRNAs regulate metabolism as well as the growth and development of plants. The authors of this study conducted a comprehensive analysis of online resources, including genome databases that incorporate ncRNA databases, to improve accessibility and understanding of ncRNA web resources for plant research.

Keywords

Noncoding RNA · Stress response · Transcriptional gene regulation · Plants

2.1 Introduction

Since they are sessile organisms, plants have evolved sophisticated regulatory systems to allow for dynamic adaptation to shifting environmental conditions. Among the coding and noncoding transcripts, ncRNAs have become important regulators that play a major role in the controlled expression of genes in plants. Plant ncRNAs are fascinating molecules that play a vital role in coordinating transcriptional events that control plant growth, development, and adaptation. This chapter explores this fascinating terrain. Abiotic stressors, such as salt, cold, heat, drought, and hypoxia, have a detrimental impact on plant development, productivity, and overall yield, leading to significant losses (Hirayama and Shinozaki 2010). Food security is threatened by a number of causes that worsen these problems, including the growing global population, urbanization that reduces arable land, changes in soil quality, and global warming (Mittler 2006). ncRNAs have emerged as important physiologically active molecules that contribute to the diversity of the genome and phenotype. Only about 2% of the eukaryotic genome's transcripts actually encode proteins, despite the fact that up to 90% of it is translated into RNA (Pauli et al. 2011; Rai et al. 2019). ncRNA transcripts lacking the ability to code for proteins and having poorly conserved sequences were once thought to be transcriptional noise, but they are now acknowledged as important parts of the residual transcriptome (Pauli et al. 2011; Ariel et al. 2015). Despite the difficulties involved, high-throughput sequencing analysis and experimental validation have demonstrated the involvement of ncRNAs in multiple levels of gene regulation, including epigenetic, transcriptional, and posttranscriptional events (Cech and Steitz 2014). Along with homology-based and experimental methods, the advent of advanced next-generation sequencing (NGS) technology has aided in the finding of ncRNAs in plants (Shimoni et al. 2007). Cultivated plants undergo a number of regulatory processes in order to maintain a constant level of productivity, and ncRNAs are frequently crucial parts of these networks. Recent developments in deep sequencing technology have facilitated the study of ncRNAs in molecular biology. The first characterized ncRNA in baker's yeast was alanine tRNA (Holley et al. 1965). The discovery of RNA's catalytic function in the 1980s offered researchers a new perspective on the intricate roles played by diverse ncRNAs (Mattick and Rinn 2015). Subsequently, the detection of regulatory ncRNA activity in *Caenorhabditis elegans* enabled the finding of several types of ncRNAs in humans, animals, and plants (Lee et al. 1993; Chen 2009; Fan et al. 2016). Technological advancements have further enabled the classification of various regulatory ncRNAs that modulate the expression of protein-coding genes through interactions with diverse molecular pathways in cellular processes. However, compared to mammals, our understanding of plant ncRNAs is relatively limited. This chapter provides an updated overview of the multifaceted regulatory functions of ncRNAs, including ncRNA variants/isoforms, circRNAs, and ncRNA derivatives in plant metabolism, stress tolerance, growth, and development. Moreover, it highlights examples of how ncRNAs govern the development of plant phenotypes with improved agronomic features, along with potential applications of this knowledge in crop enhancement programs (Xu et al. 2023).

2.1.1 Noncoding RNAs: Their Origin and Biogenesis

Nie et al. (2012) have highlighted the diverse origins and functions of ncRNAs. Several hypotheses, including duplication, evolution from pre-existing transposable elements (TEs), random hairpin structure formation, pseudogenization of protein-coding sequences, RNA virus replication, and the production of double-stranded RNA (dsRNA) from heterochromatin regions and DNA repeats, have been put forth to explain the formation of viruses. These mechanisms give rise to various types of ncRNAs, including siRNA, piRNA, and lncRNA.

The biogenesis of ncRNAs occurs through two primary pathways: canonical and noncanonical. Canonical pathways involve the synthesis of ncRNAs through conventional processes, while noncanonical pathways follow unconventional or irregular routes (Nie et al. 2012). Structural ncRNAs encompass ribosomal RNA (rRNA), transfer RNA (tRNA), small nuclear RNA (snRNA), and small nucleolar RNA (snoRNA). Regulatory ncRNAs can be broadly categorized into long ncRNAs (lncRNAs, >200 nt) and small ncRNAs (snoRNA, 18–30 nt) (Table 2.1). Other regulatory ncRNAs, such as circular RNAs (cirRNAs) and ncRNA derivatives, have also demonstrated involvement in various plant activities as revealed in multiple studies (Baev et al. 2014; Chu et al. 2018).

2.1.1.1 miRNAs

Plants have a well-defined mechanism for the synthesis of microRNAs (miRNAs), which involves numerous important proteins and processing processes (Borges and Martienssen 2015). This mechanism is essential for controlling gene expression and is largely conserved throughout plant species. Below is a summary of the key processes in plant miRNA biogenesis.

Table 2.1 Classification of noncoding RNAs: different classes and subclasses of structural and regulatory noncoding RNAs

Category	Size range	Example	Function
Housekeeping ncRNAs	Varies	• tRNAs (transfer RNA) • rRNAs (ribosomal RNA)	Essential cellular processes (protein translation, ribosome structure)
Regulatory ncRNAs			Regulate gene expression and other cellular functions
* small ncRNAs (<200 nt) *			
Short regulatory RNAs	<50 nt	• piRNAs (Piwi interacting RNAs) • siRNAs (small interfering RNAs) miRNAs (MicroRNAs)	RNA silencing, gene regulation
RNA polymerase transcripts	50–500 nt	• snRNAs (small nuclear RNAs) • snoRNAs (small nucleolar RNAs)	RNA processing, modification
* long ncRNAs (>200 nt) *			
Long noncoding RNAs (lncRNAs)	>500 nt	• Xist • HOTAIR	Diverse regulatory functions (transcription, chromatin modification)

1. **Transcription of miRNA Genes:** RNA polymerase II normally transcribes miRNA genes, resulting in lengthy primary miRNA transcripts, or pr-miRNAs. These pri-miRNAs can be found inside the introns of genes that code for proteins or they can function as separate transcriptional units.
2. **Processing of Pri-miRNAs by DCL1:** An enzyme known as DICER-LIKE 1 (DCL1), a member of the plant Dicer protein family, processes the main transcripts nuclearly. The precursor miRNA (pre-miRNA) hairpin structure is created when DCL1 cleaves the pri-miRNA (Rogers and Chen 2013).
3. **Export of Pre-miRNAs to the Cytoplasm:** The pre-miRNAs are exported from the nucleus to the cytoplasm by the exportin protein HASTY.
4. **Dicing of Pre-miRNAs by DCL1 in the Cytoplasm:** After entering the cytoplasm, DCL1 converts the pre-miRNA into a brief duplex of double-stranded RNA. The mature miRNA and its complementary strand, referred to as the miRNA* (miRNA-star), make up this duplex (Rogers and Chen 2013).
5. **Loading of miRNA into AGO1 Protein:** After that, the mature miRNA joins the AGO1 protein-containing RNA-induced silencing complex (RISC). The main effector protein in charge of mediating miRNA-mediated gene silencing is called AGO1 (Borges and Martienssen 2015; Budak et al. 2015).
6. **Target mRNA Recognition and Cleavage:** The mature miRNA is guided by the RISC complex through base pairing to its target messenger RNA (mRNA) (Borges and Martienssen 2015; Alptekin et al. 2017). Inducing cleavage or blocking translation of the target mRNA is possible when the miRNA binds to it with adequate complementarity.

7. **Degradation of miRNA* Strand:** Usually, the complementary strand of the mature miRNA, known as the miRNA* strand, is broken down. On the other hand, the miRNA* strand might occasionally be functional and aid in the regulation of genes.

It is significant to keep in mind that plants have developed specialized machinery for miRNA biogenesis and that this machinery includes elements like DCL1 and AGO1, which are essential to the process (Borges and Martienssen 2015; Alptekin et al. 2017). Furthermore, miRNAs have a variety of regulatory functions in plants and can affect a range of biological processes, such as defense mechanisms, development, and stress response. Plant miRNA biogenesis research continues to shed light on the complex regulatory systems controlling plant gene expression.

2.1.1.2 siRNAs

In contrast to miRNAs, siRNAs can originate from either exogenous or endogenous sources. Endogenous siRNAs are generated from complementary long double-stranded RNAs through the activity of RNA-DEPENDENT RNA (RDR) Pol followed by cleavage by DCL1 proteins into siRNAs. Borges and Martienssen (2015) identified several classes of endogenous siRNAs in plants, including heterochromatic siRNAs (hcsiRNAs) and secondary siRNAs such as natural antisense siRNAs (nat-siRNAs or nat-siRNAs), phased siRNAs, and transacting siRNAs (ta-siRNAs). To exert posttranscriptional gene regulation through RNA interference (RNAi), pairs of perfectly complementary transcripts called natsiRNAs are produced by the Pol II/Pol IV, RDR2, and DCL1/DCL3-dependent pathways (Borges and Martienssen 2015). The precursors of tasiRNAs and phasiRNAs are generated by Pol II transcription from noncoding loci and protein-coding genes, respectively. These precursors originate from cleavage fragments of target miRNA transcripts. RDR6 and SUPPRESSOR OF GENE SILENCING3 (SGS3) produce single-stranded RNAs (ssRNAs) and subsequently double-stranded RNAs (dsRNAs) when the miRNA-mediated AGO1/AGO7 complex cleaves the transcripts. DCL2 or DCL4 convert the dsRNAs into 21 or 24 nucleotide siRNAs, which are then loaded into AGO1/AGO4. The target is cleaved upon siRNA binding. PhasiRNAs can be activated by either a "one-hit" (one binding site per target) or "two-hit" (two binding sites per target) model (Fei et al. 2013). HcsiRNAs, ranging from 23 to 24 nucleotides, utilize the RNA-directed DNA methylation (RdDM) pathway to guide DNA and/or histone methylation for transcriptional gene silencing (TGS) (Mattick and Rinn 2015). HcsiRNAs, derived from transposable elements (TE) and hc region repeats, are produced under the control of Pol IV and CLASSY1. These elements are then processed by RDR2 and DCL3.

2.1.1.3 LncRNAs

Long noncoding RNAs (lncRNAs) in plants are produced through a variety of transcriptional and posttranscriptional mechanisms. Longer than 200 nucleotides, RNA molecules known as lncRNAs are involved in a variety of processes related to the control of gene expression but do not code for proteins. An outline of lncRNA biogenesis in plants is provided.

1. **Unconventional Genomic Regions:** lncRNAs can originate from various genomic regions, including intergenic regions, enhancers, introns, antisense strands, and pseudogenes. Unlike protein-coding genes with well-defined transcriptional start sites, lncRNA transcription may initiate from diverse locations within these regions (Fei et al. 2013).
2. **Diverse Transcriptional Mechanisms:** The transcription of lncRNAs may not strictly follow canonical promoter sequences or transcription factor binding sites. They can be transcribed by RNA polymerase II (Pol II) or even RNA polymerase III (Pol III) in some cases, and their regulation might involve unique regulatory elements (Borges and Martienssen 2015).
   (a) **Transcriptional Initiation:** RNA Polymerase II is the enzyme that transcribes the majority of plant lncRNAs. It is also the enzyme that transcribes protein-coding genes. Unique Promoter Regions: Specific transcription factors control the transcription of lncRNA genes, which frequently have unique promoter regions.
   (b) **LncRNA Precursor Processing:** Primary transcript formation (transcription) is to create primary transcripts known as precursor lncRNAs (pre-lncRNAs), RNA Polymerase II transcribes lncRNA genes.
   (c) **Capping, Splicing, and Polyadenylation:** Unlike mRNAs that typically undergo extensive splicing and polyadenylation, many lncRNAs might lack splicing or polyadenylation. Some lncRNAs may undergo alternative splicing, but many retain intronic sequences or lack a poly (A) tail (Wierzbicki et al. 2008, 2009). Alternative processing mechanisms: lncRNAs can undergo diverse processing mechanisms. Some may undergo exonucleolytic processing or alternative cleavage and polyadenylation events, resulting in noncanonical transcript maturation pathways.
   (d) **Nuclear Export: Exportin-Mediated Transport:** Through the action of exportin proteins, mature or pre-lncRNAs can be transported from the nucleus to the cytoplasm.

2.1.1.4 CircRNA

Circular RNAs (circRNAs) are generated through a unique process called back-splicing, which involves a noncanonical splicing mechanism differing from the linear splicing of canonical messenger RNAs (mRNAs) (Cai et al. 2018).

1. **Back-splicing in circRNA Biogenesis:** circRNAs are formed when a downstream 5′ splice site is joined to an upstream 3′ splice site, resulting in a covalently closed circular structure. This process, known as back-splicing, occurs when a spliceosome joins a downstream splice donor site to an upstream splice acceptor site, forming a circular transcript instead of the linear mRNA (Cai et al. 2018).
2. **Formation of Circular Structure:** The back-splicing event creates a circular RNA molecule with a covalently closed loop structure, devoid of free ends. This circular conformation makes circRNAs resistant to exonucleases and contributes to their stability compared to linear RNAs.

2.2 Influence of Noncoding RNAs on the Regulation of Plant Genes

sRNAs, such as miRNAs and siRNAs, play a crucial role in controlling gene expression through various mechanisms, including target mRNA cleavage, translational repression, and transcriptional silencing (Nie et al. 2012). In a similar manner, lncRNAs also contribute to gene expression regulation by modifying chromatin states, governing mRNA translation, stability, and interacting with proteins and miRNAs (Komiya et al. 2014).

2.2.1 Target Transcript Cleavage by siRNAs and miRNAs

The process of suppressing target mRNA cleavage is dependent on the binding of miRNAs and/or siRNAs to complementary sites on the mRNA molecules, which provides sequence-specificity (Bartel 2004). In the case of miRNAs, the DCL1, HYL1, and SE complex processes MIR genes into mature miRNA-miRNA* duplexes (Rogers and Chen 2013). The HASTY protein carries these duplexes into the cytoplasm, where they attach to the AGO1-containing RISC complex. When the target RNA transcript's sense sequence attaches to the complementary sites, the RISC-AGO1 complex cleaves the transcript. In the RISC complex, the antisense microRNA strand continues to be present. Conversely, siRNAs are designed to cleave target gene transcripts with the help of the RISC-AGO1 or AGO7 complex, with a specific focus on nucleotides 10–11 upstream of the antisense strand's 5′ ends (Liu and El-Kassaby 2017). Both the 3′ and 5′ fragments are subsequently degraded by the EXONUCLEASE 4 (XRN4) enzyme. Recent studies have shown that AGO2, AGO4, and AGO10 in plants also interact with RNA and exhibit reduced activity, indicating the complexity of the sRNA-induced gene silencing process (Maunoury and Vaucheret 2011).

2.2.2 miRNAs and siRNAs Induced Translational Inhibition of Target Genes

In plants, AGO1 and AGO10 promote the mechanism of translation inhibition when sRNAs do not properly match with the target mRNA (Brodersen et al. 2008, b). The extent of inhibition is influenced by the number of microRNA binding sites. The RISC-AGO1 complex hinders ribosome recruitment or movement by binding to the 5′ untranslated region (UTR) or open reading frame (ORF) of the target gene (Iwakawa and Tomari 2015). Additional factors such as GW-repeat protein, VARICOSE (VCS), and the microtubule enzyme KATANIN (KTN1) are involved in regulating the process of translation inhibition (Yang et al. 2012a, b). Despite the involvement of both AGO1 and AGO10, the specific roles of each AGO gene in this inhibitory mechanism remain unclear (Yang et al. 2012a, b). Further research is necessary to fully understand the underlying mechanism of sRNA-mediated translational repression and how the repressed target mRNAs avoid endonucleolytic cleavage.

2.2.3 Regulatory Roles of miRNAs and siRNAs in Suppressing Target Gene Translation

In plants, when sRNAs match the target mRNA, AGO1 and AGO10 facilitate the translation inhibition mechanism (Brodersen et al. 2008, b). The number of miRNA binding sites affects the degree of inhibition. By attaching itself to the target gene's 5′ untranslated region (UTR) or open reading frame (ORF), the RISC-AGO1 complex prevents ribosome recruitment or movement (Fei et al. 2013). Additional factors such as GW-repeat protein, VARICOSE (VCS), and the microtubule enzyme KATANIN 1 (KTN1) are involved in regulating the process of translation inhibition (Yang et al. 2012a, b).

2.2.4 miRNA and siRNA-Mediated DNA Methylation Regulation

The RdDM pathway typically utilizes 24-nt siRNAs (hcsirnas) to repress gene expression (Creasey et al. 2014). These hcsirnas, in the context of CG, CHG, and CHH sequences, facilitate cytosine methylation when targeting heterochromatic regions. AGO4 and DCL3-dependent miRNAs form a complex that leads to gene silencing through cytosine and histone methylation (Ye et al. 2012). *Arabidopsis* possesses multiple copies of the DCL family, which play a role in synthesizing mRNAs of varying lengths. DCL1 is essential for converting partly paired dsRNA precursors into mature 21 nt miRNAs. Conversely, DCL2 and DCL4 start the process of creating 20–22 nt siRNAs from precursors of fully complementary dsRNA. On the other hand, Pol IV and V, RDR, DCL, AGO, and Pol IV and V work collectively to initiate the de novo hc-siRNA-induced RdDM process by enhancing methylation at target sites and transcribed double-stranded precursors (Matzke et al. 2009). Thus, transcriptional gene silencing happens when hc-siRNA/AGO4 RISC directs DNA and H3K9 methyltransferases to the target region, whereas systemic silencing is caused by DNA methylation and methylation of lysine at the ninth position of histone H3 (H3K9).

2.3 Regulation of Gene Expression by lncRNAs

Although the specific molecular mechanisms behind lncRNAs' functions in plants are still unknown, they associate RNA with proteins and operate as transcriptional activators or repressors to control gene expression (Goodrich and Kugel 2006). This section focuses on the processes that control lncRNA genes, which have been characterized by variations in transcriptional and posttranscriptional expression levels in plants. Plant lncRNAs can be categorized as cis-acting or transacting. Cis-acting lncRNAs exert their influence directly on neighboring genes by affecting local nucleotide sequences or chromosomal regions in close proximity to their site of synthesis. On the other hand, transacting lncRNAs can act on multiple genes across different chromosomal locations, independent of their site of synthesis (Lee 2012). Moreover, lncRNAs can act as sRNA precursors. For instance, some lncRNAs have the ability to combine with natural antisense transcripts (NATs) to generate double-stranded RNA duplexes, which then act as sRNAs to perform regulatory tasks. With the help of endogenous siRNA, the pentatricopeptide repeat gene and the Rab2-like gene in *Arabidopsis* generate a NAT pair from their complementary area (Katiyar-Agarwal et al. 2006).

When lncRNAs compete with target mRNAs for miRNA binding, they disrupt miRNA activity and reduce repression of the target gene. LncRNAs have the ability to mimic miRNA targets. For example, it has been demonstrated that the Induced by Phosphate Starvation 1 (IPS1) lncRNA in *Arabidopsis* mimics miRNA399 when the plant is phosphate-deficient (Aung et al. 2006). Furthermore, epigenetic regulatory mechanisms such as DNA methylation, histone modification, and chromatin remodeling, as well as protein-protein interactions, changes, and localization, are frequently used by lncRNAs to carry out their functions. Effective regulators of DNA methylation, lncRNAs share characteristics with DNA methylation in terms of strand-specific distribution and expression patterns. Examples include the COOLAIR and COLDAIR lncRNAs, which alter chromatin during vernalization to suppress FLOWERING LOCUS C (FLC) expression (Kim et al. 2017).

2.4 Function of ncRNA in Response to Abiotic Stress in Plants

Plants have developed various systems to regulate their growth and development in response to environmental stress. Recent studies have shown that noncoding RNAs (ncRNAs) exhibit differential expression under these unfavorable conditions (Vakilian 2020). These ncRNAs can directly respond to stress or regulate gene expression within interconnected cellular networks (Helliwell et al. 2016). Therefore, this section will primarily focus on the role of miRNAs, siRNAs, and lncRNAs in plant responses to abiotic stress.

2.4.1 Involvement of miRNAs in Response to Abiotic Stress

miRNAs have emerged as promising targets for enhancing plant productivity and tolerance to abiotic stress, considering their diverse target roles (Pandita 2019). The regulation of miRNAs in response to abiotic stress plays a crucial role in their functionality. Stress-responsive miRNAs, as reported by Sunkar et al. (2007), can act as repressors of stress-sensitive genes, which may be negative regulators of stress tolerance. Furthermore, under stress conditions, miRNAs can accumulate target gene mRNAs upon downregulation, thereby positively modulating stress tolerance (Sunkar et al. 2007). This section aims to highlight the overall significance of microRNAs and the challenges involved in their systematic characterization in plants under abiotic stress. Specific abiotic stresses can either upregulate or downregulate microRNA expression.

For instance, in *Arabidopsis*, miP319c and miP389a were downregulated during dehydration stress, while miP393, miP397b, and miP402 were upregulated. Similarly, rice seedlings exhibited downregulation and upregulation of microRNAs 17 and 16, respectively, under drought stress conditions. In maize, drought stress led to the upregulation of eight miRNAs in leaves and 13 miRNAs in roots, while seven miRNAs were downregulated in roots (Liu et al. 2019). The increased expression of miR156, miR160, miR162, miR164, and miR171 in response to drought stress led to a decrease in the expression of their target genes. For instance, miR160 targets the auxin response factor (ARF) genes. ARF16 is involved in the growth of root cap cells, while ARF17 functions as an auxin gene regulator. Reduced expression of ARF16 and ARF17 inhibits plant growth and enhances maize's drought resistance. Similarly, miR169 suppression leads to the accumulation of NFY, which controls the expression of nitrate transporters AtNRT2.1 and AtNRT1.1 (Zhao et al. 2011).

Interestingly, the same miRNA can exhibit both upregulation and downregulation in response to abiotic stress within the same plant. For example, the expression of miR398a/b in *Medicago truncatula* fluctuates in response to drought stress. In two independent studies on rice subjected to drought, miR156 and miR171 showed contrasting regulatory patterns, while miR164 and miR171 displayed unique regulatory tendencies under drought stress in maize (Liu et al. 2019). Moreover, opposing miRNA expression has been observed under the same abiotic stress conditions in different plant species. While rice and wheat upregulated miR399 expression, maize and *Arabidopsis* downregulated miR159 expression in response to drought stress. *Arabidopsis* showed elevated expression of miR168, miR397, and miR188, while maize exhibited their downregulation (Liu et al. 2008). The inconsistent expression of these microRNAs, potentially linked to variations in the severity of abiotic stress and/or other plant developmental conditions, necessitates further investigation.

Certain miRNAs are evolutionarily conserved across plant species during stress adaptation. In response to nitrogen (N) deprivation, several miRNAs, including miR169, miR171, miR395, miR397, miR398, miR399, miR408, and miR827, showed elevated expression in both *Arabidopsis* and maize. Similarly, maize, *Arabidopsis*, rice, and wheat upregulated miR160, miR162, miR395, and miR827, while downregulating miR166, miR172, miR397, miR827, and miR1432 in response to drought stress (Liu et al. 2008). Although miR171 exhibits varied expression patterns under drought stress in maize and rice, the conservation of these miRNAs suggests their biological significance (Liu et al. 2019). Further investigation is required to determine the factors contributing to such variances in similar settings.

Interestingly, specific microRNAs tend to respond to different abiotic stresses. In *Arabidopsis*, miR169a and miR169c are suppressed by drought and ABA stress, while miR398 is inhibited by oxidative stress, salt, ABA, and high Cu2+ concentration (Sunkar et al. 2012). In contrast, salt, drought, cold, ABA, and UV-B stressors enhance the expression of miR393 in *Arabidopsis*. Moreover, different abiotic stresses elicit varied responses from distinct microRNAs. For instance, drought, heat, and ABA treatment inhibit miR169 expression in *Arabidopsis*, whereas salt, cold, and UV-B irradiation promote its expression, as reported by Sunkar et al. (2012).

2.4.2 Role of siRNA in Plant Response to Abiotic Stress

The involvement of siRNAs in plants' response to abiotic stress has recently received attention (Palomar et al. 2012). In *Arabidopsis*, the interaction between nat-siRNA SRO5 and pyrroline-5-carboxylate dehydrogenase (P5CDH) reduces reactive oxygen species (ROS) induced by severe salt stress (Borsani et al. 2005). Elevated levels of TasiRNAs TAS1, TAS2, and TAS3 were observed in *Arabidopsis* plants exposed to hypoxia, indicating their role in stress response. These differences in tasiRNAs levels are correlated with the levels of TAS transcripts that target miRNAs (miR173 and miR39). The downregulation of pentatricopeptide repeat (PPR) mRNAs, which are targets of TAS2, is associated with mitochondrial defense during hypoxia (Lurin et al. 2004).

In *Arabidopsis*, heat stress induces the expression of HTT1 and HTT2, which are targets of TAS1. Overexpression of TAS1 weakens thermotolerance by reducing HTT gene expression, while overexpression of HTT genes enhances thermotolerance by activating HSF genes (Li et al. 2014). Heat-resistant transgenic plants that overexpress the HSFA1a gene show increased transcription of HTT genes. HSFA1a binds to the promoters of HTT genes to initiate their expression. HTT1 interacts with HSP40 and HSP70–14 in *Arabidopsis*, suggesting that HTT1 mediates thermotolerance pathways targeting TAS1a and is activated by HSFA1a (Li et al. 2014). HTT1 acts as a cofactor for HSP70–14 complexes.

On the other hand, the ta-siRNAs transcript TAS4, regulated by miR828, plays a crucial role in anthocyanin production in *Arabidopsis* under phosphate deficiency (Luo et al. 2012). Under nitrogen-deficient conditions, induction of TAS4-siR81(−) leads to anthocyanin accumulation (Luo et al. 2012). MYB transcripts, essential for anthocyanin synthesis, are targeted by TAS4-siR81(−).

2.4.3 Role of lncRNA in Plant Response to Abiotic Stress

Recent research has highlighted the involvement of plant lncRNAs in various abiotic stress responses (Xuan et al. 2015, b). These lncRNAs exhibit distinct behaviors in response to abiotic stressors. One mechanism through which plant lncRNAs participate in abiotic stress response is target mimicry. Wang et al. (2017) demonstrated that certain plant lncRNAs act as competing endogenous RNAs (ceRNAs) that are targeted by miRNAs, thus preventing the miRNAs from interacting with their intended targets. For instance, in P-deficient *Arabidopsis*, lncRNA IPS1 mimics miRNA399, thereby preventing its interaction with its target gene PHOSPHATE 2 (PHO2), and consequently impairing miR399-mediated regulation of PHO2 (Aung et al. 2006). Additionally, Shuai et al. (2014) found that drought-sensitive lncRNAs in *Populus trichocarpa*, such as ptc-miR482a.1, ptc-miR476a, and ptc-miR156c, act as targets that resemble miRNAs in response to abiotic stress.

Plant antisense lncRNAs engage in interactions with sense mRNAs during abiotic stress, resulting in the formation of double-stranded RNA duplexes that influence the expression of genes on the opposing strand. Poplar under nitrogen-deficient conditions and *Arabidopsis* under light and heat stress, as well as maize under drought stress, have shown the presence of five antisense lncRNAs (Wunderlich et al. 2014).

Furthermore, plant lncRNAs can serve as precursors for small RNAs (sRNAs) during abiotic stress response. Chen et al. (2016) identified nine long intergenic noncoding RNAs (lincRNAs) in N-deficient *Populus* that act as precursors for 11 known miRNAs, along with five lincRNAs that serve as precursors for 14 unique miRNAs. In a study by Song et al., 9687 novel lncRNAs and 50 lncRNAs acting as miRNA precursors were discovered in *Brassica rapa* under heat and cold stress (Song et al. 2010). Similar findings were reported for *Phaeodactylum tricornutum* under phosphorus-deficient conditions (Cruz de Carvalho et al. 2016). Additionally, specific lncRNAs have been identified as miRNAs, siRNAs, and shRNAs in Zea mays under drought stress (Zhang et al. 2014) and in wheat (*Triticum aestivum*) under heat stress (Waititu et al. 2020).

These studies highlight the diverse roles of plant lncRNAs in abiotic stress responses, including target mimicry, interaction with sense mRNAs, and serving as sRNA precursors.

2.5 Conclusion

The role of ncRNAs in plant responses to abiotic stress is increasingly recognized due to advancements in sequencing technologies. Despite their shared origin and conservation, ncRNAs exhibit diverse expression patterns across plant species and in response to different stresses, suggesting complex regulatory pathways. Understanding the environmental regulation of ncRNA expression and their tissue-specificity requires further investigation. While many ncRNAs' functions remain unknown, they interact coordinately, suggesting collaboration in stress responses. Alternative pathways for miRNA and siRNA biogenesis exist beyond canonical mechanisms, indicating the complexity of posttranscriptional gene regulation. The diversity in lncRNA biogenesis contributes to their functional complexity. Deep sequencing technologies have facilitated ncRNA identification and profiling, enhancing our understanding of their regulatory roles in plant physiology. However, knowledge gaps persist regarding the proteins and enzymes controlling ncRNA expression and degradation. Functional testing and cautious use of CRISPR-Cas tools are recommended to elucidate ncRNA functions without unintended mutations. ncRNAs regulate various aspects of plant biology, and leveraging high-throughput sequencing data from diverse species can enhance ncRNA research. Developing user-friendly tools and exploring genetic engineering for ncRNA manipulation are essential for advancing plant biotechnology and stress resilience.

2.6 Future Perspective of ncRNA Research

So far, functional research on plant ncRNAs has primarily focused on miRNAs, investigating their roles in development and stress response. The complexity and functional significance of diverse ncRNAs, particularly circRNAs and RNA derivatives, require further exploration, while research on other ncRNAs is still in its early stages. It is fascinating to observe how plants coordinate the accumulation of numerous ncRNAs according to their specific needs. To harness the potential of ncRNAs for enhancing crop yield, a deeper understanding of their functional and regulatory roles in various gene regulation networks is necessary. Establishing a database of prospective ncRNAs with a specific focus on desirable traits can contribute to the development of new crop varieties. Overall, the valuable insights for researchers to better comprehend the diverse types of ncRNAs, their interplay with plant phenotypes and regulation, and their potential applications in molecular crop breeding. Despite many unresolved issues, the strategic application of ncRNA-based approaches in molecular crop breeding offers hope for addressing future food crises. Additionally, as Linux is the platform on which the majority of prediction tools and deep learning models are produced, the creation of user-friendly Windows versions would enable more researchers to examine various ncRNA types. Since ncRNAs are involved in regulation in plants, it is still unclear how to use genetic engineering to modify ncRNAs in order to control particular biological processes. Researchers will be able to use the additional ncRNAs that are discovered to their advantage in order to create plants that are resistant to abiotic stress.

References

Alptekin B, Langridge P, Budak H (2017) Abiotic stress miRNomes in the Triticeae. Funct Integr Genomics 17(2):145–170

Ariel F, Romero-Barrios N, Jégu T, Benhamed M, Crespi M (2015) Battles and hijacks: noncoding transcription in plants. Trends Plant Sci 20(6):362–371

Aung K, Lin SI, Wu CC, Huang YT, Su CL, Chiou TJ (2006) pho2, a phosphate overaccumulator, is caused by a nonsense mutation in a microRNA399 target gene. *Plant Physiol* 141(3):1000–1011

Baev V, Milev I, Naydenov M, Vachev T, Apostolova E, Mehterov N, Gozmanva M, Minkov G, Sablok G, Yahubyan G (2014) Insight into small RNA abundance and expression in high-and lowtemperature stress response using deep sequencing in *Arabidopsis*. Plant Physiol Bioch 84:105–114

Bartel DP (2004) MicroRNAs: genomics, biogenesis, mechanism, and function. Cell 116(2):281–297

Borges F, Martienssen RA (2015) The expanding world of small RNAs in plants. Nat Rev Mol Cell Biol 16(16):727–741

Borsani O, Zhu J, Verslues PE, Sunkar R, Zhu JK (2005) Endogenous siRNAs derived from a pair of natural cis-antisense transcripts regulate salt tolerance in Arabidopsis. *Cell* 123(7):1279–1291

Brodersen P, Sakvarelidze-Achard L, Bruun-Rasmussen M, Dunoyer P, Yamamoto YY, Sieburth L, Voinnet O (2008) Widespread translational inhibition by plant miRNAs and siRNAs. Science 320(5880):1185–1190

Budak H, Hussain B, Khan Z, Ozturk NZ, Ullah N (2015) From Genetics to Functional Genomics: improvement in Drought Signaling and Tolerance in Wheat. *Front Plant Sci* 6:1012

Cai Q, Qiao L, Wang M, He B, Lin FM, Palmquist J, Huang HD, Jin H (2018) Plants send small RNAs in extracellular vesicles to fungal pathogen to silence virulence genes. Science 360(6393):1126–1129

Cech TR, Steitz JA (2014) The noncoding RNA revolution-trashing old rules to forge new ones. Cell 157(1):77–94

Chen X (2009) Small RNAs and their roles in plant development. Annu Rev Cell Dev Biol 25:21–44

Chen M, Wang C, Bao H, Chen H, Wang Y (2016) Genome-wide identification and characterization of novel lncRNAs in Populus under nitrogen deficiency. Mol Genet Genomics 291:1663–1680

Chu Q, Shen E, Ye CY, Fan L, Zhu QH (2018) Emerging roles of plant circular RNAs. J Plant Cell Dev 1(1):1–14

Creasey KM, Zhai J, Borges F, Van Ex F, Regulski M, Meyers BC, Martienssen RA (2014) miRNAs trigger widespread epigenetically activated siRNAs from transposons in *Arabidopsis*. Nature 508(7496):411–415

de Carvalho C, Sun MH, Bowler HX, Chua C, N.H. (2016) Noncoding and coding transcriptome responses of a marine diatom to phosphate fluctuations. New Phytol 210(2):497–510

Fan Y, Yang J, Mathioni SM, Yu J, Shen J, Yang X, Wang L, Zhang Q, Cai Z, Xu C, Li X (2016) PMS1T, producing phased small interfering RNAs, regulates photoperiod-sensitive male sterility in rice. Proc Natl Acad Sci USA 113(52):15144–15149

Fei Q, Xia R, Meyers BC (2013) Phased, secondary, small interfering RNAs in posttranscriptional regulatory networks. Plant Cell 25(7):2400–2415

Goodrich JA, Kugel JF (2006) Non-coding-RNA regulators of RNA polymerase II transcription. Nat Rev Mol Cell Biol 7(8):612–616

Helliwell CA, Robertson M, Finnegan EJ, Buzas DM, Dennis ES (2016) Vernalization-repression of *Arabidopsis* FLC requires promoter sequences but not antisense transcripts. PLoS One 6(6):e2151

Hirayama T, Shinozaki K (2010) Research on plant abiotic stress responses in the post genome era: past, present and future. Plant J 61(6):1041–1052

Holley RW, Apgar J, Everett GA, Madison JT, Marquisee M, Merrill SH, Penswick JR, Zamir A (1965) Structure of a ribonucleic acid. Science 147(3664):1462–1465

Iwakawa HO, Tomari Y (2015) The functions of MicroRNAs: mRNA decay and translational repression. Trends Cell Biol 25(11):651–665

Katiyar-Agarwal S, Morgan R, Dahlbeck D, Borsani O, Villegas A, Zhu J-K, Staskawicz BJ, Jin H (2006) A pathogen-inducible endogenous siRNA in plant immunity. Proc Natl Acad Sci USA 103(47):18002–18007

Kim DH, Xi Y, Sung S (2017) Modular function of long noncoding RNA, COLDAIR, in the vernalization response. PLoS Genet 13(7):e1006939

Komiya R, Ohyanagi H, Niihama M, Watanabe T, Nakano M, Kurata N, Nonomura KI (2014) Rice germline-specific argonaute MEL1 protein binds to phasiRNAs generated from more than 700 lincRNAs. Plant 78(3):385–397

Lee RC, Feinbaum RL, Ambros V (1993) The C. Elegans heterochronic gene lin-4 encodes small RNAs with antisense complementarity to lin-14. Cell 75(5):843–854

Lee JT (2012) Epigenetic regulation by long noncoding RNAs. Science 338(6113):1435–1439

Li SX, Liu JX, Liu ZY, Li X, Wu F, He Y (2014) Heat-induced TAS1 target1 mediates thermotolerance via heat stress transcription factorA1a–directed pathways in *Arabidopsis*. Plant Cell 26(4):1764–1780

Liu Y, El-Kassaby Y (2017) Regulatory cross-talk between microRNAs and hormone signalling cascades controls phenotypical variations: a case study in seed dormancy modulations during seed set of *Arabidopsis* thaliana. Plant Cell Rep 36:705–717

Liu HH, Tian X, Li YJ, Wu CA, Zheng CC (2008) Microarray-based analysis of stress-regulated microRNAs in *Arabidopsis* thaliana. RNA 14(5):836–843

Liu X, Zhang X, Sun B, Hao L, Liu C, Zhang D, Tang H, Li C, Li Y, Shi Y et al (2019) Genome-wide identification and comparative analysis of drought-related microRNAs in two maize inbred lines with contrasting drought tolerance by deep sequencing. PLoS One 14(7):e0219176

Luo QJ, Mittal A, Jia F, Rock CD (2012) An autoregulatory feedback loop involving PAP1 and TAS4 in response to sugars in *Arabidopsis*. Plant Mol Biol 80:117–129

Lurin C, Andrés C, Aubourg S, Bellaoui M, Bitton F, Bruyère C, Caboche M, Debast C, Gualberto J, Hoffmann B et al (2004) Genome-wide analysis of *Arabidopsis* pentatricopeptide repeat proteins reveals their essential role in organelle biogenesis. Plant Cell 16(8):2089–2103

Mattick JS, Rinn JL (2015) Discovery and annotation of long noncoding RNAs. Nat Struct Mol Biol 22(1):5–7

Matzke M, Kanno T, Daxinger L, Huettel B, Matzke AJ (2009) RNA-mediated chromatin-based silencing in plants. Curr Opin Cell Biol 21(3):367–376

Maunoury N, Vaucheret H (2011) AGO1 and AGO2 act redundantly in miR408-mediated Plantacyanin regulation. PLoS One 6(12):e28729

Mittler R (2006) Abiotic stress, the field environment and stress combination. Trends Plant Sci 11(1):15–19

Nie L, Wu H-J, Hsu J-M, Chang S-S, Labaff AM, Li C-W, Wang Y, Hsu JL, Hung M-C (2012) Long non-coding RNAs: versatile master regulators of gene expression and crucial players in cancer. Am J Transl Res 4(2):127–150

Palomar AM, Portillo A, Santibáñez P, Mazuelas D, Arizaga J, Crespo A, Gutiérrez Ó, Cuadrado JF, Oteo JA (2012) Crimean-Congo hemorrhagic fever virus in ticks from migratory birds, Morocco. *Emerg infect diseases* 19(2):260–263

Pandita D (2019) Plant MIRnome: miRNA biogenesis and abiotic stress response. In: Plant abiotic stress response. Springer, Berlin, pp 449–474

Pauli A, Rinn JL, Schier AF (2011) Non-coding RNAs as regulators of embryogenesis. Nat Rev Genet 12(2):136–149

Rai MI, Alam M, Lightfoot DA, Gurha P, Afzal AJ (2019) Classification and experimental identification of plant long non-coding RNAs. Genomics 111(5):997–1005

Rogers K, Chen X (2013) Biogenesis, turnover, and mode of action of plant microRNAs. Plant Cell 25(7):2383–2399

Shimoni Y, Friedlander G, Hetzroni G, Niv G, Altuvia S, Biham O, Margalit H (2007) Regulation of gene expression by small-noncoding RNAs: a quantitative view. Mol Syst Biol 3(1):138

Shuai P, Liang D, Tang S, Zhang Z, Ye C-Y, Su Y, Xia X, Yin W (2014) Genome-wide identification and functional prediction of novel and drought-responsive lincRNAs in Populus trichocarpa. J Exp Bot 65(17):4975–4983

Song Y, Wu K, Dhaubhadel S, An L, Tian L (2010) *Arabidopsis* DNA methyltransferase AtDNMT2 associates with histone deacetylase AtHD2s activity. Biochem Biophys Res Commun 396(2):187–192

Sunkar R, Chinnusamy V, Zhu J, Zhu JK (2007) Small RNAs as big players in plant abiotic stress responses and nutrient deprivation. Trends Plant Sci 12(7):301–309

Sunkar R, Li YF, Jagadeeswaran G (2012) Functions of microRNAs in plant stress responses. Trends Plant Sci 17(4):196–203

Vakilian KA (2020) Machine learning improves our knowledge about miRNA functions towards plant abiotic stresses. Sci Rep 10(1):1–10

Waititu JK, Zhang C, Liu J, Wang H (2020) Plant non-coding RNAs: origin, biogenesis, mode of action and their roles in abiotic stress. Int J Mol Sci 21(21):8401

Wang J, Meng X, Dobrovolskaya OB, Orlov YL, Chen M (2017) Non-coding RNAs and their roles in stress response in plants. Genom Proteom Bioinform 15(5):301–312

Wierzbicki AT, Haag JR, Pikaard CS (2008) Noncoding transcription by RNA polymerase pol IVb/pol V mediates transcriptional silencing of overlapping and adjacent genes. Cell 135(4):635–648

Wierzbicki AT, Ream TS, Haag JR, Pikaard CS (2009) RNA polymerase V transcription guides ARGONAUTE4 to chromatin. Nature Genet 41(5):630–634

Wunderlich M, Groß-Hardt R, Schöffl F (2014) Heat shock factor HSFB2a involved in gametophyte development of *Arabidopsis* thaliana and its expression is controlled by a heat-inducible long non-coding antisense RNA. Plant Mol Biol 85:541–550

Xu D, Wenya Y, Chunjie F, Bobin L, Lu M-Z, Jin Z (2023) Opportunity and challenges of predictive approaches for the non-coding RNA in plant. Front Plant Sci 13:890663

Xuan H, Zhang L, Liu X, Han G, Li J, Li X, Zhang S (2015) PLNlncRbase: a resource for experimentally identified lncRNAs in plants. Gene 573(2):328–332

Yang L, Wu G, Poethig RS (2012a) Mutations in the GW-repeat protein SUO reveal a developmental function for microRNA-mediated translational repression in *Arabidopsis*. Proc Natl Acad Sci 109(1):315–320

Yang L, Wu G, Poethig RS (2012b) Mutations in the GW-repeat protein SUO reveal a developmental function for microRNA-mediated translational repression in *Arabidopsis*. Proc Natl Acad Sci USA 10(1):315–320

Ye R, Wang W, Iki T, Liu C, Wu Y, Ishikawa M, Qi Y (2012) Cytoplasmic assembly and selective nuclear import of *Arabidopsis* Argonaute4/siRNA complexes. Mol Cell 46(6):859–870

Zhang W, Han Z, Guo Q, Liu Y, Zheng Y, Wu F, Jin W (2014) Identification of maize long noncoding RNAs responsive to drought stress. PloS One 9(6):e98958

Zhao K, Tung CW, Eizenga GC, Wright MH, Ali ML, Price AH, Norton GJ, Islam MR, Reynolds A, Mezey J, McClung AM, Bustamante CD, McCouch SR (2011) Genome-wide association mapping reveals a rich genetic architecture of complex traits in Oryza sativa. *Nat Commun* 2:467

Posttranscriptional Gene Regulation: The Role of Noncoding RNAs

3

Jafar K. Lone and Ragini Bhardwaj

Abstract

The complex world of RNA molecules is explored with a particular emphasis on noncoding RNAs (ncRNAs) and the variety of functions they play in posttranscriptional gene regulation. This chapter covers the classification and traits of ncRNAs in eukaryotic genomes. The diverse functions of ncRNAs are highlighted, including their involvement in posttranscriptional processes such mRNA splicing, degradation, and translation, as well as transcriptional and epigenetic control. The relationships between long noncoding RNAs (lncRNAs) and microRNAs (miRNAs) are further explored in this chapter, emphasizing the function of lncRNAs as competitive endogenous RNAs (ceRNAs) and control expression of gene. The chapter also covers how lncRNAs are involved in splicing and chromatin remodeling. In conclusion, the chapter delves into the newly discovered function of circular RNAs (circRNAs) in the regulation of alternative splicing, highlighting their influence on the complexity of plant genomes and the regulation of genes.

Keywords

Alternative splicing · Circular RNAs · ncRNAs · Chromatin remodeling · Posttranscriptional gene regulation

---

J. K. Lone (✉)
ICAR-National Bureau of Plant Genetic Resources, Pusa, New Delhi, India

Department of Biochemistry, School of Bioengineering and Biosciences, Lovely Professional University, Phagwara, Punjab, India

R. Bhardwaj
ICAR-National Bureau of Plant Genetic Resources, Pusa, New Delhi, India

Department of Bioscience and Biotechnology, Banasthali Vidyapith University, Tonk, Rajasthan, India

K. Goswami et al. (eds.), *Non-coding RNAs for Crop Improvement*,
https://doi.org/10.1007/978-981-97-7213-1_3

3.1 Introduction

According to the RNA world hypothesis, life originated from RNA molecules, with DNA and RNA emerging later as carriers of genetic information. This concept forms the foundation of the core dogma of life. While it is estimated that around 90% of eukaryotic genomes are transcribed during development, only a small portion of transcribed RNAs, approximately 2%, actually code for proteins (Ariel et al. 2015; Lee et al. 2019). Consequently, the majority of RNA is classified as "noncoding RNA," lacking the ability to produce proteins. Despite the prevalent view of the noncoding genome as "junk DNA," due to its abundance of pseudogenes, simple repeats, and transposons (Comings 1972; Ohno 1972), it actually contains a significant number of ncRNAs that serve important biological functions. These ncRNAs can be categorized into two main classes based on their length: small noncoding RNAs (sncRNAs) and long noncoding RNAs (lncRNAs), with the cutoff at 200 nucleotides (nt). Both lncRNAs, such as enhancer RNAs, transposon-derived RNAs, and circular RNAs (Zhang et al. 2019); and sncRNAs such as micro RNAs (miRNAs), small interfering RNAs (siRNAs), piwi-interacting RNAs (piRNAs) (Lone et al. 2023) have well-established roles. LncRNAs exhibit a wide range of lengths, spanning from 200 bases to over 100 kb, and can function within the nucleus or cytoplasm. They are expressed in various tissues, cell types, and cellular states (Cao 2014; Quinn and Chang 2016). The continuous discovery of regulatory ncRNAs demonstrates that RNA has undergone evolutionary changes similar to proteins and DNA. Among sncRNAs, there are functional RNAs such as tRNAs, rRNAs, and snRNAs, which play crucial roles in controlling transcription and translation (Hangauer et al. 2013; Milligan and Lipovich 2015). The regulatory RNAs miRNAs, siRNAs, and piRNAs are a few examples of sRNAs that are involved in regulating gene expression (Carthew and Sontheimer 2009).

However, lncRNAs constitute a class of ncRNAs that is vast, diverse, and functionally ambiguous. The presence of introns in lncRNAs suggests the existence of splice variants, and the presence of epigenetic markers indicates variable expression, similar to coding RNA transcripts (Guttman et al. 2009; Hangauer et al. 2013). This highlights the complexity and potential multi functionality of lncRNAs. These particular lncRNAs, referred to as "bimorphic" (Hangauer et al. 2013), possess a dual characteristic as they can exist in either polyadenylated or non-polyadenylated forms. In contrast to the highly conserved secondary and tertiary structures typically observed in coding transcripts, lncRNAs exhibit a wide range of RNA types. The diverse biological effects exerted by lncRNAs stem from their ability to influence multiple stages of gene expression. This functional versatility is attributed to the capacity of ncRNAs to interact precisely with a wide array of substrates. Moreover, the expression of ncRNAs undergoes rapid changes during development and differentiation, resulting in altered translation rates, either increased or decreased (Geisler and Coller 2013). This chapter primarily focus on exploring the posttranscriptional regulatory roles of ncRNAs in various cellular processes and developmental control.

3.1.1 Small Noncoding RNAs

Over the past decade, there has been a significant research focus on sncRNAs due to their involvement in RNA interference (RNAi) pathways. These pathways play a crucial role in gene silencing and cellular defense mechanisms against viruses, mobile repetitive DNA sequences, retro-elements, and transposons (Moazed 2009). The majority of the sncRNA family consists of miRNAs and siRNAs, which are both approximately 20–30 nucleotides long. They are derived from sense and antisense DNA strands, pseudogenes, and inverted repeats (Carthew and Sontheimer 2009; Lone et al. 2023). Through mechanisms such as mRNA decay or translational repression, these molecules modulate gene expression. Remarkably, miRNAs are known to negatively regulate nearly 60% of protein-coding genes (Lewis et al. 2005). A single transcript can be influenced by one or more miRNAs, while in other cases, multiple miRNAs may be involved (Esteller 2011). miRNAs, exert regulatory control over numerous genes and play critical roles in various biological processes (Jovanovic and Hengartner 2006; Büssing et al. 2008; Schickel et al. 2008; Hrdlickova et al. 2014), including transcription. siRNAs, are short double-stranded RNA molecules typically 21–25 nucleotides long, with a 5′-phosphate group and two nucleotide overhangs at the 3′ end (Lone et al. 2023). They function as gene silencers (Elbashir et al. 2001). Gene silencing can be initiated by the introduction of sequences homologous to the target gene, which can be silenced using either short-hairpin RNA (shRNA) or long double-stranded RNA precursors (Fire 1999; Tuschl 2001). siRNAs are well-known for their role as co-transcriptional silencers of gene expression, accomplished through the regulation of chromatin (Burkhart et al. 2011). Moreover, they exert wider biological functions in plants (Carthew and Sontheimer 2009; Lone et al. 2023).

3.1.2 miRNA as Posttranscriptional Regulator

In plants, miRNAs are important modulators of posttranscriptional gene expression and are involved in a number of biological processes, such as defense systems, development, and stress responses. One well-researched example of miRNA-mediated posttranscriptional regulation in plants is miR-156 and its target gene SPL (Squamosa Promoter Binding Protein (Dong et al. 2022). SPL transcription factors control phase transition, leaf morphogenesis, blooming period, and other developmental processes; in *Arabidopsis thaliana*, miR-156 controls the expression of these transcription factors. By modifying the expression of target genes, it performs essential roles in coordinating developmental changes in plants. Furthermore, it underscores the significance of miRNA-mediated posttranscriptional regulation in molding the growth and development of plants (Jerome Jeyakumar et al. 2020).

3.1.3 Long Noncoding RNA

lncRNAs belong to a diverse group of transcripts that exceed 200 nt in length. Despite being transcribed from DNA, these lncRNAs lack the capacity to encode proteins (Derrien et al. 2012; Harrow et al. 2012; Guttman et al. 2013). Various types of lncRNAs have been identified, including intergenic, exonic, enhancer, and distal lncRNAs derived from protein-coding genes (Wang and Chang 2011; Geisler and Coller 2013).

The noncoding nature of lncRNAs arises when they are located within pseudogenes or mutated copies of coding genes (Milligan and Lipovich 2015). Extensive evidence demonstrates that numerous lncRNAs overlap with coding genes (Chen et al. 2004). Depending on their genetic origins, whether from the sense or antisense strand, lncRNAs may fully or partially encompass a gene (Pandey et al. 2008; Pastori et al. 2014). Different names are assigned to lncRNAs based on their characteristics, including "intergenic" (lincRNA), "natural antisense transcripts (NATs)," "enhancer" (eRNA), "circle" (circRNA), and "promoter associated long" (pRNA), among others. Compared to protein-coding RNAs, lncRNAs generally possess fewer exons, and those exons tend to be longer (Cabili et al. 2011; Harrow et al. 2012). Additionally, lncRNAs exhibit distinctive features, such as (a) highly conserved promoter regions among vertebrates, (b) specific promoters for each lncRNA with unique DNA-binding motifs and preferred transcription factors (TFs), (c) less conserved exons across species, and (iv) expression profiles that vary across different tissues. The expression levels of lncRNAs are typically much lower than those of protein-coding genes (Derrien et al. 2012; Djebali et al. 2012), with only 11–29% of lncRNAs being expressed in all organs. Computational analysis of RNA-Seq data suggests that lncRNA transcription is autonomous and may influence the transcription of nearby protein-coding genes (Derrien et al. 2012; Popadin et al. 2013). In plants, lncRNAs are primarily associated with the regulation of gene chromatin environments, modulating modifications, RNA Polymerase processing, transcription efficiency, and facilitating the formation of activating and repressive chromatin loops with neighboring genes under epigenetic regulation (Hamid et al. 2020; Palos et al. 2023; Tremblay and Qüesta 2023).

The origin of lncRNAs remains a topic of debate. A study (Kapusta et al. 2013), revealed the presence of transposable elements (TEs) in over two-thirds of mature lncRNA transcripts. This discovery has led to the hypothesis that TE insertion was crucial to the development of most lncRNAs (Nekrutenko and Li 2001). LncRNAs have the ability to control the expression of genes by a variety of methods, such as transcription, posttranscriptional gene regulation, chromatin remodeling, *cis* and *trans* gene expression, and other phenomena. In overview, the Fig. 3.1 elaborates the functioning of posttranscriptional roles by lncRNA in plants.

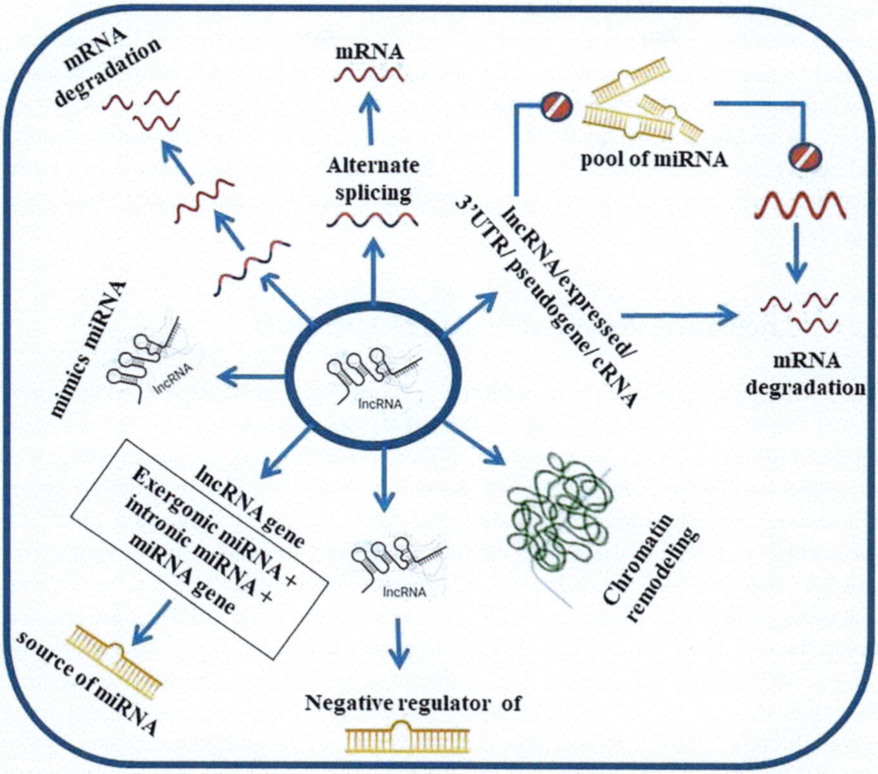

Fig. 3.1 Posttranscriptional gene regulation by lncRNA

3.1.4 LncRNAs as Posttranscriptional Regulators

When it comes to posttranscriptional regulation, lncRNAs play a key role as competing endogenous RNAs that affect the abundance of miRNAs, which in turn modify mRNA levels through modifications to mRNA stability, mRNA degradation, and translation (Dykes and Emanueli 2017). lncRNAs are pivotal in various aspects of plant vegetative growth, including germination, leaf development, root growth, photomorphogenesis, and the regulation of phytohormones. Recent research has unveiled the posttranscriptional gene expression regulation mediated by select lncRNAs. For instance, in rice, the lncRNA TWISTED LEAF (TL) maintains a flat leaf shape by regulating the expression of an R2R3-MYB transcription factor gene (Liu et al. 2018). Similarly, in Arabidopsis, the NAT lncRNA, MASm controls flowering time by enhancing histone 3 lysine 4 trimethylation (H3K4me3) at the MADS AFFECTING FLOWERING4 (MAF4) locus (Zhao et al. 2018). In apple, the photo-induced lncRNA MdLNC499 modulates anthocyanin accumulation by regulating the transcription levels of MdERF109 (Ma et al. 2021). In tomato plants, lncRNA16397 induces SlGRX22 expression, boosting tomato resistance to late blight (Zhang et al. 2015). Furthermore, some lncRNAs, unlike conventional ones, can encode open reading frames (ORFs), generating short peptides (SEPs) that are pivotal in plant development and responses, regulating processes such as growth, development, and resistance responses at posttranscriptional stage (Zhao et al. 2022). In addition to the well-established functions of lncRNA in transcriptional and epigenetic gene regulation, they also have the potential to promote or inhibit posttranscriptional processes such as mRNA splicing, degradation, and translation (Fig. 3.1).

3.2 LncRNA: A Potential Source of MicroRNA

It is possible to classify primitive miRNAs(pri-miRNAs) as lncRNAs because their typical length is more than 1 kb (Kim et al. 2009). pri-miRNAs come from two main places in the genome: (a) genes in which they are embedded, whose expression is often but not always linked to that of the parent transcript, and (b) miRNA genes themselves, which contain promoters that are regulated primarily by RNA polymerase II (RNA Pol II) in a manner analogous to that of messenger RNA (mRNA) (Dykes and Emanueli 2017). Around 50% of miRNAs are derived from noncoding transcripts (Saini et al. 2007). Indeed, numerous miRNAs are located within the introns of noncoding genes (Dykes and Emanueli 2017), where they exist within both coding and noncoding genomic regions. The genetic arrangement illustrates that the host long noncoding RNA (lncRNA) serves additional roles beyond merely hosting pri-miRNAs, functions that are presumably encoded by the exons. This suggests that the host lncRNA likely carries out diverse functions beyond those associated solely with pri-miRNA processing and regulation.

3.3 Interactions Between lncRNAs and miRNAs

The interactions between miRNAs and lncRNAs play crucial roles in regulating various biological processes, including development, nutrient absorption, and responses to biotic and abiotic stresses (Zhou et al. 2020; Jiang et al. 2020). Target mimicry stands out as one of the most significant mechanisms in the interplay between miRNAs and lncRNAs. In this mechanism, lncRNAs containing endogenous target mimic (eTM) sites sequester miRNAs through sequence complementarity, thereby inhibiting their effects on target mRNA. Target mimicry is also referred to as miRNA decoy, miRNA sponge, or competing endogenous RNA (ceRNA) in animals (Salmena et al. 2011).

Arabidopsis IPS1 (INDUCED BY PHOSPHATE STARVATION 1) represents the first identified lncRNA known to pair with miR399 (Franco-Zorrilla et al. 2007). LncRNA acts as an endogenous miRNA eTM to regulate miRNA expression and function. During pollen development in turnip (*Brassica rapa* subsp. rapa), lncRNAs function as targets to regulate the function of miRNAs (eTMs), of which bra-eTM160-1 and bra-eTM160-2, are predicted to be eTMs of bra-mir160-5p. By overexpressing the eTMs in *B. rapa*, bra-eTM160-1 and bra-eTM160-2 had inhibitory ability on bra-miR160-5p and participated in pollen formation and male fertility through upregulating ARF genes, especially BrARF17. Although transgenic plants have a normal flower shape, half of the pollen grains in the anthers are small and wrinkled, are non-viable, and lack the nucleus and pollen lining (Huang et al. 2018).

miRNAs frequently attach to miRNA response elements (MREs) found in the 3′ untranslated regions (UTRs) of target transcripts to suppress gene expression. Usually, this is accomplished by the short seed sequence of the miRNA, (usually 6–8 nucleotides). Computational predictions have identified the presence of miRNA binding sites in numerous lncRNAs using miRNA seed sequences. This intriguingly suggests that many lncRNAs may regulate gene expression by sequestering miRNAs, thereby reducing the concentration of miRNAs and limiting their accessibility within the cell. In this beneficial mechanism, the lncRNA acts as a modulator of gene expression by inhibiting miRNA activity. This phenomenon is described by the "ceRNA hypothesis" (Thomas et al. 2010; Salmena et al. 2011) (Fig. 3.1).

To compete with miRNAs, cells produce a specific class of lncRNAs known as competitive endogenous RNAs (ceRNAs), which can exist in linear or circular forms, and function by sequestering miRNAs or suppressing miRNA function. Linear lncRNAs bind to miRNAs and facilitate their degradation, thus controlling gene expression (Dempsey and Cui 2017). Circular RNAs (circRNAs) are generated through a back-splicing process that forms a covalent bond between the 3′ and 5′ ends of the lncRNA, resulting in a ring-shaped molecule (Ashwal-Fluss et al. 2014; Ebbesen et al. 2016). Additionally, lncRNAs can suppress mRNA translation or degradation by engaging in base-pairing interactions, facilitated by Staufen proteins, between their Alu elements and the 3' UTR sequences of mRNAs (Gong and Maquat 2011).

Pseudogenes, which are noncoding transcripts sharing a high degree of sequence similarity with coding genes, are considered strong contenders for acting as ceRNAs (Poliseno et al. 2010; Milligan and Lipovich 2015; An et al. 2017), as they are expected to possess many MREs similar to the coding gene. One such lncRNA is a pseudogene with similarity to the gene encoding the tumor suppressor phosphatase and tensin homolog (PTEN). This lncRNA shares the same 3' UTR as the coding gene and contains numerous MREs (Poliseno et al. 2010).

Phased small interfering RNAs (phasiRNAs) can be produced by miRNAs targeting lncRNA transcripts (Jin et al. 2021). In plants, a subset of lncRNAs generating a class of phasiRNAs named trans-acting siRNAs (tasiRNAs) was first identified in Arabidopsis (Yoshikawa et al. 2005; Axtell et al. 2006; Howell et al. 2007). These lncRNAs are targeted by miRNAs, including miR173 and miR390, to produce tasiRNAs (Yoshikawa et al. 2005; Axtell et al. 2006; Howell et al. 2007). Recently, certain lncRNAs from reproductive organs have been demonstrated to generate reproductive phasiRNAs (Fei et al. 2016; Xia et al. 2019).

Co-expression assays in *N. benthamiana* and maize protoplasts have demonstrated that PILNCR1 can mitigate the effects of miR399 in maize. This reduction depends on complementary regions present in both PILNCR1 and miR399 (Du et al. 2018). In maize, a novel lncRNA target of miR399, PILNCR1, is also essential for low phosphate (Pi) tolerance (Du et al. 2018), indicating that the lncRNA-miR399-PHO2 regulatory module may be a widespread mechanism in plant responses to Pi deficiency. Furthermore, lncRNA39026 binds miR168a and inhibits its function, consequently enhancing tomato resistance to *Phytophthora infestans* (Hou et al. 2020). These findings suggest that lncRNAs may modulate various biological processes by sequestering miRNAs.

Wu et al. developed a computational method and identified 36 and 189 potential eTMs in Arabidopsis and rice, respectively (Wu et al. 2013). Since then, additional lncRNAs potentially acting as decoys for miRNAs have been identified in maize (Fan et al. 2015), cassava (Li et al. 2017), tomato (Yang et al. 2019; Cui et al. 2020), and melon (Gao et al. 2020). However, the functional significance of these lncRNA-miRNA interactions still requires further analysis.

3.4 Degradation of Messenger RNA (mRNA) Through miRNA-Independent lncRNAs

In addition to their influence on gene expression through interaction with microRNAs, certain lncRNAs also mRNA for degradation. One mechanism involved in mRNA degradation is known as nonsense-mediated mRNA decay (NMD) (Park and Maquat 2013). Among the proteins involved in this process is Staufen 1 (STAU1). The 3' UTR of ARF1 mRNA contains a specific double-stranded RNA motif that can be recognized and bound by STAU1, leading to the formation of a stem-loop structure. However, the 3' UTR of other mRNAs, such as SERPINE1 (serpin peptidase inhibitor-clade E member 1), targeted for degradation by Staufen lacks this stem-loop structure and instead contains only a single-stranded binding site. Interestingly, in the case of these mRNAs, STAU1 is recruited through an interaction with a specific lncRNA that carries a corresponding single-stranded binding site. The weak binding of this lncRNA to the mRNA results in the formation of a double-stranded RNA-binding motif. These lncRNAs are referred to as "half STAU1 binding site RNAs" (Gong and Maquat 2011).

3.5 Interactions of LncRNAs with Splicing Factor

The plant hormone auxin plays crucial roles in development, including the enhancement of protein-coding capacity and the promotion of developmental flexibility (Yan et al. 2012; Palusa and Reddy 2015). In *Arabidopsis*, a lincRNA called Alternative Splicing Competitor (ASCO) has been identified to interact with nuclear speckle-localized splicing factors (SFs), specifically the nuclear speckle RNA-binding proteins NSRa and NSRb, in vivo (Bardou et al. 2014). These NSR proteins are involved in the regulation of auxin's molecular and growth responses. Notably, in the NSRa/b double mutant, more than 2200 genes exhibited differential regulation compared to wild-type plants after auxin treatment, leading to reduced sensitivity to auxin and fewer lateral root development. Analysis of RNA processing events in the NSRa/b mutant revealed a significant occurrence of intron retention and differential 5′ or 3′ ends in a subset of genes, including a high proportion of auxin-related genes that exhibited similar patterns as observed in ASCO overexpressing lines (Bardou et al. 2014; Tran et al. 2016). Intriguingly, in vitro binding assays demonstrated that ASCO competes with mRNA targets for NSR binding, suggesting that ASCO modulates the alternative splicing of pre-mRNAs in response to auxin by interacting with NSRs (Bardou et al. 2014).

In *A. baumannii*, a recent study employed an NSRa-directed RNA immunoprecipitation (RIP)-Seq strategy, revealing that NSRa primarily targets lncRNAs, similar to the findings in Arabidopsis (Bazin et al. 2018). Given that the majority of NSRa targets are enriched for genes involved in biotic stress responses, it was proposed that NSR-containing complexes, involving lncRNAs and alternative splicing mRNAs, integrate the auxin and immune response pathways (Bazin et al. 2018). The knockdown and overexpression of ASCO resulted in the dysregulation of gene expression and splicing in numerous genes associated with biotic stress and the response to flagellin in Arabidopsis (Rigo et al. 2020). This observation supports the hypothesis that ASCO controls alternative splicing in an NSR-independent manner. Furthermore, additional protein partners of ASCO, such as the pre-mRNA processing splicing factor 8A (PRP8a) and the spliceosome-core component SmD1b, were identified through mass spectrometry and RIP assays in conjunction with chromatin isolation by RNA purification (ChIRP) (Grainger and Beggs 2005; Elvira-Matelot et al. 2016; Rigo et al. 2020). Similar to NSRs, the binding of PRP8a to specific mRNA targets is inhibited upon ASCO overexpression (Rigo et al. 2020).

3.6 Regulation of Splicing by lncRNAs through Chromatin Remodeling

It has been found in recent research on *A. thaliana*, circRNAs are important regulators of alternative splicing (AS) (Conn et al. 2017). Specifically, a naturally occurring AS variation known as SEP3.3, which involves the skipping of exon 6, is found to be more prevalent when a circRNA encompassing the entire exon 6 of the SEPALLATA 3 (SEP3) gene is overexpressed (Conn et al. 2017). Alterations in the splicing of SEP3 have been observed to result in distinct flower development phenotypes (Severing et al. 2012; Conn et al. 2017). SEP3 belongs to the MADS (MCM1-AGAMOUS-DEFICIENS-SRF)-box transcription factor superfamily and is known to play a crucial role in flower development.

Remarkably, the circRNA derived from exon 6 of SEP3 can form a hybrid structure known as an RNA: DNA R-loop with its corresponding DNA locus. This R-loop creation affects alternative splicing, which causes transcriptional pause, and it occurs concurrently with the recruitment of splicing factors. The significance of

R-loop formation and chromatin conformation as critical variables influencing splicing patterns is highlighted by this finding (Conn et al. 2017). Furthermore, it suggests that circRNAs expressed from distant loci might enhance the efficiency of splicing for their associated messenger RNAs that undergo exon skipping. In plants, miniature inverted-repeat transposable elements (MITES) discovered in the flanking introns of rice exonic circRNAs (Lu et al. 2015), and reverse complementary pairs of LINE1-like elements (LLECRPs) in *Zea mays* (Chu et al. 2020), play pivotal roles in circRNA biogenesis. Exploring the involvement of repetitive sequences in the flanking regions of circRNAs in polyploid species is essential for understanding genome complexity (Chu et al. 2018). Longer flanking introns containing repetitive or reverse complementary sequences only cover a limited amount of exonic circRNA production in plants (Ye et al. 2015; Chu et al. 2020). Consequently, Adenosine (A) to Inosine (I) editing by adenosine deaminase acting on RNA1 (ADAR1) occurs frequently in intronic sequences, which are crucial for circularizing intervening exons (Ivanov et al. 2015).

Approximately 6% and 1% of circRNAs in *A. thaliana* originate from chloroplast and mitochondrial genes, respectively, indicating their presence and regulation in subcellular organelles (Sun et al. 2016). Trans back-splicing is also notable in plants, with reported occurrences of 13% of circRNAs in *A. thaliana* and 34% in *O. sativa* (Ye et al. 2015; Chu et al. 2020), besides a single circRNA in *N. benthamiana* identified by the MDA-cloning method (Guria et al. 2019). However, the impact of these circRNAs on gene regulation and potential physiological changes requires thorough evaluation. Most exonic circRNAs formed in *O. sativa*, *A. thaliana*, and *G. max* consist of 1–4 exons (Lu et al. 2015; Ye et al. 2015; Zhao et al. 2017), likely generated post-transcriptionally after removing intervening introns, unlike single exonic circRNA, which is derived through the co-transcriptional pathway (Chu et al. 2020).

3.7 Alternative Splicing Regulation by LncRNA-RNA Duplexes

Transcriptional data obtained from *A. thaliana* revealed the existence of numerous convergent overlapping pairs (COPs) of genes, which possess the capability to generate dsRNAs (Jen et al. 2005). Within these COPs, there is a notable concentration of genes that contain introns and/or exhibit AS of transcripts. Furthermore, polyadenylation and/or alternative splicing variations typically happen at locations where sense transcript introns and antisense transcripts overlap. This implies that RNA molecules and NATs lncRNA pairing may control alternative splicing in protein-coding genes (Jen et al. 2005). Through a comprehensive genome-wide scan in Arabidopsis, a total of 1320 potential trans-NAT couples were identified (Wang et al. 2006).

Most of these transcripts generate extended double-stranded RNA duplexes, which can indirectly regulate alternative splicing and silence genes, especially when sense and antisense molecules are expressed in the same subcellular compartment

(Wang et al. 2006). All of these findings show that lncRNAs are involved in regulating alternative splicing, which reprogrammed the transcriptome and integrates it into a dynamic splicing network.

3.8 Conclusions

The versatility of RNA molecules within cells is evident in the wide range of biological functions performed by lncRNAs. Research on various types of noncoding RNAs has revealed their complexity and their ability to operate as interconnected and regulated networks. The interaction of lncRNAs with transcriptional regulation roles of noncoding RNAs demonstrates the diverse mechanisms through which they regulate gene expression after transcription. It is increasingly apparent that lncRNAs employ multiple strategies to control gene expression. In addition to their recognized role as transcriptional regulators, lncRNAs also exert significant influence over pre-mRNA splicing, mRNA degradation, and protein synthesis. These characteristics establish lncRNAs, alongside RNA-binding proteins and microRNAs, as bonafide regulators of posttranscriptional gene expression. Moreover, future discoveries are expected to reveal additional instances where lncRNAs compete or collaborate with RNA-binding proteins, microRNAs, and possibly other lncRNAs to exert posttranscriptional activities. Unlike microRNAs, which rely primarily on sequence, or RNA-binding proteins, which depend on structure, lncRNAs, particularly larger ones, possess a distinct ability to function through both sequence homology/complementarity with other nucleic acids and structural features. This enables them to serve as molecular frameworks and scaffolds for the assembly of large macromolecular complexes. The conserved functions of diverse lncRNAs across species are believed to be determined by specific short sequences and their secondary structures. The discovery of an increasing amount of cold-responsive lncRNAs, such as SVALK, is intriguing because it suggests that the noncoding fraction of the transcriptome, which controls transcriptional and posttranscriptional levels of gene expression, is essential for environmental reactions. Modern methods such as SHAPE-seq (selective 2′-hydroxyl acylation analyzed by primer extension sequencing) may be able to provide light on whether plant lncRNAs change their structure in response to temperature and become new regulators that precisely modify the protein-coding genome in response to climate change.

3.9 Future Perspective

The complex roles that lncRNAs play in gene regulation and cellular processes will be further explored. Particular attention will be paid to the elucidation of intricate regulatory networks involving interactions with regulatory molecules. Furthermore, the focus will be on investigating new functional activities of lncRNAs that go beyond transcriptional regulation, such as their participation in protein synthesis, mRNA degradation, and pre-mRNA splicing. The study of lncRNA structures and their dynamic alterations in response to environmental stimuli will be made easier by advancements in structural biology tools, which will also shed light on their regulator. A comparative genomics study will provide further insight into the evolutionary conservation of lncRNAs and aid in the identification of conserved sequences and structural motifs between species. Moreover, lncRNAs have the potential to be biomarkers for diagnosis and prognosis as well as prospective targets for therapeutic intervention in illnesses, promoting the use of precision medicine techniques.

References

An Y, Furber KL, Ji S (2017) Pseudogenes regulate parental gene expression via ce RNA network. J Cell Mol Med 21:185–192

Ariel F, Romero-Barrios N, Jégu T, Benhamed M, Crespi M (2015) Battles and hijacks: noncoding transcription in plants. Trends Plant Sci 20:362–371

Ashwal-Fluss R, Meyer M, Pamudurti NR, Ivanov A, Bartok O, Hanan M, Evantal N, Memczak S, Rajewsky N, Kadener S (2014) circRNA biogenesis competes with pre-mRNA splicing. Mol Cell 56:55–66

Axtell MJ, Jan C, Rajagopalan R, Bartel DP (2006) A two-hit trigger for siRNA biogenesis in plants. Cell 127:565–577

Bardou F, Ariel F, Simpson CG, Romero-Barrios N, Laporte P, Balzergue S, Brown JW, Crespi M (2014) Long noncoding RNA modulates alternative splicing regulators in Arabidopsis. Dev Cell 30:166–176

Bazin J, Romero N, Rigo R, Charon C, Blein T, Ariel F, Crespi M (2018) Nuclear speckle RNA binding proteins remodel alternative splicing and the non-coding Arabidopsis transcriptome to regulate a cross-talk between auxin and immune responses. Front Plant Sci 9:1209

Burkhart KB, Guang S, Buckley BA, Wong L, Bochner AF, Kennedy S (2011) A pre-mRNA–associating factor links endogenous siRNAs to chromatin regulation. PLoS Genet 7:e1002249

Büssing I, Slack FJ, Großhans H (2008) let-7 microRNAs in development, stem cells and cancer. Trends Mol Med 14:400–409

Cabili MN, Trapnell C, Goff L, Koziol M, Tazon-Vega B, Regev A, Rinn JL (2011) Integrative annotation of human large intergenic noncoding RNAs reveals global properties and specific subclasses. Genes Dev 25:1915–1927

Cao J (2014) The functional role of long non-coding RNAs and epigenetics. Biol Proced Online 16:1–13

Carthew RW, Sontheimer EJ (2009) Origins and mechanisms of miRNAs and siRNAs. Cell 136:642–655

Chen J, Sun M, Kent WJ, Huang X, Xie H, Wang W, Zhou G, Shi RZ, Rowley JD (2004) Over 20% of human transcripts might form sense–antisense pairs. Nucleic Acids Res 32:4812–4820

Chu Q, Shen E, Ye C-Y, Fan L, Zhu Q-H (2018) Emerging roles of plant circular RNAs. J Plant Cell Dev 1:1–14

Chu Q, Bai P, Zhu X, Zhang X, Mao L, Zhu Q-H, Fan L, Ye C-Y (2020) Characteristics of plant circular RNAs. Brief Bioinform 21:135–143

Comings DE (1972) The structure and function of chromatin. Adv Human Genet:237–431

Conn VM, Hugouvieux V, Nayak A, Conos SA, Capovilla G, Cildir G, Jourdain A, Tergaonkar V, Schmid M, Zubieta C (2017) AcircRNA from SEPALLATA3 regulates splicing of its cognate mRNA through R-loop formation. Nat Plants 3:1–5

Cui J, Jiang N, Hou X, Wu S, Zhang Q, Meng J, Luan Y (2020) Genome-wide identification of lncRNAs and analysis of ceRNA networks during tomato resistance to *Phytophthorainfestans*. Phytopathology 110(2):456–464

Dempsey JL, Cui JY (2017) Long non-coding RNAs: a novel paradigm for toxicology. Toxicol Sci 155:3–21
Derrien T, Johnson R, Bussotti G, Tanzer A, Djebali S, Tilgner H, Guernec G, Martin D, Merkel A, Knowles DG (2012) The GENCODE v7 catalog of human long noncoding RNAs: analysis of their gene structure, evolution, and expression. Genome Res 22:1775–1789
Djebali S, Davis CA, Merkel A, Dobin A, Lassmann T, Mortazavi A, Tanzer A, Lagarde J, Lin W, Schlesinger F (2012) Landscape of transcription in human cells. Nature 489:101–108
Dong Q, Hu B, Zhang C (2022) microRNAs and their roles in plant development. Front Plant Sci 13:824240
Du Q, Wang K, Zou C, Xu C, Li W-X (2018) The PILNCR1-miR399 regulatory module is important for low phosphate tolerance in maize. Plant Physiol 177:1743–1753
Dykes IM, Emanueli C (2017) Transcriptional and post-transcriptional gene regulation by long non-coding RNA. Genomics Proteomics Bioinformatics 15:177–186
Ebbesen KK, Kjems J, Hansen TB (2016) Circular RNAs: identification, biogenesis and function. Biochim Biophys Acta 1859:163–168
Elbashir SM, Lendeckel W, Tuschl T (2001) RNA interference is mediated by 21-and 22-nucleotide RNAs. Genes Dev 15:188–200
Elvira-Matelot E, Bardou F, Ariel F, Jauvion V, Bouteiller N, Le Masson I, Cao J, Crespi MD, Vaucheret H (2016) The nuclear ribonucleoprotein SmD1 interplays with splicing, RNA quality control, and posttranscriptional gene silencing in Arabidopsis. Plant Cell 28:426–438
Esteller M (2011) Non-coding RNAs in human disease. Nat Rev Genet 12:861–874
Fan C, Hao Z, Yan J, Li G (2015) Genome-wide identification and functional analysis of lincRNAs acting as miRNA targets or decoys in maize. BMC Genomics 16:793. https://doi.org/10.1186/s12864-015-2024-0
Fei Q, Yang L, Liang W, Zhang D, Meyers BC (2016) Dynamic changes of small RNAs in rice spikelet development reveal specialized reproductive phasiRNA pathways. J Exp Bot 67:6037–6049
Fire A (1999) RNA-triggered gene silencing. Trends Genet 15:358–363
Franco-Zorrilla JM, Valli A, Todesco M, Mateos I, Puga MI, Rubio-Somoza I, Leyva A, Weigel D, García JA, Paz-Ares J (2007) Target mimicry provides a new mechanism for regulation of microRNA activity. Nat Genet 39:1033–1037
Gao C, Sun J, Dong Y, Wang C, Xiao S, Mo L, Jiao Z (2020) Comparative transcriptome analysis uncovers regulatory roles of long non-coding RNAs involved in resistance to powdery mildew in melon. BMC Genomics 21:1–16
Geisler S, Coller J (2013) RNA in unexpected places: long non-coding RNA functions in diverse cellular contexts. Nat Rev Mol Cell Biol 14:699–712
Gong C, Maquat LE (2011) lncRNAstransactivate STAU1-mediated mRNA decay by duplexing with 3′ UTRs via Alu elements. Nature 470:284–288
Grainger RJ, Beggs JD (2005) Prp8 protein: at the heart of the spliceosome. RNA 11:533–557
Guria A, VelayudhaVimala Kumar K, Srikakulam N, Krishnamma A, Chanda S, Sharma S, Fan X, Pandi G (2019) Circular RNA profiling by Illumina sequencing via template-dependent multiple displacement amplification. Biomed Res Int 2019:2756516
Guttman M, Amit I, Garber M, French C, Lin MF, Feldser D, Huarte M, Zuk O, Carey BW, Cassady JP (2009) Chromatin signature reveals over a thousand highly conserved large non-coding RNAs in mammals. Nature 458:223–227
Guttman M, Russell P, Ingolia NT, Weissman JS, Lander ES (2013) Ribosome profiling provides evidence that large noncoding RNAs do not encode proteins. Cell 154:240–251
Hamid R, Jacob F, Marashi H, Rathod V, Tomar RS (2020) Uncloaking lncRNA-meditated gene expression as a potential regulator of CMS in cotton (Gossypiumhirsutum L.). Genomics 112:3354–3364
Hangauer MJ, Vaughn IW, McManus MT (2013) Pervasive transcription of the human genome produces thousands of previously unidentified long intergenic noncoding RNAs. PLoS Genet 9:e1003569

Harrow J, Frankish A, Gonzalez JM, Tapanari E, Diekhans M, Kokocinski F, Aken BL, Barrell D, Zadissa A, Searle S (2012) GENCODE: the reference human genome annotation for The ENCODE Project. Genome Res 22:1760–1774

Hou X, Cui J, Liu W, Jiang N, Zhou X, Qi H, Meng J, Luan Y (2020) LncRNA39026 enhances tomato resistance to *Phytophthorainfestans* by decoying miR168a and inducing *PR* gene expression. Phytopathology 110(4):873–880

Howell MD, Fahlgren N, Chapman EJ, Cumbie JS, Sullivan CM, Givan SA, Kasschau KD, Carrington JC (2007) Genome-wide analysis of the RNA-DEPENDENT RNA POLYMERASE6/DICER-LIKE4 pathway in Arabidopsis reveals dependency on miRNA-and tasiRNA-directed targeting. Plant Cell 19:926–942

Hrdlickova B, de Almeida RC, Borek Z, Withoff S (2014) Genetic variation in the non-coding genome: involvement of micro-RNAs and long non-coding RNAs in disease. Biochim Biophys Acta 1842:1910–1922

Huang L, Dong H, Zhou D, Li M, Liu Y, Zhang F, Feng Y, Yu D, Lin S, Cao J (2018) Systematic identification of long non-coding RNA s during pollen development and fertilization in *Brassica rapa*. Plant J 96(1):203–222

Ivanov A, Memczak S, Wyler E, Torti F, Porath HT, Orejuela MR, Piechotta M, Levanon EY, Landthaler M, Dieterich C (2015) Analysis of intron sequences reveals hallmarks of circular RNA biogenesis in animals. Cell Rep 10:170–177

Jen C-H, Michalopoulos I, Westhead DR, Meyer P (2005) Natural antisense transcripts with coding capacity in Arabidopsis may have a regulatory role that is not linked to double-stranded RNA degradation. Genome Biol 6:1–10

Jerome Jeyakumar JM, Ali A, Wang W-M, Thiruvengadam M (2020) Characterizing the role of the miR156-SPL network in plant development and stress response. Plan Theory 9:1206

Jiang N, Cui J, Hou X, Yang G, Xiao Y, Han L, Meng J, Luan Y (2020) Sl-lncRNA15492 interacts with Sl-miR482a and affects *Solanumlycopersicum* immunity against *Phytophthorainfestans*. Plant J 103:1561–1574

Jin J, Lu P, Xu Y, Li Z, Yu S, Liu J, Wang H, Chua N-H, Cao P (2021) PLncDB V2. 0: a comprehensive encyclopedia of plant long noncoding RNAs. Nucleic Acids Res 49:D1489–D1495

Jovanovic M, Hengartner MO (2006) miRNAs and apoptosis: RNAs to die for. Oncogene 25:6176–6187

Kapusta A, Kronenberg Z, Lynch VJ, Zhuo X, Ramsay L, Bourque G, Yandell M, Feschotte C (2013) Transposable elements are major contributors to the origin, diversification, and regulation of vertebrate long noncoding RNAs. PLoS Genet 9:e1003470

Kim VN, Han J, Siomi MC (2009) Biogenesis of small RNAs in animals. Nat Rev Mol Cell Biol 10:126–139

Lee H, Zhang Z, Krause HM (2019) Long noncoding RNAs and repetitive elements: junk or intimate evolutionary partners? Trends Genet 35:892–902

Lewis BP, Burge CB, Bartel DP (2005) Conserved seed pairing, often flanked by adenosines, indicates that thousands of human genes are microRNA targets. Cell 120:15–20

Li S, Yu X, Lei N, Cheng Z, Zhao P, He Y, Wang W, Peng M (2017) Genome-wide identification and functional prediction of cold and/or drought-responsive lncRNAs in cassava. Sci Rep 7:45981

Liu X, Li D, Zhang D, Yin D, Zhao Y, Ji C, Zhao X, Li X, He Q, Chen R, Hu S, Zhu L (2018) A novel antisense long noncoding RNA , TWISTED LEAF, maintains LEAF blade flattening by regulating its associated sense R2R3- MYB gene in rice. New Phytol 218:774–788

Lone JK, Mushtaq M, Gupta OP (2023) Future prospective of small RNA molecules: food crop improvement and agricultural sustainability. In: Plant Small RNA in Food Crops. Elsevier, pp 571–600

Lu T, Cui L, Zhou Y, Zhu C, Fan D, Gong H, Zhao Q, Zhou C, Zhao Y, Lu D (2015) Transcriptome-wide investigation of circular RNAs in rice. RNA 21:2076–2087

Ma H, Yang T, Li Y, Zhang J, Wu T, Song T, Yao Y, Tian J (2021) The long noncoding RNA MdLNC499 bridges MdWRKY1 and MdERF109 function to regulate early-stage light-induced anthocyanin accumulation in apple fruit. Plant Cell 33:3309–3330

Milligan MJ, Lipovich L (2015) Pseudogene-derived lncRNA s: emerging regulators of gene expression. Front Genet 5:476

Moazed D (2009) Small RNA s in transcriptional gene silencing and genome defence. Nature 457:413–420

Nekrutenko A, Li W-H (2001) Transposable elements are found in a large number of human protein-coding genes. Trends Genet 17:619–621

Ohno S (1972) So much" junk" DNA in our genome.In" Evolution of Genetic Systems". In: Brookhaven Symposium in Biology. pp 366–370

Palos K, Yu L, Railey CE, Nelson Dittrich AC, Nelson AD (2023) Linking discoveries, mechanisms, and technologies to develop a clearer perspective on plant long noncoding RNA s. Plant Cell 35:1762–1786

Palusa SG, Reddy AS (2015) Differential recruitment of splice variants from SR pre-mRNA s to polysomes during development and in response to stresses. Plant Cell Physiol 56:421–427

Pandey RR, Mondal T, Mohammad F, Enroth S, Redrup L, Komorowski J, Nagano T, Mancini-DiNardo D, Kanduri C (2008) Kcnq1ot1 antisense noncoding RNA mediates lineage-specific transcriptional silencing through chromatin-level regulation. Mol Cell 32:232–246

Park E, Maquat LE (2013) Staufen-mediated mRNA decay. Wiley Interdiscip Rev RNA 4:423–435

Pastori C, Peschansky VJ, Barbouth D, Mehta A, Silva JP, Wahlestedt C (2014) Comprehensive analysis of the transcriptional landscape of the human FMR1 gene reveals two new long noncoding RNA s differentially expressed in Fragile X syndrome and Fragile X-associated tremor/ataxia syndrome. Hum Genet 133:59–67

Poliseno L, Salmena L, Zhang J, Carver B, Haveman WJ, Pandolfi PP (2010) A coding-independent function of gene and pseudogene mRNA s regulates tumour biology. Nature 465:1033–1038

Popadin K, Gutierrez-Arcelus M, Dermitzakis ET, Antonarakis SE (2013) Genetic and epigenetic regulation of human lincRNA gene expression. Am J Hum Genet 93:1015–1026

Quinn JJ, Chang HY (2016) Unique features of long non-coding RNA biogenesis and function. Nat Rev Genet 17:47–62

Rigo R, Bazin J, Romero-Barrios N, Moison M, Lucero L, Christ A, Benhamed M, Blein T, Huguet S, Charon C (2020) The Arabidopsis lnc RNA ASCO modulates the transcriptome through interaction with splicing factors. EMBO Rep 21:e48977

Saini HK, Griffiths-Jones S, Enright AJ (2007) Genomic analysis of human microRNA transcripts. Proc Natl Acad Sci 104:17719–17724

Salmena L, Poliseno L, Tay Y, Kats L, Pandolfi PP (2011) AceRNA hypothesis: the Rosetta Stone of a hidden RNA language? Cell 146:353–358

Schickel R, Boyerinas B, Park SM, Peter ME (2008) MicroRNA s: key players in the immune system, differentiation, tumorigenesis and cell death. Oncogene 27:5959–5974

Severing EI, van Dijk AD, Morabito G, Busscher-Lange J, Immink RG, van Ham RC (2012) Predicting the impact of alteRNA tive splicing on plant MADS domain protein function. PLoS One 7:e30524

Sun X, Wang L, Ding J, Wang Y, Wang J, Zhang X, Che Y, Liu Z, Zhang X, Ye J, Wang J, Sablok G, Deng Z, Zhao H (2016) Integrative analysis of *Arabidopsis thaliana*transcriptomics reveals intuitive splicing mechanism for circular RNA. FEBS Lett 590:3510–3516

Thomas M, Lieberman J, Lal A (2010) Desperately seeking microRNA targets. Nat Struct Mol Biol 17:1169–1174

Tran VDT, Souiai O, Romero-Barrios N, Crespi M, Gautheret D (2016) Detection of generic differential RNA processing events from RNA -seq data. RNA Biol 13:59–67

Tremblay BJM, Qüesta JI (2023) Mechanisms of epigenetic regulation of transcription by LNCRNA s in plants. IUBMB Life 75:427–439

Tuschl T (2001) RNA interference and small interfering RNAs. Chembiochem 2:239–245

Wang KC, Chang HY (2011) Molecular mechanisms of long noncoding RNAs. Mol Cell 43:904–914

Wang H, Chua N-H, Wang X-J (2006) Prediction of trans-antisense transcripts in Arabidopsis thaliana. Genome Biol 7:1–11

Wu H-J, Wang Z-M, Wang M, Wang X-J (2013) Widespread long noncoding RNA s as endogenous target mimics for microRNA s in plants. Plant Physiol 161:1875–1884
Xia R, Chen C, Pokhrel S, Ma W, Huang K, Patel P, Wang F, Xu J, Liu Z, Li J (2019) 24-nt reproductive phasiRNA s are broadly present in angiosperms. Nat Commun 10:627
Yan K, Liu P, Wu C-A, Yang G-D, Xu R, Guo Q-H, Huang J-G, Zheng C-C (2012) Stress-induced alteRNA tive splicing provides a mechanism for the regulation of microRNA processing in Arabidopsis thaliana. Mol Cell 48:521–531
Yang Y, Liu T, Shen D, Wang J, Ling X, Hu Z, Chen T, Hu J, Huang J, Yu W (2019) Tomato yellow leaf curl virus intergenic siRNA s target a host long noncoding RNA to modulate disease symptoms. PLoS Pathog 15:e1007534
Ye C, Chen L, Liu C, Zhu Q, Fan L (2015) Widespread noncoding circular RNAs in plants. New Phytol 208:88–95
Yoshikawa M, Peragine A, Park MY, Poethig RS (2005) A pathway for the biogenesis of trans-acting siRNA s in Arabidopsis. Genes Dev 19:2164–2175
Zhang Y, Wang W, Chen J, Liu J, Xia M, Shen F (2015) Identification of miRNA s and their targets in cotton inoculated with Verticilliumdahliae by high-throughput sequencing and degradome analysis. Int J Mol Sci 16:14749–14768
Zhang P, Wu W, Chen Q, Chen M (2019) Non-coding RNA s and their integrated networks. J Integr Bioinform 16:20190027
Zhao W, Cheng Y, Zhang C, You Q, Shen X, Guo W, Jiao Y (2017) Genome-wide identification and characterization of circular RNA s by high throughput sequencing in soybean. Sci Rep 7:5636
Zhao X, Li J, Lian B, Gu H, Li Y, Qi Y (2018) Global identification of Arabidopsis lncRNA s reveals the regulation of MAF4 by a natural antisense RNA. Nat Commun 9:5056
Zhao S, Meng J, Luan Y (2022) LncRNA -encoded short peptides identification using feature subset recombination and ensemble learning. Interdisciplinary Sciences: Computational Life Sciences 1–12
Zhou X, Cui J, Meng J, Luan Y (2020) Interactions and links among the noncoding RNA s in plants under stresses. Theor Appl Genet 133:3235–3248

Interaction Between Small and Long Noncoding RNAs

4

Shivani Singh, Sharad Kumar Jaiswal, and Budhayash Gautam

Abstract

Noncoding RNAs (ncRNAs) have become crucial components of gene regulatory networks in numerous plant species. Short noncoding RNAs (sRNAs), specifically 20–30-nucleotide ones, play significant roles in modulating developmental processes and responses to biological stress. Various classes of sRNAs have been identified, and their distinct functions are being characterized based on differences in their biogenesis mechanisms. This article examines the current understanding of major classes of sRNAs, such as microRNAs, small nuclear RNAs (snRNAs), and secondary classes. In eukaryotic genomes, a wide range of ncRNAs, including long ncRNAs (lncRNAs) and microRNAs (miRNAs) are crucial for transcriptional and posttranscriptional control of gene expression. Evidence suggests that these ncRNAs play important regulatory roles in plant stress responses. High-throughput sequencing has revealed that the majority of RNAs lack the ability to code for proteins. lncRNAs are transcripts longer than 200 nucleotides that do not encode proteins. Computational methods have uncovered numerous lncRNAs in various plant species. LncRNAs interact with DNA, RNA, and protein molecules to regulate the expression of their target genes at the

S. Singh
Department of Biological Sciences, Faculty of Science, Sam Higginbottom University of Agriculture, Technology & Sciences, Prayagraj, Uttar Pradesh, India

S. K. Jaiswal · B. Gautam (✉)
Department of Computational Biology and Bioinformatics, Jacob Institute of Biotechnology & Bioengineering, Sam Higginbottom University of Agriculture, Technology & Sciences, Prayagraj, Uttar Pradesh, India
e-mail: budhayash.gautam@shiats.edu.in

K. Goswami et al. (eds.), *Non-coding RNAs for Crop Improvement*,
https://doi.org/10.1007/978-981-97-7213-1_4

epigenetic, transcriptional, posttranscriptional, or translational levels. This chapter provides examples of miRNA and lncRNA-mediated plant stress tolerance mechanisms, discusses lncRNA biogenesis and regulatory processes, and summarizes ongoing research on the identification, characterization, bioinformatics tools, and resources for plant miRNAs and lncRNAs.

Keywords

Noncoding RNA · Small RNA · Long noncoding RNA · Databases · Gene regulation

4.1 Introduction

Noncoding RNAs (ncRNAs) are RNA molecules transcribed from DNA that do not encode proteins but play important roles in various epigenetic regulatory processes. Their regulatory effects are primarily mediated by complementary base pairing with messenger RNAs (mRNAs) (Phillips 2008). Despite their significance, ncRNAs were often overlooked in genomic data due to their small size and were considered background noise or junk DNA. However, recent advancements in next-generation sequencing (NGS) technology have enabled high-throughput experimental investigations on ncRNAs, providing valuable information about their coverage, diversity, and potential target gene annotations (Eddy 2001; Shimoni et al. 2007).

Over time, ncRNAs have been categorized into different classes based on their biogenesis routes, with some classes having distinct roles but overlapping functions (Van Bonnet et al. 2006). The broadest class of ncRNAs is small noncoding RNAs (snRNAs) and Piwi-interacting RNAs (piRNAs) (Budak and Akpinar 2015; Borges and Martienssen 2015). Additionally, smaller classes of species-specific small RNAs (sRNAs) are emerging. These, collectively called RNA interfering (RNAi), are important in the posttranscriptional regulation of plant genes by targeting mRNA for sRNA degradation or translational silencing (Pattanayak et al. 2012).

RNAi, including ncRNAs, has been found to significantly impact various stress responses, including disease resistance, which is particularly important in plants (Kantar et al. 2011; Budak and Akpinar 2015). Plants constantly engage in co-evolutionary conflicts with pests and pathogens, resulting in substantial crop losses. Plant diseases alone have led to global crop losses of 10–16% despite current disease management practices (Kandan et al. 2016). Crop species such as wheat (*Triticum aestivum*) and rice (*Oryza sativa*) have been severely affected. With climate change favoring invasive species, the impact of diseases and invasions is expected to increase (Boyd et al. 2013; Gautam et al. 2013). RNAi plays a crucial role in regulating resistance genes in host plants and targeting genes of invading pathogens or pests, making it a key player in this ongoing battle (Ghag 2017; Islam et al. 2018). Host-induced gene silencing (HIGS) has been observed, offering new avenues for crop disease management and the potential transfer of mRNA between species (Baulcombe 2015).

ncRNAs can be classified based on their length into small ncRNAs (18–30 nucleotides), medium-sized ncRNAs (31–200 nucleotides), and lncRNAs (greater than 200 nucleotides). Among them, plant miRNAs are a well-known class of small ncRNAs that are typically 21–24 nucleotides in length. They play a role in suppressing the expression of target genes by directing the degradation or inhibition of translation of target mRNAs (Rajagopalan et al. 2006). In addition to miRNAs, a large number of lncRNAs have been discovered in plants. Initially, lncRNAs were considered transcriptional "noise" due to their low expression levels and sequence conservation compared to protein-coding mRNAs (Liu et al. 2015). However, growing evidence indicates that certain lncRNAs regulate gene expression through cis- or trans-regulation (Liu et al. 2015).

Plants face various stressors, including viral infections, salinity, drought, cold, and heat, which hinder growth and productivity. To cope with these unfavorable conditions, plants have evolved multiple gene regulatory systems to maintain cellular homeostasis. Recent research has highlighted the essential role of ncRNAs, particularly siRNAs and lncRNAs, in controlling gene expression in response to stress (Xu et al. 2016). The advent of NGS technology and bioinformatics approaches has greatly contributed to the discovery and characterization of numerous ncRNAs in plants.

4.1.1 Discovery and Classification of LncRNAs in Plants

The identification of lncRNAs in plants has greatly expanded our knowledge of molecular biology and gene regulation. Longer than 200 nucleotides, or LncRNAs, are transcripts that are essential to many cellular functions but do not code for proteins. Plant molecular biology was initially primarily concerned with genes that coded for proteins. However, as high-throughput sequencing technologies and bioinformatics advancements became more common, scientists started delving deeper into the noncoding region of the genome.

According to studies, lncRNAs play significant roles in the growth, development, and reaction of plants to abiotic stress. Figure 4.1 represents the timelines of lncRNAs discovery.

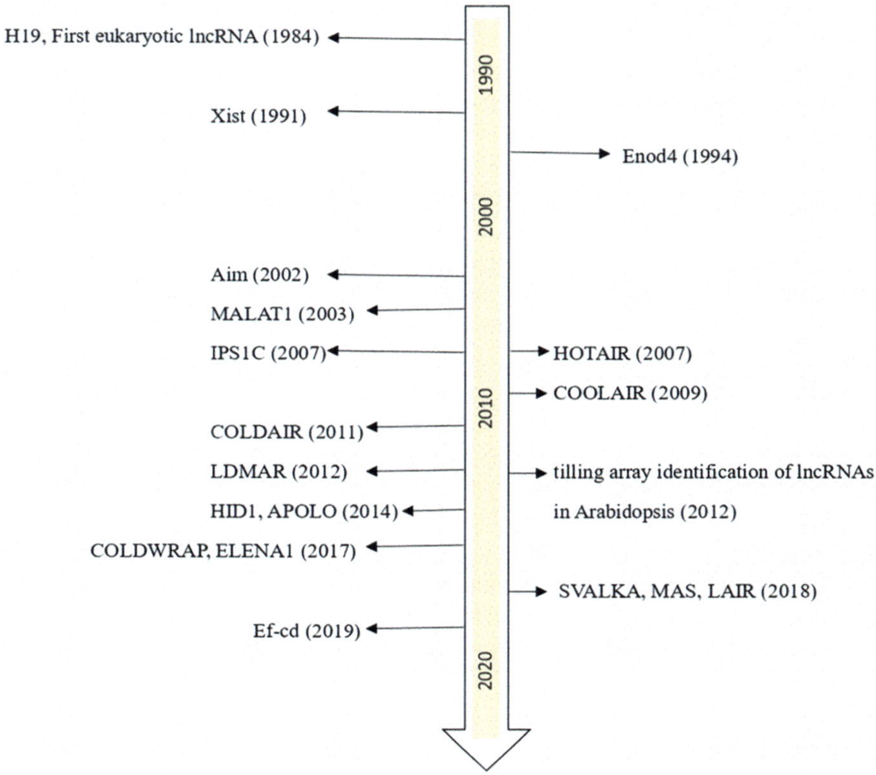

Fig. 4.1 Discovery and classification of lncRNAs

4.1.2 Characteristics of lncRNAs

4.1.2.1 Number and Size of lncRNA Transcripts

Numerous plant species, including Arabidopsis, rice, and maize, have been found to possess lncRNAs and have lengths of more than 200 nucleotides. In Arabidopsis, a comprehensive examination uncovered approximately 6480 (Liu et al. 2012) to 6510 (Zhao et al. 2018) lncRNAs (Table 4.1). LncRNAs generally exhibit fewer exons and shorter lengths compared to protein-coding mRNAs. Some lncRNAs contain open reading frames (ORFs) capable of generating short peptides, although it is unclear whether these peptides have functional roles (Yin et al. 2020).

4.1.2.2 Expression Specification and Functionality

Expression of lncRNAs is often more tissue-specific compared to mRNAs of protein-coding genes. Approximately 32% of Arabidopsis lncRNAs exhibit organ-specific expression patterns, which can be validated using techniques like qRT-PCR (Liu et al. 2012). Due to their specificity, lncRNAs can serve as valuable markers for

Table 4.1 Summary list of key databases for lncRNA identification and annotation in plants

Name	Description	Weblink	Reference
GREENC	A wiki-based database of 37 plant species' lncRNAs	http://greenc.sciencedesigners.com/	Paytuví Gallart et al. (2016)
lncRNAdb	Literature outlining lncRNA functions	http://lncRNAdb.org	Quek et al. (2015)
PLncDB	lncRNAs from Arabidopsis	http://chualab.rockefeller.edu/gbrowse2/homepage.html	Jin et al. (2013)
PNRD	lncRNAs primarily from four plant species: Poplar, maize, rice, and Arabidopsis	http://structuralbiology.cau.edu.cn/PNRD/	Yi et al. (2015)
CANTATAdb	Ten model plant species' lncRNAs	http://yeti.amu.edu.pl/CANTATA/	Szcześniak et al. (2019)
PLNlncRbase	Ten model plant species' lncRNAs	http://bioinformatics.ahau.edu.cn/PLNlncRbase/pcsb	Xuan et al. (2015)

tissues and developmental stages. The generally low expression levels of lncRNAs and the limitations of traditional mRNA sequencing methods in terms of detection could contribute to their apparent selectivity.

4.1.2.3 Salient Features of lncRNAs

Biotypes

Based on the position of lncRNAs on chromosomes relative to nearby genes, various lncRNA biotypes have been identified (Budak et al. 2020). There are five main categories of lncRNAs: (1) intergenic, located between two genes; (2) intronic, derived from introns; (3) sense-semantic, spanning exons on the same strand; (4) antisense, spanning exons on the opposite strand; and (5) bidirectional, where the expression of the lncRNA and the adjacent coding transcript on the opposite strand is initiated in close genomic proximity (Boller and He 2009).

Similarity to mRNAs

Similarities exist between lncRNAs and mRNAs in terms of the RNA polymerases responsible for transcription, polyadenylation, capping, 5′ modifications, and alternative splicing models (Derrien et al. 2012). However, lncRNAs typically lack the ability to code for proteins (Karapetyan et al. 2013). During splicing, exon/intron lengths of most lncRNAs are comparable to those of mRNA-coding genes (Guttman et al. 2009). However, certain long intergenic ncRNAs (lincRNAs) in plants like chickpea and cotton have occasionally been found to have much longer average exon lengths compared to mRNAs (Dinger et al. 2008).

Tissue-Specificity

Tissue-specific expression levels of lncRNAs have been consistently observed in mammals, suggesting their involvement in differentiation, development, repair, maintenance, and other activities (Dinger et al. 2008). Recent investigations in plants have also identified tissue-specific lncRNAs expressed at different developmental stages. These studies have shown the involvement of lncRNAs in fiber development in cotton, flower development in chickpea, strawberry, and mulberry, as well as floral organs and roots in mulberry (Kang and Liu 2015; Khemka et al. 2016; Song et al. 2016; Dinger et al. 2008).

Cell-Type-Specificity

In studies analyzing tissue specificity of lncRNAs, there is a general trend of lower expression levels compared to protein-coding genes. This may be due to chronically low expression in all cells or restricted expression in specific cell types or subpopulations (Gao et al. 2016). Cell type-specific lncRNAs have been extensively discovered in both animals and humans. Single-cell transcriptomic has revealed distinct and abundant expression of lncRNAs, such as LOC646329 in radial glia subpopulations of the human neocortex during development (Wang et al. 2016). Similarly, cell type-specific lincRNAs have been found in Arabidopsis roots, although their expression levels were lower than those of mRNAs even in specialized cell types (Wang et al. 2016).

Mechanisms of Action

lncRNAs have been implicated in various mechanisms of action beyond their role in protein-coding genes. They are involved in pre-transcriptional regulation, transcriptional control, and posttranscriptional regulation, and serve as scaffolds, targets for endogenous miRNA mimics, and precursors for other noncoding RNAs (Zhu and Wang 2012; Quek et al. 2015).

4.2 Regulatory Interactions Between Small and Long Noncoding RNAs

A complex and dynamic network in cellular processes is formed by the regulatory interactions between lncRNAs and sncRNAs, such as miRNAs and siRNAs. At different levels, these interactions play a major role in fine-tuning gene expression (Bardou et al. 2014).

4.2.1 Processing of lncRNAs into Small RNAs

LncRNAs can sometimes be converted into smaller, useful RNAs such as miRNAs, siRNAs competing ceRNA. Dicer-like enzymes, for instance, have the ability to break certain long noncoding RNAs (lncRNAs) into smaller pieces, which helps to produce small RNAs (Karapetyan et al. 2013).

4.2.2 miRNA-Mediated Regulation of lncRNAs

Certain miRNAs have the ability to target and control the expression of particular lncRNAs. When the miRNA identifies complementary sequences in the lncRNA, translational repression or lncRNA degradation results (Zhai et al. 2011; Zhang et al. 2019). For example, It has been demonstrated that the lncRNA COOLAIR in *Arabidopsis thaliana* recruits the PRC2 complex to control the expression of the FLC gene. This control affects flowering time by interaction with miR-782, which targets COOLAIR (Csorba et al. 2014).

4.2.3 Competitive Endogenous RNA (ceRNA) Mechanism

By severing miRNAs from their target mRNAs, lncRNAs can function as competitive endogenous RNAs. The availability of miRNAs is influenced by this ceRNA crosstalk, which subsequently affects the expression of genes that code for proteins (Henriques et al. 2017).

4.2.4 siRNA-Mediated Regulation of lncRNAs

Under certain conditions, siRNA-triggered lncRNA degradation. siRNAs produced through RNAi pathways can direct the degradation of particular lncRNAs. The regulation of lncRNA levels in response to viral infections or environmental stimuli is facilitated by this process (Baldrich et al. 2014). For example, the lncRNA lncRNA16397 interacts with the siRNA ZMR4 and the miRNA miR2118 in maize (*Zea mays*) to produce a complex that regulates genes linked to plant protection against diseases.

4.2.5 Target Mimicry

An lncRNA may include sequences complementary to miRNAs in this method. LncRNAs can behave as decoys or sponges by attaching to miRNAs and sequestering them from their target mRNAs, preventing them from carrying out their regulatory actions. Through this process, miRNAs' availability and activity are efficiently modulated, which indirectly affects the expression of miRNA-target genes (Franco-Zorrilla et al. 2007), an example of target mimicry involving the lncRNA IPS1.

4.2.6 RNA-RNA Duplex Formation

In addition to acting as miRNA sponges, lncRNAs can form RNA-RNA duplexes with miRNAs through partial sequence complementarity. This interaction can lead to the degradation or sequestration of miRNAs, thereby influencing miRNA availability and activity (Baldrich et al. 2014).

The example of miR156 and its target mRNA SPL (Squamosa Promoter-Binding Protein-Like) in plants:

miR156: miR156 is a conserved miRNA in plants that plays a crucial role in regulating various aspects of plant development, including the timing of vegetative phase change, leaf morphology, and flowering time.

SPL: SPL genes encode transcription factors that are involved in the regulation of plant developmental processes. SPL mRNAs contain binding sites for miR156 in their 3′ untranslated regions (UTRs) (Jung et al. 2012).

4.3 LncRNA and sRNA Cooperative Pathway

In plants, the regulated path that involves sRNAs and lncRNAs is a synchronized network of regulatory proteins that is important for cellular activities, gene expression control, and other biological processes. lncRNAs, siRNAs and miRNAs, are important regulators of posttranscriptional gene expression and have the ability to interact with each other (Franco-Zorrilla et al. 2007). Here are some examples of cooperative actions where sRNAs and lncRNAs work together to regulate specific pathways in plants:

4.3.1 Plant Development and Stress Responses

Several studies have highlighted the collaboration between miRNAs and lncRNAs in regulating plant growth and stress responses (Le et al. 2014). For instance, during abiotic stresses like drought or high salinity, specific miRNAs target transcripts encoding stress-responsive proteins, while lncRNAs can modulate the expression of these miRNAs or function as precursors for small RNAs involved in stress adaptation (Khraiwesh et al. 2012). This cooperative action assists in fine-tuning gene expression patterns to help plants cope with environmental stresses while maintaining growth and development. For example, a long noncoding RNA, ALEX1 (Antisense to Long day-specific Expression 1) has been identified in the well-studied model plant *Arabidopsis thaliana*. For the miR166, which is involved in leaf development, ALEX1 functions as a molecular sponge (Yu et al. 2019). ALEX1 affects the expression of its target genes, such as the homeobox transcription factor gene, which is involved in stress responses and leaf development, by sequestering miR166 (Wang et al. 2017). Another example shows how lncRNA Tt40179 interacts with a circular RNA (circTt40179) in wheat under drought stress. The expression of genes linked to the drought stress response, such as those involved in ABA signaling, is influenced by both ncRNAs. The plant's capacity to withstand drought is enhanced by this cooperative effort (Sunkar et al. 2012).

4.3.2 ncRNA-miRNA-mRNA Axis in Flowering Regulation

In plants, flowering time is a crucial developmental process regulated by multiple signaling pathways. Both miRNAs and lncRNAs participate in the regulation of flowering time genes (Castaings et al. 2014). Certain miRNAs target mRNAs encoding transcription factors involved in flowering time control, while lncRNAs might act as molecular decoys or guides to modulate the activity of these miRNAs. This cooperative regulation ensures proper timing for flowering, which is essential for reproductive success. For example, in maize, miR399, which targets the PHO2 gene involved in phosphate homeostasis, interacts with the long noncoding RNA INDUCED BY PHOSPHATE STARVATION 1 (IPS1). By acting as a semblance of miR399, IPS1 decoys it from binding to PHO2 and permit its expression. Plants' ability to adapt to phosphate deficiency is influenced by this regulatory axis (Michaels and Amasino 1999).

4.3.3 The Role of miRNA-circRNA in Root Development

It has been discovered that rice contains a circular RNA (circLCT1) that absorbs miR1432. Many target genes, including a receptor-like kinase involved in root development, are influenced by this interaction in terms of their expression. In rice plants, the cooperative activity of miR1432 and circLCT1 helps to regulate root architecture.

4.3.4 The Role of lncRNA and miRNAin Nutrient Homeostasis and Signaling

Noncoding RNAs, including miRNAs and lncRNAs, participate in the regulation of nutrient uptake, transport, and signaling pathways in plants. miRNAs can target transcripts involved in nutrient uptake and homeostasis, while lncRNAs may interact with these miRNAs to modulate their expression or processing (Zhao et al. 2018). This cooperation helps in maintaining nutrient balance and proper plant growth. For example, lncRNA IPS1 (INDUCED BY PHOSPHATE STARVATION 1) in *Arabidopsis thaliana* functions as a molecular sponge for miR399. miR399 typically targets the PHO2 gene, which regulates the expression of phosphate transporters and is involved in maintaining phosphate homeostasis. In the event of phosphate deficiency, IPS1 binds to miR399 and inhibits its interaction with PHO2, increasing PHO2 expression to enhance phosphate uptake.

4.3.5 The Role of lncRNA and miRNA in Defense Responses against Pathogens

During pathogen infection, plants employ intricate defense mechanisms involving various regulatory pathways. Both miRNAs and lncRNAs contribute to the regulation of defense-related genes (Lagarde et al. 2016). Certain miRNAs target transcripts encoding defense proteins, and lncRNAs might interact with these miRNAs to influence their activity or stability. This collaboration aids in mounting an effective defense response against pathogens. For example-Tomato nucleotide-binding site leucine-rich repeat (NBS-LRR) resistance genes are regulated by the interaction between the lncRNA SlNET3 and miR482b (Liu et al. 2015). This network modifies the expression of genes related to defense, which affects tomatoes' resistance to disease. These examples illustrate how sncRNAs and lncRNAs in plants collaboratively regulate diverse pathways involved in development, stress responses, nutrient signaling, and defense mechanisms. The cooperative actions between these noncoding RNAs involve modulation of gene expression, RNA stability, and interactions within regulatory networks (Zhang et al. 2013). Understanding the synergistic roles of sncRNAs and lncRNAs provides insights into the complexity of plant regulatory mechanisms and their adaptation to changing environmental conditions. Examples of small and long noncoding RNAs working in concert to regulate gene expression in plant.

4.4 Bioinformatics Tools to Study the Interaction between lncRNAs and Small RNAs

4.4.1 Plant lncRNAs Databases

To facilitate future lncRNA research, several databases have been developed to organize the identified lncRNAs (Quek et al. 2015) and their biological characteristics (Table 4.1).

4.4.2 lncRNA-miRNA Interaction Prediction

To predict the interactions between lncRNAs and sRNAs in plants, a variety of datasets and techniques are available (Table 4.2). These tools make use of a variety of computational techniques and experimental data to deduce possible interactions between sRNAs and lncRNAs.

Table 4.2 Summary list of key databases/tool for lncRNA-sRNA interaction in plants

Name	Description	Web link	Reference
LncTar	The capacity to accurately and widely forecast the RNA targets of long noncoding RNAs	https://www.cuilab.cn/lnctar	Li et al. (2015)
StarBase	(sRNA target Base), Investigation of miRNA-target interaction maps derived from Degradome-Seq and CLIP-Seq data	https://bio.tools/starbase	Yang et al. (2011)
miRcode	The GENCODE gene annotation, which includes long noncoding RNA genes, is used to predict "whole transcriptome" human microRNA targets	http://www.mircode.org/	Jeggari et al. (2012)
LncRNA2Target	An extensive database on mouse and human lncRNA target genes	https://ngdc.cncb.ac.cn/databasecommons/database/id/27	Cheng et al. (2019)
LncBase v2	Computationally estimated and experimentally confirmed microRNA sites on lengthy noncoding RNAs	https://rnacentral.org/expert-database/lncbase	Paraskevopoulou et al. (2016)
PlantNATsDB	Plant Natural Antisense Transcript Database	https://fairsharing.org/FAIRsharing.rn1pxg	Chen et al. (2012)

4.5 Summary

The diversity and significance of lncRNA molecules in plants are rapidly expanding, but studying them presents challenges. Research primarily focuses on lncRNA evolution in plants, particularly in angiosperms, to understand their role in environmental adaptation. However, exploration of lncRNAs in a wider range of plant species is needed for a comprehensive understanding. Most studies on stress-responsive lncRNAs have focused on single stress types, neglecting simultaneous or recurrent stresses. To grasp the molecular pathways of lncRNAs, additional research is essential. miRNAs and lncRNAs are crucial ncRNA categories in plants, with bioinformatics tools aiding in their prediction and identification.

Single-cell and single-molecule sequencing technologies offer new opportunities to expand the understanding of lncRNAs. ceRNA networks in plants influence various biological processes, with dysregulation potentially affecting plant resilience and productivity. The dual functionality of lncRNAs as precursors for small RNAs complicates gene regulation in plants, necessitating a deeper understanding of their role in regulatory networks.

Interaction between miRNAs and lncRNAs, as well as lncRNAs acting as precursors of miRNAs and competing endogenous RNAs (ceRNAs), are primary mechanisms of lncRNA-mediated stress response. Understanding these mechanisms can reveal how gene regulatory networks respond to stress. LncRNAs and small RNAs modulate chromatin structure and epigenetic modifications, significantly influencing gene expression and plant responses to environmental stimuli. Experimental techniques like RNA-Seq, CLIP-Seq, and ChIP-Seq, along with bioinformatics tools, have advanced understanding of RNA-mediated regulatory networks. Continued innovation in experimental and computational approaches will further illuminate these intricate mechanisms in plants and other organisms.

4.6 Future Perspective

The study of sncRNAs and lncRNAs in plants represents a burgeoning field with significant implications for understanding gene regulation, plant development, stress responses, and potential applications in agriculture. Considering future perspectives and emerging trends, along with therapeutic implications targeting the interplay between these ncRNAs in plants, leads to several potential directions.

The continual development of bioinformatics tools and databases devoted to the study of ncRNA interactions and functions will provide researchers with better means of investigating the intricacies of ncRNA-mediated gene regulation in plants, leading to discoveries with broad applications in biotechnology and agriculture.

References

Baldrich P, Kakar K, Sire C, Moreno AB, Berger A, Garcı'aChapa M et al (2014) Small RNA profiling reveals regulation of Arabidopsis miR168 and heterochromatic siRNA415 in response to fungal elicitors. BMC Genomics 15:1083

Bardou F, Ariel F, Simpson CG et al (2014) Long noncoding RNA modulates alternative splicing regulators in Arabidopsis. Dev Cell 30(2):166–176

Baulcombe DC (2015) VIGS, HIGS and FIGS: small RNA silencing in the interactions of viruses or filamentous organisms with their plant hosts. Curr Opin Plant Biol 26:141–146

Boller T, He SY (2009) Innate immunity in plants: an arms race between pattern recognition receptors in plants and effectors in microbial pathogens. Science 324(5928):742–744

Borges F, Martienssen RA (2015) The expanding world of small RNAs in plants. Nat Rev Mol Cell Biol 16(12):727–741

Boyd LA, Ridout C, O'Sullivan DM, Leach JE, Leung H (2013) Plantpathogen interactions: disease resistance in modern agriculture. Trends Genet 29(4):233–240

Budak H, Akpinar BA (2015) Plant miRNAs: biogenesis, organization and origins. Funct Integr Genomics 15:523–531

Budak H, Kaya SB, Cagirici HB (2020) Long non-coding RNA in plants in the era of reference sequences. Front Plant Sci 11:276

Castaings L, Bergonzi S, Albani MC et al (2014) Evolutionary conservation of cold-induced antisense RNAs of FLOWERING LOCUS C in Arabidopsis thaliana perennial relatives. Nat Commun 5(1):4457

Chen D, Yuan C, Zhang J, Zhang Z, Bai L, Meng Y, Chen LL, Chen M (2012) PlantNATsDB: a comprehensive database of plant natural antisense transcripts. Nucleic Acids Res 40(D1):D1187–D1193

Cheng L, Wang P, Tian R, Wang S, Guo Q, Luo M, Zhou W, Liu G, Jiang H, Jiang Q (2019) LncRNA2Target v2.0: a comprehensive database for target genes of lncRNAs in human and mouse. Nucleic Acids Res 47(D1):D140–D144

Csorba T, Questa JI, Sun Q, Dean C (2014) Antisense COOLAIR mediates the coordinated switching of chromatin states at FLC during vernalization. Proc Natl Acad Sci 111(45):16160–16165

Derrien T, Johnson R, Bussotti G, Tanzer A, Djebali S, Tilgner H, Guernec G, Martin D, Merkel A, Knowles DG, Lagarde J (2012) The GENCODE v7 catalog of human long noncoding RNAs: analysis of their gene structure, evolution, and expression. Genome Res 22(9):1775–1789

Dinger ME, Amaral PP, Mercer TR, Pang KC, Bruce SJ, Gardiner BB, Askarian-Amiri ME, Ru K, Soldà G, Simons C, Sunkin SM (2008) Long noncoding RNAs in mouse embryonic stem cell pluripotency and differentiation. Genome Res 18(9):1433–1445

Eddy SR (2001) Non-coding RNA genes and the modern RNA world. Nat Rev Genet 2(2):919–929

Franco-Zorrilla JM, Valli A, Todesco M, Mateos I, Puga MI, Rubio-Somoza I et al (2007) Target mimicry provides a new mechanism for regulation of microRNA activity. Nat Genet 39(98):1033–1037

Gao R, Liu P, Irwanto N et al (2016) Upregulation of LINC-AP2 is negatively correlated with AP2 gene expression with Turnip crinkle virus infection in Arabidopsis thaliana. Plant Cell Rep 35:2257–2267

Gautam HR, Bhardwaj ML, Kumar R (2013) Climate change and its impact on plant diseases. Curr Sci 105:1685–1691

Ghag SB (2017) Host induced gene silencing, an emerging science to engineer crop resistance against harmful plant pathogens. Physiol Mol Plant Pathol 100:242–254

Guttman M, Amit I, Garber M, French C, Lin MF, Feldser D, Huarte M, Zuk O, Carey BW, Cassady JP et al (2009) Chromatin signature reveals over a thousand highly conserved large non-coding RNAs in mammals. Nature 458(7235):223–227

Henriques R, Wang H, Liu J et al (2017) The antiphasic regulatory module comprising CDF5 and its antisense RNA flore links the circadian clock to photoperiodic flowering. New Phytol 216(3):854–867

Islam W, Noman A, Qasim M, Wang L (2018) Plant responses to pathogen attack: small RNAs in focus. Int J Mol Sci 19(2):E515

Jeggari A, Marks DS, Larsson E (2012) miRcode: a map of putative microRNA target sites in the long non-coding transcriptome. Bioinformatics (Oxford, England) 28(15):2062–2063

Jin J, Liu J, Wang H, Wong L, Chua NH (2013) PLncDB: plant long non-coding RNA database. Bioinformatics (Oxford, England). 29(8):1068–1071

Jung JH, Seo PJ, Ahn JH (2012) Oryza sativa Jumonji C (OsJMJ705) is a potential regulator for root development in rice. Plant Signal Behav 7(9):1162–1164

Kandan A, Akhtar J, Singh B, Kham Z, Pedapati A, Agarwal PC (2016) Dynamics of crop protection and climate change. Studera Press, Delhi

Kang C, Liu Z (2015) Global identification and analysis of long non-coding RNAs in diploid strawberry Fragaria vesca during flower and fruit development. BMC Genomics 16:815

Kantar M, Lucas SJ, Budak H (2011) miRNA expression patterns of Triticum dicoccides in response to shock drought stress. Planta 233:471–484

Karapetyan AR, Buiting C, Kuiper RA, Coolen MW (2013) Regulatory roles for long ncRNA and mRNA. Cancers 5(2):462–490

Khemka N, Singh VK, Garg R, Jain M (2016) Genome-wide analysis of long intergenic non-coding RNAs in chickpea and their potential role in flower development. Sci Rep 6(1):33297

Khraiwesh B, Zhu JK, Zhu J (2012) Role of miRNAs and siRNAs in biotic and abiotic stress responses of plants. Biochim Biophys Acta 1819 2:137–148

Lagarde J, Uszczynska-Ratajczak B, Santoyo-Lopez J et al (2016) Extension of human lncRNA transcripts by RACE coupled with long-read high-throughput sequencing (RACE-Seq). Nat Commun 7(1):1–11

Le TN, Schumann U, Smith NA, Tiwari S, Au PC, Zhu QH et al (2014) DNA demethylases target promoter transposable elements to positively regulate stress responsive genes in Arabidopsis. Genome Biol 15:1–18

Li J, Ma W, Zeng P, Wang J, Geng B, Yang J, Cui Q (2015) LncTar: a tool for predicting the RNA targets of long noncoding RNAs. Brief Bioinform 16(5):806–812

Liu H, Jin T, Liao R, Wan L, Xu B, Zhou S et al (2012) miRFANs: an integrated database for Arabidopsis thaliana microRNA function annotations. BMC Plant Biol 12:1–18

Liu X, Hao L, Li D, Zhu L, Hu S (2015) Long non-coding RNAs and their biological roles in plants. Genomics Proteomics Bioinformatics 13(3):137–147

Michaels SD, Amasino RM (1999) FLOWERING LOCUS C encodes a novel MADS domain protein that acts as a repressor of flowering. Plant Cell 11(5):949–956

Paraskevopoulou MD, Vlachos IS, Karagkouni D, Georgakilas G, Kanellos I, Vergoulis T, Zagganas K, Tsanakas P, Floros E, Dalamagas T, Hatzigeorgiou AG (2016) DIANA-LncBase v2: indexing microRNA targets on non-coding transcripts. Nucleic Acids Res 44(D1):D231–D238

Pattanayak D, Solanke AU, Kumar PA (2012) Plant RNA interference pathways: diversity in function, similarity in action. Plant Mol Biol Report 31:493–506

Paytuví Gallart A, Hermoso Pulido A, Martínez A, de Lagrán I, Sanseverino W, Aiese Cigliano R (2016) GREENC: a Wiki-based database of plant lncRNAs. Nucleic Acids Res 44(D1):D1161–D1166

Phillips T (2008) Small non-coding RNA and gene expression. Nat Educ 1:115

Quek XC, Thomson DW, Maag JL, Bartonicek N, Signal B, Clark MB, Gloss BS, Dinger ME (2015) lncRNAdb v2.0: expanding the reference database for functional long noncoding RNAs. Nucleic Acids Res 43(D1):D168–D173

Rajagopalan R, Vaucheret H, Trejo J, Bartel DP (2006) A diverse and evolutionarily fluid set of micrornas in Arabidopsis thaliana. Genes Dev 20(24):3407–3425

Shimoni Y, Friedlander G, Hetzroni G, Niv G, Altuvia S, Biham O et al (2007) Regulation of gene expression by small-noncoding RNAs: a quantitative view. Mol Syst Biol 3(1):138

Song X, Sun L, Luo H, Ma Q, Zhao Y, Pei D (2016) Genome-wide identification and characterization of long non-coding RNAs from mulberry (Morus notabilis) RNA-seq data. Genes 7(3):11

Sunkar R, Li YF, Jagadeeswaran G (2012) Functions of microRNAs in plant stress responses. Trends Plant Sci 17(4):196–203

Szcześniak MW, Bryzghalov O, Ciomborowska-Basheer J, Makałowska I (2019) CANTATAdb 2.0: expanding the collection of plant long noncoding RNAs. Methods Protoc:415–429

Van Bonnet E, de Peer Y, Rouzé P (2006) The small RNA world of plants. New Phytol 171(3):451–468

Wang Y, Li H, Sun Q, Yao Y (2016) Characterization of small RNAs derived from tRNAs, rRNAs and snoRNAs and their response to heat stress in wheat seedlings. PLoS One 11(3):e0150933

Wang J, Meng X, Dobrovolskaya OB, Orlov YL, Chen M (2017) Non-coding RNAs and their roles in stress response in plants. Genom Proteom Bioinform 15(5):301–312

Xu XW, Zhou XH, Wang RR, Peng WL, An Y, Chen LL (2016) Functional analysis of long intergenic non-coding RNAs in phosphate-starved rice using competing endogenous RNA network. Sci Rep 6(1):20715

Xuan H, Zhang L, Liu X, Han G, Li J, Li X, Liu A, Liao M, Zhang S (2015) PLNlncRbase: a resource for experimentally identified lncRNAs in plants. Gene 573(2):328–332

Yang JH, Li JH, Shao P, Zhou H, Chen YQ, Qu LH (2011) starBase: a database for exploring microRNA-mRNA interaction maps from Argonaute CLIP-Seq and Degradome-Seq data. Nucleic Acids Res 39(D1):D202–D209

Yi X, Zhang Z, Ling Y, Xu W, Su Z (2015) PNRD: a plant non-coding RNA database. Nucleic Acids Res 43:982–989

Yin Y, Lu JY, Zhang X (2020) U1 snRNP regulates chromatin retention of noncoding RNAs. Nature 580(7801):147–150

Yu Y, Zhang Y, Chen X, Chen Y (2019) Plant noncoding RNAs: hidden players in development and stress responses. Annu Rev Cell Dev Biol 35:407–431

Zhai J, Jeong DH, Paoli ED, Park S, Rosen BD, Li Y et al (2011) MicroRNAs as master regulators of the plant NB-LRR defense gene family via the production of phased, transacting siRNAs. Genes Dev 25(23):2540–2553

Zhang J, Mujahid H, Hou Y, Nallamilli BR, Peng Z (2013) Plant long ncRNAs: a new frontier for gene regulatory control. Am J Plant Sci 4:1038–1045

Zhang P, Du H, Wang J (2019) Multiplex CRISPR/Cas9-mediated metabolic engineering increases soya bean isofavone content and resistance to soya bean mosaic virus. Plant Biotechnol J 18(6):1384–1395

Zhao X, Li J, Lian B (2018) Global identifcation of Arabidopsis lncRNAs reveals the regulation of MAF4 by a natural antisense RNA. Nat Commun 9(1):5056

Zhu QH, Wang MB (2012) Molecular functions of long non-coding RNAs in plants. Genes 3(1):176–190

Exploring Noncding RNAs Through Next-Generation Sequencing and Bioinformatics

5

Anita Tripathi and Kavita Goswami

Abstract

The field of genomics has undergone a revolution with the introduction of next-generation sequencing (NGS) technologies, which have allowed researchers to explore the complexities of the genome in greater detail. Our study focuses on the investigation of noncoding RNAs (ncRNAs) using cutting-edge bioinformatics tools and NGS techniques. Previously regarded as genomic "dark matter," noncoding RNAs have become important participants in a variety of biological processes, influencing the development of disease, cellular homeostasis, and gene regulation. The large volume of data produced by NGS platforms is analysed and interpreted using state-of-the-art bioinformatics techniques, which shed light on the intricate realm of ncRNAs. Current developments in computational technologies have created a prodigious momentum for the discovery and functional annotations of ncRNAs. Due to their sequence-specific mechanism, they are effectively being discovered using different bioinformatics approaches. These bioinformatics approaches play a crucial role in reducing the large pool of potential ncRNAs by pointing out to the high-confidence candidates. In this chapter, we addressed the advances in ncRNA research coupled with NGS and computational tools.

Keywords

NGS · ncRNA · Bioinformatics · Databases · Tools

A. Tripathi (✉)
NerdGene Healthcare Pvt. Ltd, Pune, Maharashtra, India

K. Goswami (✉)
Plant RNAi Biology, International Centre for Genetic Engineering and Biotechnology, New Delhi, India

K. Goswami et al. (eds.), *Non-coding RNAs for Crop Improvement*,
https://doi.org/10.1007/978-981-97-7213-1_5

5.1 Introduction

To withstand the rapidly changing environmental conditions, plants employ a variety of response mechanisms that provide broad phenotypic malleability to permit the molecular-genetic changes to respond environment-related reactions. Plants have adopt a mechanism of evasion and/or tolerance for their survival when exposed to stress (Jackson et al. 2009). ncRNAs mainly microRNAs (miRNAs), short interfering RNAs (siRNAs), and long noncoding RNAs (lncRNAs), have emerged as indispensable regulators of plant response under various environmental conditions (Hombach and Kretz 2016). ncRNAs were initially considered to be completely junk transcripts having no functional role but have now been well established as regulators of gene expression. To define precisely, an ncRNA is a functional RNA that is transcribed from DNA but does not encode a protein (Palazzo and Lee 2015). Plant ncRNAs have mainly have three key characteristics; they provide no (or very little) protein-coding function, act as regulatory elements and are primarily located in the intergenic regions of the genome (Bhogireddy et al. 2021). In view of their large numbers, it is expected that the list of biologically validated ncRNAs, as recorded in the databases (Quek et al. 2015), will continue to rise exponentially. Till now only 10% of known ncRNAs have been validated to be functional, yet it has signified affluence of new biology (Palazzo and Lee 2015).

Computational biology is playing a critical role in supporting the growing body of data analyzing the significance of ncRNAs in cellular activities and disease development during the past several decades (Asim et al. 2021). The process of identifying and selecting important ncRNAs from a large pool inside a biological system to investigate and evaluate their functional implications can be speeded up by creating computational tools to predict ncRNAs and their roles (Rincón-Riveros et al. 2021). It is now easier to get global ncRNA profiles and comprehend their mechanism of action in plants because of the development of high-throughput sequencing (HTS) technology (Pundhir et al. 2015). The field of studying ncRNAs has undergone a revolution due to the development of bioinformatics databases, tools, and automation (Raza and Ahmad 2019).

Plant ncRNAs are classified as small RNAs (18–30 nt), medium-sized ncRNAs (31–200 nt), and lncRNAs (lncRNAs) (>200 nt) based on the length of their sequences. These sizes are taken into consideration as a first filter when determining which type of ncRNAs to identify (Jannesar et al. 2020). The importance of bioinformatics techniques in the discovery and investigation on the regulatory roles in the plant's adaptive response to varied situations is discussed below.

5.2 Sequencing Technologies

Gilbert and Maxam created the chemical sequencing technique (Gilbert 1981). The purification of DNA and radioactive labelling at the 5′ end of the DNA fragment to be sequenced were essential for this technique. In four distinct reactions (G, A + G,

C, and C + T), chemical treatment was applied to produce breakage at a tiny portion of one or two of the four nucleotide bases. The broad implementation of this technology was hindered by the technical difficulties of employing radioactivity. Around the same time, Fredric Sanger and others developed a different method for "sequence-by-synthesis" using chain termination method (Jay et al. 1974).

The discovery of a luminous technique for monitoring the generation of pyrophosphate throughout DNA synthesis was the next ground-breaking development in sequencing technology (Hyman 1988; Nyrén and Lundin 1985). This method utilises two steps in an enzymatic reaction wherein pyrophosphate is transformed into adenosine triphosphate (ATP) by ATP sulfurylase. ATP was then used as the substrate for luciferase, which produced light in direct proportion to the quantity of pyrophosphate. The utilisation of natural nucleotides, as opposed to the highly modified deoxy nucleotide diphosphate (dNTPs) employed in chain termination protocols, and the ability to obtain real-time outcomes instead of requiring drawn-out electrophoresis made this pyrosequencing technique beneficial (Ronaghi et al. 1998). After being licenced to 454 Life Sciences (which Roche eventually acquired), pyrosequencing was developed into the first significant and profitable NGS technology. By conducting enormous numbers of parallel sequencing processes on a micrometre scale, different NGS machines created a paradigm shift and significantly amplified the quantity of DNA that could be sequenced in a single run (Margulies et al. 2005) (Fig. 5.1).

A brief overview of the first, second, and third generations of sequencing technology is provided in the following sections, where concentrating on the platforms and sequencing techniques that define every sequencing generation method (Table 5.1).

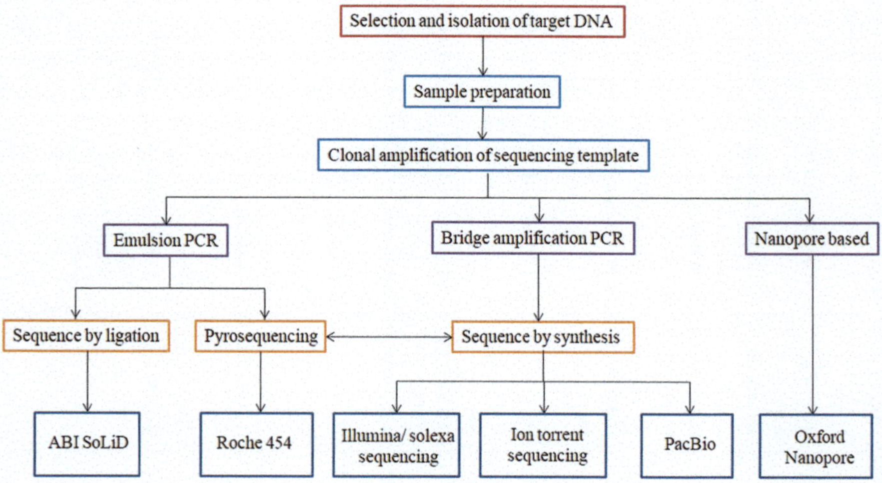

Fig. 5.1 Schematic representation of the various next-generation sequencing platforms

Table 5.1 Sequencing methods and platforms developed to study nucleotides

Generations	First-generation	Second-generation				Third-generation	
	Sanger sequencing	Illumina sequencing	Ion torrent	SoliD ABI	Roche	PacBio	Oxford Nanopore
Sequencing principle	Chain termination method	Sequencing by synthesis	Sequencing by synthesis	Sequencing by ligation	Pyrosequencing	Sequencing by synthesis	Nanopore based
Technology	Single DNA fragment at a time	Polonies cleavable dye terminators	emPCR, H+ detection	emPCR, ligation with cleavable dye terminators	emPCR, pyrosequencing	Real-time, single-molecule template	Not required
Read per run	96	25 M–400 M	2 M–60 M	3B–6B	200–1 M	432–660	100
Read type	SE	SE/PE	SE	SE	SE/PE	SE	1D,2D
Output range	0.00069 to 0.0021 Gb	600 Gb	1–10 Gb	120 Gb	0.7 Gb	0.5 to 1Gb	2 Tb to 4 Tb
Avg read length	400–900*	150–300	200–400	75	100–700	1300–13,500	9545–9846
Accuracy		99.90%	99%	99.94%	99%		
Primary error	NA	Mismatch	Indel	Mismatch	Indel	Indel	Indel/ mismatch
Error rate (%)	0.30%	0.1–1	1	–0.1	1	15	12
Year	2002	2007	2011	2007	2005	2011	2014
Instruments	3730xl	Genome analyzer, MiSeq, HiSeq,MiniSeq, NextSeq, HiSeqX	PGM 314 chip v2, PGM 316 chip v2, PGM 318 chip v2, ion proton, ion S5/S5XL 520 3, ion S5/S5XL 530, ion S5/S5XL 540	5500 W, 5500xl W	GS20, GS FLX, GS FLX titanium, GS FLX, titanium+, GS junior, GS junior+	RS C1, RS C2, RS C2 XL, RS II C2 XL, RS II P5 C3, RS II P6 C4, sequel	MinION Mk, PromethION

5.2.1 First-Generation Sequencing

The First-Generation Sequencing Technologies, which included Sanger and Maxam-Gilbert, were credited with launching the field of DNA sequencing with their publications in 1977 (Metzker 2010; Thudi et al. 2012).

5.2.1.1 Sanger Sequencing

Frederick Sanger and others created the classical DNA sequencing technique known as "Sanger sequencing," or "chain termination sequencing," in the 1970s. It is a typically used method that has historical significance for figuring out the nucleotide base sequence of a DNA molecule. The process produces a collection of fragments with a single nucleotide length difference because it incorporates chain-terminating dideoxynucleotides (ddNTPs) selectively during DNA synthesis (Sanger et al. 1977).

Producing a DNA template for sequencing is the first step in the Sanger sequencing process. A combination of normal deoxynucleotides (dNTPs) and a tiny amount of chain-terminating ddNTPs (Guo et al. 2008; Sanger et al. 1977), each labelled with a distinct fluorescent dye, are added to the template along with a DNA primer and DNA polymerase enzyme. Premature termination of the developing DNA strand results from the DNA polymerase occasionally incorporating a ddNTP instead of a standard dNTP as it stretches the primer along the template. This produces a sequence of DNA fragments, each labelled with a specific fluorescent colour that corresponds to the type of ddNTP that was employed, and concludes with a ddNTP (Sanger et al. 1977; Guo et al. 2008). The order of nucleotides in the original DNA template is then ascertained by examine the arrangement of colored peaks, each peak denoting distinct nucleotide after the fragments have been size-separated using gel electrophoresis or capillary electrophoresis. While newer methods have essentially replaced Sanger sequencing for large-scale, high-throughput sequencing projects, Sanger sequencing has been instrumental in many scientific breakthroughs and is still important for specialised sequencing applications.

5.2.1.2 Maxam-Gilbert Sequencing

This is one of the initial techniques of DNA sequencing, created in tandem with Sanger sequencing by Allan Maxam and Walter Gilbert in the late 1970s (Gilbert 1981). The methodology of chemical sequencing is based on the selective breakage of DNA at particular nucleotides by chemical reactions, which permits the identification of the DNA sequence. Using the Maxam-Gilbert approach, the DNA fragment is first labelled at one end and the sample is split into four reactions, each with a distinct chemical reagent. Dimethyl sulphate (DMS), piperidine, hydrazine, and formic acid are the substances that are used (Eckert and McConnell-Ginet 1992). When these chemicals encounter certain bases, the DNA backbone cleaves at the locations of the targeted bases. Following the chemical processes, gel electrophoresis is used to sort the fragments based on size (Eckert and McConnell-Ginet 1992). After that, the sequence is inferred by examining the gel's band pattern. Although Maxam-Gilbert sequencing was popular in the early days of DNA sequencing, Sanger sequencing and, more recently, high-throughput sequencing methods have essentially replaced it because of their efficiency and automation. Maxam-Gilbert sequencing made a substantial contribution to our understanding of DNA sequences and structure, though its utilisation has declined.

5.2.2 Second-Generation Sequencing

5.2.2.1 Roche 454

454 Life Sciences introduced and promoted the initial NGS technique in 2005. It uses base calling to extract the sequence from recorded signals (Margulies et al. 2007). First, beads are connected to a library of DNA molecules using a known adaptor sequence consisting of four nucleotides. Next, each bead is coated using a water-in-oil emulsion polymerase chain reaction (PCR) (Griffiths and Tawfik 2006), which amplifies each DNA molecule in its own droplet inside the emulsion on average. After coating the beads with DNA, they are rinsed over a picoliter reaction plate that can hold one bead per well. After washing the plate with the smaller bead-linked enzymes and dNTPs, the pyrophosphate release was measured using a charged couple device sensor located beneath the wells to initiate sequencing. The fragment's sequence may be ascertained using the adapter sequence, which is included at the beginning of the flow gram. The number of nucleotides in the sequence is directly correlated with the flow gram's peak intensity (Margulies et al. 2005). This method has a cheap capital and per-experiment cost and is quick, efficient, and appropriate for sequencing read lengths ranging from 450 to 900 bases. Low throughput, high cost per megabyte of data, and error-prone (polybase >6) sequencing are its main drawbacks (Goswami and Sanan-Mishra 2022).

5.2.2.2 ABI SoLid

Life Technologies invented this technique for sequencing by Oligonucleotide Ligation and Detection (SOLiD) in 2006 (Ansorge 2009). This technique, which builds on ideas previously established in George Church's group, uses a two-base fluorescent probe and obtains sequences via ligation (Shendure and Ji 2008). It makes use of DNA ligase's mismatch sensitivity to identify the nucleotide at a certain location in the sequence. To create clonal bead populations and ensure that each magnetic bead has only one species of fragment on its surface, emulsion PCR is carried out using a library of DNA fragments that have been produced from the sample that will be sequenced. The fragments that are affixed to the magnetic beads possess a universal adaptor sequence, guaranteeing that each fragment has an identical and known beginning sequence. Four di-base probes with fluorescent labels compete with one another to ligate to the sequencing primer. To obtain di-base probe specificity, each ligation reaction's first and second bases must be questioned. The final read length is determined by the number of cycles that are completed for ligation, detection, and cleavage. The specificity and precision of the template strand may be effectively increased by using several rounds and cycles, which allow for the individual inspection of each base in two distinct ligation processes and two distinct probes (Margulies et al. 2005; Massingham and Goldman 2012). The technique's drawbacks include a significant number of gaps in the assembly, a high capital cost, and a restriction of 35–75 bases per read.

5.2.2.3 ION Torrent

Like previous methods, electrochemical detection is employed here rather than luminescence-based camera scanning or fluorescence detection, which both need for the amplification of sequence. Emulsion PCR is used to create beads containing clonal populations of DNA fragments in a way like the one previously reported. After washing these over a pico well plate, the addition of each nucleotide is determined by measuring the pH change brought on by the polymerisation process' release of protons (H^+ ions). The complementary metal-oxide-semiconductor technology utilised in the production of microprocessor chips makes this feasible (Merriman et al. 2012). Although this method is less sensitive to interpreting homopolymer sequences, it can sequence 150–200 bases per read, enabling highly quick sequencing during the actual detection phase.

5.2.2.4 Illumina

Most popular and successful NGS technologies, with a vast array of transcriptome, regulome, microbiome and genome-related applications. For both single-read and paired-end (3′ and 5′ ends) reads of long or short inserts, the sequencing method is beneficial. Bridge amplification is used in Illumina sequencing to clonally construct the fragments, which may then be sequenced using synthesis. Out of the four commonly used sequencing methods, it has the largest throughput of 360–500 Gb per day and can sequence read lengths between 100 and 150. With a 48 Gb RAM, a run may take five or 6 days (Liu et al. 2012). Illumina utilises a variety of sequencing systems, including NextSeq 500, HiSeq X Ten, GAIIx (Genome analyzer), Hiseq 2000/HiSeq 2500, and MiSeq (Goswami et al. 2019).

Small RNA-sequencing has benefited greatly from this technology. Applications that work well include differential expression analysis, single-base resolution characterisation of isomers, and the discovery of novel miRNAs. Sequencing, cluster creation, and library preparation are necessary for the procedure. To create adapter-ligated small RNA libraries from total RNA, specialised kits are offered. A tiny cluster of RNA molecules with the same sequence is produced by solid-phase bridge amplification once the adapter-ligated short RNAs are connected to the flow cell. Primers, polymerase, and reversible terminator nucleotides labelled with fluorescent dye are added to the flow cell to initiate the first sequencing cycle (Ju et al. 2006).

The polymerase inserts the labelled terminator nucleotide (chemically cleavable moiety) at the 3′ OH position of the new strand after the primer binds to the segment that is being sequenced. For simple identification, every nucleotide has a distinct fluorescently labelled terminator group. The device makes use of dual surface imaging technology, which lowers signal interference and improves sequencing accuracy. At the conclusion of each sequencing cycle, the fluorescent label is separated from the integrated base, and the procedure is repeated, one base at a time, to decode the sequence (Ju et al. 2006).

5.2.3 Third-Generation Sequencing

5.2.3.1 Pacific Biosciences SMRT Sequencing

Utilising the SMRT technique, Pacific Biosciences created the first genomic sequencer, which is now the most used third-generation sequencing technology. Using the same fluorescent labelling as the other methods, Pacific Biosciences detects the signals as they are released during incorporations, as opposed to running cycles of amplification nucleotide. It employs a device made up of several SMRT cells, each of which has zero-mode waveguides (ZMWs)-microfabricated nanostructures that are wells tens of nanometres in diameter-microfabricated in a metal sheet and then coated onto a glass substrate (McCoy et al. 2014; Rhoads and Au 2015). These ZMWs take use of the fact that light cannot travel through apertures larger than its wavelength because of the characteristics of light flowing through them. The light intensity drops down the wells' short diameter, leaving the bottoms lighted. On the bottom of every ZMW is a DNA polymerase, along with the target DNA fragment that must be sequenced. Fluorescently labelled nucleotides (of varying colours) are integrated into the DNA fragment by DNA polymerase during the sequencing step. Every time a nucleotide is integrated, sensors capture the bright signal that is released. The DNA sequence may be ascertained by the identification of the labelled nucleotides.

5.2.3.2 Oxford Nanopore Sequencing

The Oxford Nanopore sequencing (ONT) method was created to ascertain the nucleotide order of a DNA sequence. With the debut of the MinION device in 2014, Oxford Nanopore Technologies intended to improve resolution of structural genomic variations and repetitive content by producing longer readings (Laehnemann et al. 2016; Mikheyev and Tin 2014). The four-inch-long, transportable single-molecule Nanopore sequencing device is connected to a laptop computer via a USB 3.0 interface. In order to assess the effectiveness of the MinION sequencer, this device has been made available for community testing as part of the MinION Access Programme (MAP) (Laver et al. 2015). A hairpin connects a DNA molecule's initial strand to its complementary strand in this sequencing process. The DNA fragment is routed via a protein Nanopore, which is a nanoscale aperture, composed of synthetic materials or proteins (Heo 2015).

A change in the ionic current is produced when the DNA fragment is translated through the pore by a motor protein that is connected to the pore. This is due to variations in the moving nucleotides that occupy the pore. The sequence is identified by first interpreting the sequential recording of this ionic current fluctuation on a visual model. The direct strand is sequenced to produce the "template read," which is followed by the inverse strand to produce the "complement read" and the hairpin structure to produce what are known as "1D" reads. The consensus sequence that results from combining the "temple" and "complement" reads is known as a "two-direction read," or "2D" (Jain et al. 2016; Lu et al. 2016).

5.3 Bioinformatics Approaches to Study ncRNAs

Computer-based methods are being applied in the study of ncRNAs majorly in seven different aspects (Fig. 5.2).

1. Recognition of new ncRNAs from genomic sequence or by amalgamating computational analysis with experimental data (e.g. RNA-seq, tiling array data).
2. Target gene identification.
3. Secondary and tertiary structure prediction of ncRNAs.
4. Conservation and evolution studies of ncRNA genes or their target genes.
5. Computational prediction of ncRNA functions, such as "guilt by association".
6. Establishment of integrated regulatory networks comprising all regulatory layers of ncRNA.
7. Development of databases and web servers to accelerate and simplify noncoding RNA studies.

To achieve these goals, several strategies are being adopted by bioinformaticians worldwide which are mainly based on,

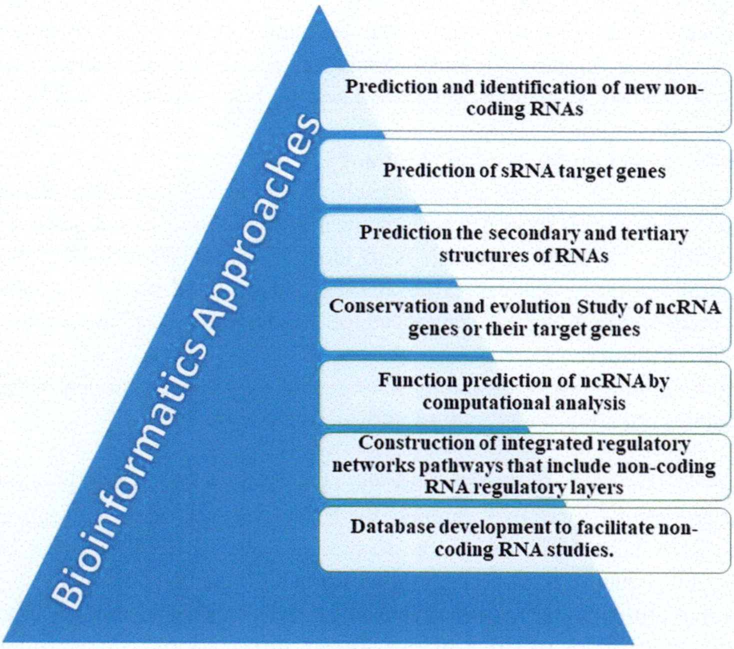

Fig. 5.2 Computer-based methods being applied in studying ncRNAs

1. Homology based ncRNA identification.
2. De-novo methods for ncRNA identification.
3. Assembly-based methods for ncRNA identification.
4. Family-based method for ncRNA identification.

5.3.1 Homology-Based Methods for ncRNA Identification

Homology-based approach is based on the sequence and structural similarity between RNAs evolving from a common ancestor and thus share function similarities. These methods generally compare the unknown sequence of interest with already reported ncRNAs in public databases and based on the similarity at both sequence and secondary structure are predicted and assigned to a specific category or ncRNAs family. There are a number of ncRNA databases one vastly utilised databases with 4108 families of ncRNA catalogued (version 14.9) (Kalvari et al. 2021).

1. **Sequence-based Homology Search:**
   It depends completely on sequence similarities deduced by sequence alignment methods. These algorithms are programmed to identify short significant stretches of similarity between sequences, irrespective of where they are in the sequences. And these alignments are evaluated based on three parameters, lowest E-Value, highest alignment length/coverage, and percentage similarity between aligned sequences, then assigned to best matching ncRNA family (Biegert and Söding 2009).
2. **Methods Based on Structural Homology:**
   The sequences are less conserved when compare to structures of ncRNAs, hence relying just on sequence alignments may result in incorrect ncRNA family assessments, the better approach is to identify conserved sequences for their structural similarity (Panwar et al. 2014b). This method predicts structures of both query and homologous sequences followed by structure base alignments.
3. **Hybrid Approach:**
   Much reliable and gives much robust similarity results, obtained by incorporating both sequence and structure information (Rocha et al. 2004).

Nevertheless, homology approaches are the fastest and widely used, however they have few limitations. First, they compare the RNA sequence of interest with already known ncRNA families hence unable to predict novel ncRNA families. Second, they count on sequence or structure conservation and thus are unsuitable to predict ncRNAs less similarity in sequence and structure. Sticking to this approach will narrow down our scope of study only to known information. As a result, de-novo methods are put further to solve such dilemma.

5.3.2 De-Novo Methods of ncRNA Identification

De-novo methods, instead of looking for homology, predicts ncRNA from primary sequences or structures based on universal principles that direct ncRNA folding energetics and/or statistical tendencies that native ncRNA sequences and structures acquire (Tran et al. 2009). Based on the source of shared features, de-novo methods can be divided into sequence-based, structure-based, and hybrid feature methods which use both features.

1. **De-novo Methods Based on Sequence Features:** Looks for nucleotide composition features of ncRNA families such as GC content, (Wang et al. 2013), the presence of several *k*-mers (specific nucleotide sequence of length k) for ncRNAs in a specific family usually sharing comparable probability distribution. Profiling of 6-mer to distinguish between protein-coding and non-coding sequences, independent of known annotations (Panwar et al. 2014a), in addition to predicting lncRNAs and messenger RNAs based on a combination of 1–5 mers (Sun et al. 2013). A single sequence feature is incompatible to reliably identify ncRNAs; hence mostly tools and algorithms incorporate more than one sequence features, to characterise ncRNAs.
2. **De-novo Methods based on Structural Features:** As secondary structures of several functional RNA are more preserved, such as miRNAs have hairpin loop whereas tRNAs share cloverleaf structure (Washietl et al. 2005), and the structures with lowest Minimum Free Energy (MFE) are usually best possible fold structure, hence MFE is being widely used for ncRNA secondary structure, but it is not enough to use MEF alone, morestructure features like thermodynamic stability are also being utilised for predicting ncRNA structures (Tran et al. 2009).
3. **Hybrid Method:** Merging sequence and structural characteristics, gives higher confidence on predicted ncRNAs (Nawrocki et al. 2009; Zhang et al. 2017).

With de-novo approaches the prediction of new ncRNA families with lacking in sequence conservation, they are more sensitive but less accurate when compared to homology approaches.

5.3.3 Assembly-Based Methods for ncRNA Identification

In genomics, assembly-based techniques are essential, and their effectiveness is frequently increased by utilising different characteristics present in DNA or RNA sequences (Eng et al. 1999). These features fall into three general categories: sequence features, structural features, and hybrid approaches that combine the two in a beneficial way.

1. **Assembly-based Methods Based on Sequence Features***:* Assembly techniques based on sequence features make use of data that is encoded in the primary nucleotide sequence. Short sub-sequences (k-mers) are used in methods such as k-mer-based assembly to find overlaps and reconstruct longer transcriptomic or genomic sequences (Eng et al. 1999). Seed-and-extend algorithms capture subtle variations and repetitive elements by focusing on exact or near matches between sequences. Rebuilding sequences with greater reliability is made possible by quality score-aware assembly, which considers the quality scores linked to sequencing reads.
2. **Assembly-based Methods Based on Structural Features:** Nucleotide properties and their three-dimensional arrangement are considered by structural feature-based assembly techniques. These techniques are essential for capturing intricate genomic architectures (Bhattacharya 2016). For example, the assembly process incorporates structural features like secondary structures, conserved motifs, or repetitive elements. This is especially important when discussing RNA-sequencing, since transcript assembly is influenced by the secondary structures of RNA molecules.
3. **Hybrid Method:** Hybrid assembly techniques combine structural and sequence elements to maximise the benefits of each strategy (Zhang et al. 2017). These techniques are meant to minimise the drawbacks of sequence-based and structurally-based strategies while simultaneously leveraging their respective advantages.

To accurately and thoroughly reconstruct genomic or transcriptomic information, assembly-based techniques that make use of sequence features, structural features, or a combination of both are crucial.

5.3.4 Family-Specific Methods

ncRNA annotation and analysis techniques are family-specific and take into account the distinctive traits and evolutionary histories of particular ncRNA families. It is more accurate and sensitive to identify different classes of ncRNAs using family-specific methods because they have different features, such as transfer RNAs (tRNAs), lncRNAs, and miRNAs (Emamjomeh et al. 2019; Kalvari et al. 2018). Here are a few instances of ncRNA analysis techniques that are specific to families:

5.3.4.1 Small ncRNAs

The increasing advancements of sequencing technologies became a crucial turning point in the discovery of new small ncRNA specifically for miRNAs and siRNA (ta-siRNAs, nat-siRNAs, hc-siRNAs, ra-siRNAs), accelerating biological research and significantly cuts down the cost enabling access to the full complexity of sRNAs in plants in addition to these, provides quantitative information of the expression profiles (Goswami et al. 2019).

small RNA sequencing (sRNA-Seq) has been supporting many compelling research avenues, including pre-profiling known and novel transcripts and expression quantifications of small RNA, predicting cleaved target genes of miRNA candidates, identifying specific biomarkers for diagnostics and classification of infection or stress conditions in plants and discovering regulatory networks and understanding the mechanisms of tissue or organism development based on transcript profiles. Amongst all sRNAs, the most studied is miRNAs because of their tremendously significant role in regulation of gene expression.

5.3.4.2 Long ncRNAs

Long noncoding RNAs (lncRNAs), longer than 200 nt, comprise both long intronic and intergenic ncRNA and are reported to regulate gene expression. The prediction of lncRNAs is very challenging, because of less sequence and structure conservation. General approaches of lncRNA identification are to first identify and filter out the coding sequences, then remove sequences <200 nt to filter out other small ncRNAs, and remaining once are taken further as candidate lncRNAs. Finally, to predict the novel lncRNA, the assembled transcripts are assessed based on distinctive features such as absence of open reading frame (ORF) or start and stop codons (Kung et al. 2013). Other characteristics include secondary structures features, protein-coding potential (Mercer et al. 2009), conserved splicing sites, DGE + RNA-Seq, and conserved promoters (Derrien et al. 2012). Another approach adopted was through specific chromatin modification signatures, such as the K4-K36 methylation marks (Guttman et al. 2009), only those passing certain significance levels are inferred to be lncRNAs. Initially, this approach was used in mammals, where the occurrence of such tags abetted the identification of lncRNAs present within the intergenic regions. Later this was successfully applied to identify lncRNAs in *A.thaliana* (Liu et al. 2015).

The progress of high-throughput sequencing methods became a resource of quantifying the transcriptome by performing whole transcriptome sequencing or RNA-sequencing (RNA-Seq). RNA-seq by ribosomal RNA (rRNA) depletion, which retains the complete transcriptional information present in a sample (Both poly-A+ and poly-A− transcripts) is routinely used for lncRNA discovery.

Bioinformatics techniques and databases are essential for the prediction and comprehension of ncRNAs, including miRNAs and lncRNAs, in the field of plant molecular biology (Table 5.2). A notable example of such a resource is miRBase (Kozomara et al. 2019), which provides an extensive database of miRNA sequences, annotations, and projected targets, providing insightful information on regulatory

Table 5.2 List of ncRNA prediction tools

Tool	Description	Link	Reference
miRNA prediction tools			
miRDeep2	miRDeep2 is able to recognise miRNAs	http://www.australianprostatecentre.org/research/software/mirdeep-star	Friedländer et al. (2012), An et al. (2013)
Mircat2	Identification and characterisation of new miRNAs provide separate modules for plant and animal miRNA	http://srna-workbench.cmp.uea.ac.uk/	Paicu et al. (2017)
miRNAFinder	A pre-microRNA classifier for plants and analysis of feature impact	http://mirnafinder.shyaman.me/	Ihalagedara et al. (2020)
miRPlant	The purpose of miRPlant is to identify plant miRNAs by combining secondary structure and sequence features	http://sourceforge.net/projects/mirplant/	An et al. (2014)
lncRNA identification tools			
PLEK	PLEK uses machine learning in conjunction with sequence features. For lncRNA prediction, it takes into account k-mer frequencies and secondary structure data	https://sourceforge.net/projects/plek/files/	Li et al. (2014)
CPC1 and CPC2	Accurate coding potential calculator based on sequence intrinsic features, especially for long noncoding transcripts	http://cpc2.cbi.pku.edu.cn	Kang et al. (2017)
LncFinder	R package that applies ORF, secondary structure, and physicochemical property-based features to identify lncRNAs	http://bmbl.sdstate.edu/lncfinder/	Han et al. (2019)
LncDC	A machine learning-based tool for long noncoding RNA detection from RNA-Seq data	https://github.com/lim74/LncDC	Li and Liang (2022)

networks and LncDC, a machine learning-based tool for lncRNA detection from RNA-Seq data (Li and Liang 2022).

These RNA family-specific techniques highlight the variety of ncRNA analysis methodologies by considering the unique traits shared by each RNA family. Hybrid approaches that combine structural and sequence features improve the precision and dependability of ncRNA annotation and advance our knowledge of the various functional functions of ncRNA classes.

5.4 Secondary and Tertiary Structures Prediction of ncRNAs

To predict the secondary and tertiary structures of noncoding RNAs, the following general methods can be used (Table 5.3),

Table 5.3 List of secondary and tertiary structure prediction tools

Name	Description	Link	Reference
Mfold	ARNA secondary structures prediction tool	http://www.bioinfo.rpi.edu/applications/mfold	Zuker (2003)
RNAfold (ViennaRNA package)	Web server to predict secondary structures of single-stranded RNA or DNA sequences	http://www.tbi.univie.ac.at/RNA	Lorenz et al. (2011)
CMfinder	A covariance model based RNA motif finding algorithm	http://bio.cs.washington.edu/yzizhen/CMfinder/	Yao et al. (2006)
LocARNA	Accurate boundary prediction and improved detection of structural RNAs	http://www.bioinf.uni-freiburg.de/Supplements/LocARNA-P/	Will et al. (2012)
RNAComposer	RNA 3D structure prediction methods	http://rnacomposer.cs.put.poznan.pl	Biesiada et al. (2016)

5.5 Databases and Web Servers

The discovery of small ncRNA molecules has revolutionised after the introduction of NGS technologies. In conjunction with computational tools, it facilitates the identification and distinction of small ncRNA species and offers insight into their biological roles. The NGS data can be processed and analysed using a variety of software programs and databases (Table 5.4). Preparing the raw sequencing data for adapter trimming and quality filtering comes first in the analysis process, and then known and novel small RNAs are predicted and identified.

Table 5.4 List of database and server of ncRNAs

Name	Description	Link	Reference
Databases for NGS data analysis			
NCBI sequence read archive (SRA)	A vast collection of raw NGS data, encompassing datasets from RNA and DNA sequencing	https://www.ncbi.nlm.nih.gov/sra	Leinonen et al. (2010b)
European nucleotide archive (ENA)	ENA is a repository for exchanging and storing nucleotide sequence data, much like SRA	https://www.ebi.ac.uk/ena/browser/home	Leinonen et al. (2010a)
Genome browser (UCSC and Ensembl)	Interactive platforms for the analysis and visualisation of genomic data, such as NGS datasets	https://plants.ensembl.org/index.html	Navarro Gonzalez et al. (2021)
1000 genomes project	A comprehensive attempt to classify genetic variations in humans, offering a useful resource for variant analysis and population genomics	https://www.internationalgenome.org/	Siva (2008)
Databases and web servers for ncRNA analysis			
NONCODE	An extensive database dedicated to the annotation and prediction of noncoding RNAs, such as microRNAs and long noncoding RNAs (lncRNAs)	http://www.noncode.org/	Zhao et al. (2016)
miRBase	A database of microRNA sequences and annotations with functional analysis, target, and miRNA prediction tools	https://mirbase.org/	Kozomara et al. (2019)
Rfam	A collection of RNA families, comprising RNAs with and without codes	https://rfam.org/	Kalvari et al. (2021)
lncRNAdb	A carefully curated database with details on the structures, functions, and sequences of long noncoding RNAs that have been experimentally confirmed	https://ngdc.cncb.ac.cn/databasecommons/database/id/23	Quek et al. (2015)

5.6 Conclusion

All things considered, the study of ncRNAs in plant biology is a developing field with significant ramifications. These mysterious compounds, which were formerly dismissed as genomic "dark matter," are now understood to be crucial regulators of gene expression and are involved in plant development, stress responses, and environmental adaptability. Researchers have shown the intricacy of ncRNA-mediated regulation networks in plants, thanks to the development of next-generation sequencing and sophisticated bioinformatics tools.

In addition to advancing our knowledge of basic biological processes, this current study has enormous potential for agricultural applications and could lead to new ways to improve crop sustainability, productivity, and resilience in the face of changing environmental conditions. We are setting out on a quest to fully utilise plant ncRNAs for the good of both the earth and humankind as we continue to delve into their mysteries.

5.7 Future Perspectives

Next-generation ncRNA research in plant biology holds great potential for a revolutionary understanding of the complex systems underlying stress responses, environmental adaptation, and plant development. Researchers will investigate the regulatory networks and functional annotation of various kinds of ncRNAs in plants in greater detail. This comprises, siRNAs, lncRNAs, and miRNAs. Researchers will decipher the intricate interactions between ncRNAs and other molecular components in plant systems by combining ncRNA data with multi-omics datasets and using systems biology techniques. This will provide a comprehensive understanding of plant biology. Additionally, using ncRNA-based tactics could improve agricultural sustainability, production, and resilience to changing environmental conditions, opening the door for more advancement in agriculture and food security.

References

An J, Lai J, Lehman ML, Nelson CC (2013) miRDeep*: an integrated application tool for miRNA identification from RNA sequencing data. Nucleic Acids Res 41(2):727–737

An J, Lai J, Sajjanhar A, Lehman ML, Nelson CC (2014) miRPlant: an integrated tool for identification of plant miRNA from RNA sequencing data. BMC Bioinform 15:1–4

Ansorge WJ (2009) Next-generation DNA sequencing techniques. New Biotechnol 25(4):195–203

Asim MN, Ibrahim MA, Imran Malik M, Dengel A, Ahmed S (2021) Advances in computational methodologies for classification and sub-cellular locality prediction of non-coding RNAs. Int J Mol Sci 22(16):8719

Bhattacharya D (2016) Probabilistic graphical models for protein structure prediction. University of Missouri, Columbia

Bhogireddy S, Mangrauthia SK, Kumar R, Pandey AK, Singh S, Jain A, Budak H, Varshney RK, Kudapa H (2021) Regulatory non-coding RNAs: a new frontier in regulation of plant biology. Funct Integr Genomics 21(3):313–330

Biegert A, Söding J (2009) Sequence context-specific profiles for homology searching. Proc Natl Acad Sci 106(10):3770–3775

Biesiada M, Pachulska-Wieczorek K, Adamiak RW, Purzycka KJ (2016) RNAComposer and RNA 3D structure prediction for nanotechnology. Methods 103:120–127

Derrien T, Johnson R, Bussotti G, Tanzer A, Djebali S, Tilgner H, Guernec G, Martin D, Merkel A, Knowles DG (2012) The GENCODE v7 catalog of human long noncoding RNAs: analysis of their gene structure, evolution, and expression. Genome Res 22(9):1775–1789

Eckert P, McConnell-Ginet S (1992) Think practically and look locally: language and gender as community-based practice. Annu Rev Anthropol 21(1):461–488

Emamjomeh A, Zahiri J, Asadian M, Behmanesh M, Fakheri BA, Mahdevar G (2019) Identification, prediction and data analysis of noncoding RNAs: a review. Med Chem 15(3):216–230

Eng T-H, Ling Z-K, Olson W, McLean C (1999) Feature-based assembly modeling and sequence generation. Comput Ind Eng 36(1):17–33

Friedländer MR, Mackowiak SD, Li N, Chen W, Rajewsky N (2012) miRDeep2 accurately identifies known and hundreds of novel microRNA genes in seven animal clades. Nucleic Acids Res 40(1):37–52

Gilbert FS (1981) Foraging ecology of hoverflies: morphology of the mouthparts in relation to feeding on nectar and pollen in some common urban species. Ecol Entomol 6(3):245–262

Goswami K, Sanan-Mishra N (2022) RNA-Seq for revealing the function of the transcriptome. In: Bioinformatics. Academic Press, New York, pp 105–129

Goswami K, Tripathi A, Gautam B, Sanan-Mishra N (2019) Impact of next-generation sequencing in elucidating the role of microRNA related to multiple abiotic stresses. In: Molecular plant abiotic stress: biology and biotechnology. Wiley, New York, pp 389–426

Griffiths AD, Tawfik DS (2006) Miniaturising the laboratory in emulsion droplets. Trends Biotechnol 24(9):395–402

Guo J, Xu N, Li Z, Zhang S, Wu J, Kim DH, Sano Marma M, Meng Q, Cao H, Li X (2008) Four-color DNA sequencing with 3′-O-modified nucleotide reversible terminators and chemically cleavable fluorescent dideoxynucleotides. Proc Natl Acad Sci 105(27):9145–9150

Guttman M, Amit I, Garber M, French C, Lin MF, Feldser D, Huarte M, Zuk O, Carey BW, Cassady JP (2009) Chromatin signature reveals over a thousand highly conserved large non-coding RNAs in mammals. Nature 458(7235):223–227

Han S, Liang Y, Ma Q, Xu Y, Zhang Y, Du W, Wang C, Li Y (2019) LncFinder: an integrated platform for long non-coding RNA identification utilizing sequence intrinsic composition, structural information and physicochemical property. Brief Bioinform 20(6):2009–2027

Heo Y (2015) Improving quality of high-throughput sequencing reads. University of Illinois at Urbana-Champaign, Urbana

Hombach S, Kretz M (2016) Non-coding RNAs: classification, biology and functioning. Adv Exp Med Biol 937:3–17

Hyman ED (1988) A new method of sequencing DNA. Anal Biochem 174(2):423–436

Ihalagedara P, Lokuge S, Jayasundara S, Herath D, Kahanda I (2020) Mirnafinder: a pre-microrna classifier for plants and analysis of feature impact. In: 2020 IEEE conference on computational intelligence in bioinformatics and computational biology (CIBCB). IEEE, New York

Jackson MB, Ishizawa K, Ito O (2009) Evolution and mechanisms of plant tolerance to flooding stress. Ann Bot 103(2):137–142

Jain M, Olsen HE, Paten B, Akeson M (2016) The Oxford Nanopore MinION: delivery of nanopore sequencing to the genomics community. Genome Biol 17:1–11

Jannesar M, Seyedi SM, Moazzam Jazi M, Niknam V, Ebrahimzadeh H, Botanga C (2020) A genome-wide identification, characterization and functional analysis of salt-related long non-coding RNAs in non-model plant Pistacia vera L. using transcriptome high throughput sequencing. Sci Rep 10(1):5585

Jay E, Bambara R, Padmanabhan R, Wu R (1974) DNA sequence analysis: a general, simple and rapid method for sequencing large oligodeoxyribonucleotide fragments by mapping. Nucleic Acids Res 1(3):331–354

Ju J, Kim DH, Bi L, Meng Q, Bai X, Li Z, Li X, Marma MS, Shi S, Wu J (2006) Four-color DNA sequencing by synthesis using cleavable fluorescent nucleotide reversible terminators. Proc Natl Acad Sci 103(52):19635–19640

Kalvari I, Nawrocki EP, Argasinska J, Quinones-Olvera N, Finn RD, Bateman A, Petrov AI (2018) Non-coding RNA analysis using the Rfam database. Curr Protoc Bioinform 62(1):e51

Kalvari I, Nawrocki EP, Ontiveros-Palacios N, Argasinska J, Lamkiewicz K, Marz M, Griffiths-Jones S, Toffano-Nioche C, Gautheret D, Weinberg Z (2021) Rfam 14: expanded coverage of metagenomic, viral and microRNA families. Nucleic Acids Res 49(D1):D192–D200

Kang Y-J, Yang D-C, Kong L, Hou M, Meng Y-Q, Wei L, Gao G (2017) CPC2: a fast and accurate coding potential calculator based on sequence intrinsic features. Nucleic Acids Res 45(W1):W12–W16

Kozomara A, Birgaoanu M, Griffiths-Jones S (2019) miRBase: from microRNA sequences to function. Nucleic Acids Res 47(D1):D155–D162

Kung JT, Colognori D, Lee JT (2013) Long noncoding RNAs: past, present, and future. Genetics 193(3):651–669

Laehnemann D, Borkhardt A, McHardy AC (2016) Denoising DNA deep sequencing data—high-throughput sequencing errors and their correction. Brief Bioinform 17(1):154–179

Laver T, Harrison J, O'Neill P, Moore K, Farbos A, Paszkiewicz K, Studholme DJ (2015) Assessing the performance of the oxford nanopore technologies minion. Biomol Detect Quant 3:1–8

Leinonen R, Sugawara H, Shumway M, Collaboration, I. N. S. D (2010a) The sequence read archive. Nucleic Acids Res 39(suppl_1):D19–D21

Leinonen R, Akhtar R, Birney E, Bower L, Cerdeno-Tárraga A, Cheng Y, Cleland I, Faruque N, Goodgame N, Gibson R (2010b) The European nucleotide archive. Nucleic Acids Res 39(suppl_1):D28–D31

Li M, Liang C (2022) LncDC: a machine learning-based tool for long non-coding RNA detection from RNA-Seq data. Sci Rep 12(1):19083

Li A, Zhang J, Zhou Z (2014) PLEK: a tool for predicting long non-coding RNAs and messenger RNAs based on an improved k-mer scheme. BMC Bioinform 15:1–10

Liu L, Li Y, Li S, Hu N, He Y, Pong R, Lin D, Lu L, Law M (2012) (2012) comparison of next-generation sequencing systems. J Biomed Biotechnol 251364:251364

Liu X, Hao L, Li D, Zhu L, Hu S (2015) Long non-coding RNAs and their biological roles in plants. Genom Proteom Bioinform 13(3):137–147

Lorenz R, Bernhart SH, Höner Zu Siederdissen C, Tafer H, Flamm C, Stadler PF, Hofacker IL (2011) ViennaRNA Package 2.0. Algorithms Mol Biol 6:1–14

Lu H, Giordano F, Ning Z (2016) Oxford Nanopore MinION sequencing and genome assembly. Genom Proteom Bioinform 14(5):265–279

Margulies M, Egholm M, Altman WE, Attiya S, Bader JS, Bemben LA, Berka J, Braverman MS, Chen Y-J, Chen Z (2005) Genome sequencing in microfabricated high-density picolitre reactors. Nature 437(7057):376–380

Margulies M, Jarvie TP, Knight JR, Simons JF (2007) The 454 life sciences picoliter sequencing system. Perspect Bioanal 2:153–186

Massingham T, Goldman N (2012) Error-correcting properties of the SOLiD exact call chemistry. BMC Bioinform 13:1–11

McCoy RC, Taylor RW, Blauwkamp TA, Kelley JL, Kertesz M, Pushkarev D, Petrov DA, Fiston-Lavier A-S (2014) Illumina TruSeq synthetic long-reads empower de novo assembly and resolve complex, highly-repetitive transposable elements. PLoS One 9(9):e106689

Mercer TR, Dinger ME, Mattick JS (2009) Long non-coding RNAs: insights into functions. Nat Rev Genet 10(3):155–159
Merriman B, Ion Torrent R&D Team, Rothberg JM (2012) Progress in ion torrent semiconductor chip based sequencing. Electrophoresis 33(23):3397–3417
Metzker ML (2010) Sequencing technologies—the next generation. Nat Rev Genet 11(1):31–46
Mikheyev AS, Tin MM (2014) A first look at the Oxford Nanopore MinION sequencer. Mol Ecol Resour 14(6):1097–1102
Navarro Gonzalez J, Zweig AS, Speir ML, Schmelter D, Rosenbloom KR, Raney BJ, Powell CC, Nassar LR, Maulding ND, Lee CM (2021) The UCSC genome browser database: 2021 update. Nucleic Acids Res 49(D1):D1046–D1057
Nawrocki EP, Kolbe DL, Eddy SR (2009) Infernal 1.0: inference of RNA alignments. Bioinformatics 25(10):1335–1337
Nyrén P, Lundin A (1985) Enzymatic method for continuous monitoring of inorganic pyrophosphate synthesis. Anal Biochem 151(2):504–509
Paicu C, Mohorianu I, Stocks M, Xu P, Coince A, Billmeier M, Dalmay T, Moulton V, Moxon S (2017) miRCat2: accurate prediction of plant and animal microRNAs from next-generation sequencing datasets. Bioinformatics 33(16):2446–2454
Palazzo AF, Lee ES (2015) Non-coding RNA: what is functional and what is junk? Front Genet 6:127231
Panwar B, Arora A, Raghava GP (2014a) Prediction and classification of ncRNAs using structural information. BMC Genomics 15:1–13
Panwar R, Paul K, Nybakk E, Hansen E, Thompson D (2014b) The legitimacy of CSR actions of publicly traded companies versus family-owned companies. J Bus Ethics 125:481–496
Pundhir S, Poirazi P, Gorodkin J (2015) Emerging applications of read profiles towards the functional annotation of the genome. Front Genet 6:146125
Quek XC, Thomson DW, Maag JL, Bartonicek N, Signal B, Clark MB, Gloss BS, Dinger ME (2015) lncRNAdb v2. 0: expanding the reference database for functional long noncoding RNAs. Nucleic Acids Res 43(D1):D168–D173
Raza K, Ahmad S (2019) Recent advancement in next-generation sequencing techniques and its computational analysis. Int J Bioinforma Res Appl 15(3):191–220
Rhoads A, Au KF (2015) PacBio sequencing and its applications. Genom Proteom Bioinform 13(5):278–289
Rincón-Riveros A, Morales D, Rodríguez JA, Villegas VE, López-Kleine L (2021) Bioinformatic tools for the analysis and prediction of ncRNA interactions. Int J Mol Sci 22(21):11397
Rocha A, Ruiz S, Estepa A, Coll J (2004) Application of inducible and targeted gene strategies to produce transgenic fish: a review. Mar Biotechnol 6:118–127
Ronaghi M, Uhlén M, Nyrén P (1998) A sequencing method based on real-time pyrophosphate. Science 281(5375):363–365
Sanger F, Nicklen S, Coulson AR (1977) DNA sequencing with chain-terminating inhibitors. Proc Natl Acad Sci 74(12):5463–5467
Shendure J, Ji H (2008) Next-generation DNA sequencing. Nat Biotechnol 26(10):1135–1145
Siva N (2008) 1000 Genomes project. Nat Biotechnol 26(3):256–257
Sun K, Chen X, Jiang P, Song X, Wang H, Sun H (2013) iSeeRNA: identification of long intergenic non-coding RNA transcripts from transcriptome sequencing data. BMC Genomics 14:1–10
Thudi M, Li Y, Jackson SA, May GD, Varshney RK (2012) Current state-of-art of sequencing technologies for plant genomics research. Brief Funct Genomics 11(1):3–11
Tran TT, Zhou F, Marshburn S, Stead M, Kushner SR, Xu Y (2009) De novo computational prediction of non-coding RNA genes in prokaryotic genomes. Bioinformatics 25(22):2897–2905
Wang C, Wei L, Guo M, Zou Q (2013) Computational approaches in detecting non-coding RNA. Curr Genomics 14(6):371–377
Washietl S, Hofacker IL, Stadler PF (2005) Fast and reliable prediction of noncoding RNAs. Proc Natl Acad Sci 102(7):2454–2459

Will S, Joshi T, Hofacker IL, Stadler PF, Backofen R (2012) LocARNA-P: accurate boundary prediction and improved detection of structural RNAs. RNA 18(5):900–914

Yao Z, Weinberg Z, Ruzzo WL (2006) CMfinder—a covariance model based RNA motif finding algorithm. Bioinformatics 22(4):445–452

Zhang Y, Huang H, Zhang D, Qiu J, Yang J, Wang K, Zhu L, Fan J, Yang J (2017) A review on recent computational methods for predicting noncoding RNAs. BioMed Res Int 2017:9139504

Zhao Y, Li H, Fang S, Kang Y, Wu W, Hao Y, Li Z, Bu D, Sun N, Zhang MQ (2016) NONCODE 2016: an informative and valuable data source of long non-coding RNAs. Nucleic Acids Res 44(D1):D203–D208

Zuker M (2003) Mfold web server for nucleic acid folding and hybridization prediction. Nucleic Acids Res 31(13):3406–3415

Functional Characterization Methods for Non-coding RNAs

6

Adeeb Rahman and Neeti Sanan-Mishra

Abstract

Discovery of non-coding RNAs (ncRNAs) has led to a drastic change in the understanding of the central dogma of molecular biology. These molecules play a complex regulativory role in genome organization and gene expression. Particularly, the small ncRNAs, are responsible for controlling fundamental processes such as development, physiology, and various defence responses in plants and animals. Multiple *in silico* approaches have been designed to predict and characterize different types of ncRNAs. To validate them experimentally, traditional methods such as quantitative-real time PCR (qPCR) and northern blot are most commonly utilized; but these have been limited to some extent. Advanced methods such as RNA seq, *in-situ* hybridization, RLM-RACE, PARE, luciferase assays, and pulldown approaches have been of great importance. These methods are quite efficient in identifying the numerous transcripts, their interactions and functional complexes, which can give a detailed view of their roles in biological processes. Functional analysis of ncRNAs adds a new paradigm in the understanding of RNAi machinery to draw some useful solutions for the ongoing challenges in food production, medical sciences, and industrial applications. This chapter focuses on the methods employed for the characterization of known and novel ncRNAs.

Keywords

Northern blot · RACE · miRNA · Stem-loop PCR · Techniques

A. Rahman · N. Sanan-Mishra (✉)
Plant RNAi Biology, International Centre for Genetic Engineering and Biotechnology, New Delhi, India
e-mail: neeti@icgeb.res.in

K. Goswami et al. (eds.), *Non-coding RNAs for Crop Improvement*,
https://doi.org/10.1007/978-981-97-7213-1_6

6.1 Introduction

Non-coding RNAs (ncRNAs) constitute a class of RNA molecules that do not code for any protein. These include housekeeping RNAs (tRNA, rRNA, snRNA, shRNA) and regulatory RNAs such as small ncRNAs (miRNA, siRNA, circRNAs), and long ncRNAs (lncRNAs). Among them, the regulatory RNAs have gained tremendous attention in plant and animal science research. miRNAs and siRNAs are the most common type of small ncRNAs synthesized endogenously or exogenously, respectively, and have a typical length of 21–24 nt. They regulate gene expression by binding to their complementary target transcripts, leading to their degradation or translational repression. miRNAs are involved in various biological processes, including development, cell differentiation, metabolism, and response to environmental cues (Adjei et al. 2021). To understand the endogenous RNAi process and their significance in different life processes, the functional analysis of miRNAs and their target is important. Each miRNA can target number of mRNA transcripts, it means a single miRNA can regulate multiple genes at a time (Fischer et al. 2015). miRNAs can bind to the exons or 3′ untranslated region (UTR) of target mRNAs, in plants and animals, respectively, with the help of 2–8 nt seed sequence present at the 5′ end. Computational tools predict target sequences within 3′UTR of mRNA and used these sequences for function prediction by looking their evolutionary conservation in related species (Dash et al. 2022). Emergence of next generation sequencing (NGS) based high throughput methods contributed in the discoveries of miRNA pool but their biological functions still need to be explored. To get functional insights on the large amount of RNA information generated through sequencing and computational methods, several experimental and characterization-based tools were designed. Quantification of small ncRNAs can be achieved by classical methods such as quantitative real-time PCR (RT-qPCR), northern hybridization, and microarray (Sects. 6.2.1 and 6.2.2). To gain insights into miRNA-target interaction, the RLM-RACE, degradome analysis, and pull-down methods are used. For evaluating their expression, the activities of the miRNAs and different promoters can be checked through Luciferase and GFP reporter assays (Fig. 6.1).

Other than that, there are circular RNAs that possess closed-loop structures that are formed with the help of covalent bonds between 5′ and 3′ ends via back splicing (Jeck et al. 2013; Memczak et al. 2013). Due to their closed structure, they are protected from ribonuclease R (Suzuki et al. 2006). These are validated through qPCR and northern blot methods (Kristensen et al. 2019) (Pandey et al. 2020). For circRNA amplification, divergent primers are designed in which a forward primer is specific to circRNA region while reverse primers are specific to complementary region of back-splice junction sequence (Panda and Gorospe 2018). Studies have reported that circRNAs can act as miRNA sponge because they have miRNA binding sites, which indicates that circRNAs regulate miRNAs through different mechanisms (Kulcheski et al. 2016).

Another category of ncRNAs is lncRNAs, which are cis-or trans-acting and are also involved in multiple biological processes (Yuan et al. 2016). lncRNAs can act as precursor for miRNAs and also act as target mimics of miRNAs (Yan et al. 2020).

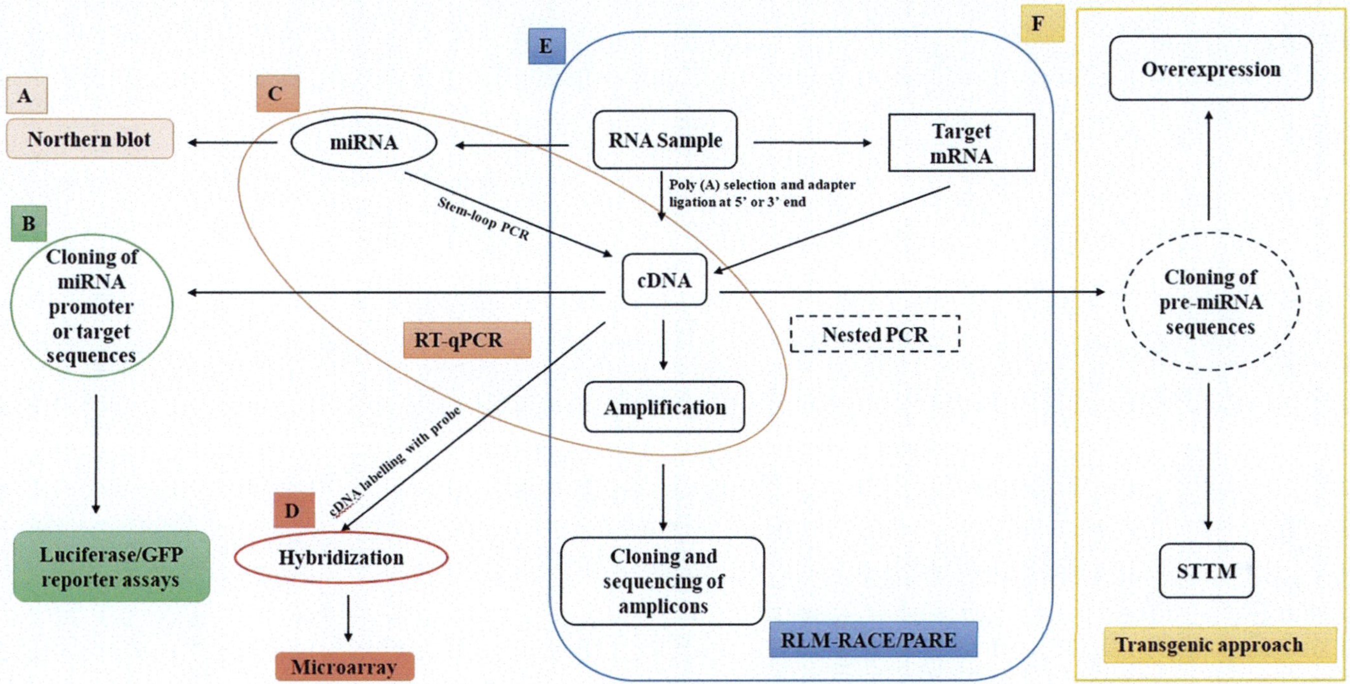

Fig. 6.1 Schematic representation of common methods that are used for the characterization of non-coding RNAs. (**a**) Northern blotting: used to separate and detect specific RNAs based on their size, (**b**) Reporter assay: used to depict promoter activity and subcellular localization, (**c**) RT-qPCR: used to get the quantified expression data of specific transcripts, (**d**) Microarray: used for the analysis of multiple targets and samples in a single go, (**e**) RLM-RACE: used for target validation via ligation of adapter sequences with cleaved transcripts, PARE: also used for target confirmation but is suitable for genome wide study, (**f**) Transgenic approach: used for overexpression and knockdown studies of miRNAs in plants

This suggests that the lncRNA and miRNA can interact in several ways. The majority of lncRNAs can be validated through RT-qPCR, northern blotting, SAGE (serial analysis of gene expression) (Robinson et al. 2004), PARE (parallel analysis of RNA ends) (German et al. 2008), RNA seq (Nagalakshmi et al. 2008), and microarray (Yazaki et al. 2007).

Noncoding RNAs, such as miRNA and lncRNAs, can also be cloned into desired vectors for overexpression or knockdown/knockout purposes. For overexpression, amplified ncRNA is cloned under the constitutive promoter while for knockdown, short tandem target mimics (STTM) is used to target miRNAs (Tang et al. 2012). The artificial miRNAs (amiRNAs) can be used to target lncRNA. amiRNA can be designed using WMD3 software (Schwab et al. 2006). After designing and synthesizing the sequence, desired ncRNA is cloned in a vector having a 35S promoter followed by mobilization in *Agrobacterium tumefaciens*. Transgenic plants having overexpression and knockdown cassettes are generated through agrobacterium-mediated plant transformation and are selected using the resistance marker gene present in the binary vector. By raising transgenics, the effects of ncRNAs on plant phenotype and characterize their specific functions in plant systems can be checked. The first part of this chapter focuses on the common techniques which can be preferably used for ncRNA validation while the second part discusses the methods used for their functional characterization in plants.

6.2 Common Techniques for Validation of Non-coding RNAs

Progression of reverse genetics and sequencing techniques widened the discovery of a variety of ncRNAs in plants and animals. The use of bioinformatics tools shared a major contribution to ncRNA prediction. *In silico* methods proved to be cost-effective, more efficient, comparatively faster, and versatile in nature. Majority of them utilised the knowledge of homologous sequences of known miRNAs in or across genomes that served as the important fundamental for ncRNA prediction (Lagos-Quintana et al. 2001) (Lee and Ambros 2001). To characterize and explore the function of each ncRNA experimental validation becomes a crucial aspect. For validation purposes, a variety of classical as well as modern methods were developed. Each method has its added advantages and some limitations (Table 6.1). This section includes the details of routinely used techniques for ncRNA validation.

6.2.1 Stem-Loop PCR

Most of the ncRNAs, such as miRNAs and siRNAs, have very short lengths (21–24 nt) that becomes very challenging for PCR. To overcome this, research groups started to lengthen small RNAs by polyadenylation (Ro and Yan 2010) or increased the amplicon size by stem-loop PCR. Stem-loop PCR uses specialized primers in which the forward primers have a fixed sequence of 44 plus 6 extra nt at 3′ end. These extra nucleotides are specific to mature miRNA sequence. Stem loop

Table 6.1 List of some common techniques involved in non-coding RNA characterization and validation.

Technique	Advantages	Limitations	References
RT-qPCR	Targeted quantification of small RNAs, highly sensitive and specific	Low to moderate throughput	Kang et al. (2012)
Microarray	High throughput	Low sensitivity and specificity, expensive	Siddika and Heinemann (2021)
Northern blot	Detect precursors and mature forms of small RNAs based on their sizes, cost-effective	Poor sensitivity, time-consuming, low throughput, requires higher amount of sample as compared to RT-qPCR	Dong et al. (2013)
RLM-RACE	Maps RNA ends, useful in miRNA-target validation	Low throughput, time consuming, low ligation efficiency	Song et al. (2017)
PARE (Degradome sequencing)	miRNA-target validation, high throughput	Low ligation efficiency, possibilities of secondary structure, requires efficient data interpretation	Song et al. (2017)
NGS	Gives wider coverage and depth, unbiased data acquisition	Requires computational analysis, expensive	Siddika and Heinemann (2021)
Reporter assays	Identifies miRNA binding sites on their targets, monitors promoter activity (luciferase and GFP-based)	Low throughput, laborious, time consuming, less suitable for time kinetics studies	Repele (2019)
miRNA-mRNA pull down assay	Identifies miRNA targets	Sometimes gives non-specific target	Dash et al. (2018)
RNA immunoprecipitation assays	Validates RNA-protein interactions	Needs lot of optimization at every step, antibody-dependent, chances of getting non-specific signals	Marmisollé et al. (2018)

structure having complementarity with those extra 6 nt can now anneal with the small RNA-specific forward primer. The reverse primer is universal in all cases because it shares the complementarity with the 44 nt stem-loop sequence (Chen et al. 2005). In stem-loop PCR, forward primer is known as stem-loop (SL) RT primer. Sometimes, RT primers can non-specifically bind with the pri−/pre-miRNAs if they are present in excess. To avoid this, it is better to isolate small RNAs from total RNAs. Studies also used hydrolysis probe which are complementary to the 3′ end of the 44 nt stem-loop sequence to maximize the length and chances of better amplification (Kramer 2011).

6.2.2 Quantitative Reverse Transcription PCR

RT-qPCR is the most common, sensitive, and reliable technique for quantification of ncRNAs (Kang et al. 2012). RT-qPCR approach can be used for almost all types of transcripts. This method measures the changes in the relative expression of a gene in different samples. The first step includes the synthesis of cDNA through reverse transcription PCR and then it can be quantified. In the reaction mixture, cDNA is used as a template. For real-time monitoring, kits like SYBR Green and TaqMan are used. The fluorescence signal emitted during initial cycles is too weak but during the exponential phase, it becomes almost double. The saturation point is reached after 35 rounds of cycles while the Ct gives the initial amount of the target in a sample (Ponchel et al. 2003). There are two quantification strategies, one is absolute, which determines the desired gene amount in the sample with the help of standard curve and the other is relative, which evaluates changes in gene expression with respect to the reference sample (Bustin 2002). A reference sample, basically an untreated control is used to correct the differences due to the PCR set-up and amount of cDNA added. In plant samples, generally, 18SrRNA, actin, and glyceraldehyde-3-phosphate dehydrogenase (GAPDH) are used as a reference because their expression levels are quite stable irrespective of the treatment (Livak and Schmittgen 2001). $2^{\wedge-}\Delta\Delta$Ct is the extensively used method for result interpretation. It uses threshold cycle (Ct) values which are generated at the end of RT-qPCR. Ct values help to calculate ΔCt that reflects the threshold difference between treated and untreated samples. Relative gene expression (fold change) for the reference sample is equivalent to 1 due to ΔΔCt value equal to 0 (Rao et al. 2013). This method can work with a small amount of templates and provides accurate results.

6.2.3 Northern Blot

One of the classical methods that can track pre- and mature miRNAs and give information about size changes in miRNA is northern blotting (Lee et al. 2002; Schmittgen et al. 2004). Briefly, the extracted RNA is heated to denature it, so that the secondary structure could be avoided. Gel electrophoresis is used to separate the strands according to the size of the RNA molecules. Usually, polyacrylamide or denaturing agarose gels are used to run small RNAs isolated from total RNA pool through enrichment analysis. After that, the gel is transferred onto a nylon membrane followed by crosslinking, hybridization with probes and detection (Lau et al. 2001; Li and Ruan 2009). Crosslinking of membranes helps in fixation of RNAs to allow long-time storage and better hybridization. The drawback of using traditional DNA probes is the poor detection of miRNAs having low levels or having similar sequences with fewer mismatches. With the use of Locked nucleic acid (LNA) probes, the sensitivity and specificity have been improved. These probes have chemically locked furanose rings in an N-type (C3′-endo) conformation via 2′-O-4′-C methylene bridge which provides better base stacking. In DNA nucleotide probes,

LNA is introduced at every third nucleotide position which enhances their mismatch discrimination power (Válóczi et al. 2004). No signal or weak signal might be due to poor RNA transfer, inefficient probe labelling or maybe higher hybridization temperature. Sometimes, there are nonspecific background signals which can be due to non-stringent hybridization temperature and washing. After hybridization, the blot can be used for stripping in which the first probe and detecting reagent are wiped off so that the same blot can re-probe for different miRNAs. For result interpretation, positive and negative controls must be used. Positive control can be the sample already known for miRNA of interest while negative control is the sample in which miRNA of interest is not present. Positive control ensures the efficiency of hybridization whereas the negative control reflects the non-specific background (Várallyay et al. 2008).

6.2.4 Microarray

Microarray is a high throughput technique where we can simultaneously analyse thousands of RNA sequences. Microarrays have several added advantages in the identification of novel ncRNAs, genome-wide profiling, and expression profiling under different conditions and samples for comparative studies. This technique is based on hybridization between nucleic acids and their complementary probes. The probes designed for miRNA have amine-modified 5′ termini and are immobilized on glass surface via covalent crosslinking. The procedure starts with the RNA isolation followed by small RNA enrichment analysis. Denaturing polyacrylamide gel electrophoresis (PAGE) separates several small RNA precursor molecules to eliminate the noise. Isolated miRNAs were subjected to fluorescent labelling and then hybridized with their corresponding probes present on glass slides. The fluorescent dye helps in detection of signals emitted during hybridization and the interpretation of signal data to analyse the relative expression of miRNAs in given samples. In microarrays, the preparation of small RNA samples, probe designing, and labelling of small RNAs are the most crucial steps (Li and Ruan 2009). Linker and capture sequences are the two main regions of any miRNA probe. In which linker sequences have poly d(T) or poly d(A) with a modified amine terminus and are close to the glass surface while the capture sequences have miRNA complementary sequences. Another important factor is the normalization of Tm of the probes which should be in a broader range so that hundreds/thousands of probes can hybridize within this range. For instance, if the Tm of the probe is less than the hybridization temperature then lower binding will be achieved while if Tm of probe is greater than the hybridization temperature will give a better binding but can distort the fluorescence signals. By adjusting the sequences or length of the probe we can normalize the Tm but not for the small RNAs which are smaller in size. To overcome this problem, various modified probes such as LNAs are used to maximize the Tm (Castoldi et al. 2006). These LNAs can be subjected to adjusting the sequence and length for Tm normalization. The effective labelling of miRNA is another important part of microarray which is divided into two categories: direct and indirect labelling. In direct labelling, miRNAs are directly conjugated with the fluorescent dye while in indirect labelling the reverse transcript of miRNA is labelled rather than miRNA. For direct conjugation, the labelling methods such as guanine labelling reagent (Babak et al. 2004), T4 RNA ligase (Thomson et al. 2004), poly(A) polymerase (Shingara et al. 2005), Biotinylation (Liang et al. 2005), RNA-primed array-based Klenowenzyme (RAKE) assay (Nelson et al. 2004) are used while indirect methods include labelling of RT-PCR products by adding adaptor sequences at miRNA ends and biotin/fluorescence modified primer (Miska et al. 2004), *in-vitro* transcript labelling by introduction of promoter sequences in adaptor region followed by fluorescent-modified labelling during in vitro transcription (Barad et al. 2004). Microarray is the best-suited method to study global transcript expression levels. This technique has huge applications in plants as well as in clinical research. In clinical research, it is used to diagnose biomarkers, detect gene expression in various cancers, drug discovery, and understand mechanisms involved in disease progression (Fahmideh et al. 2016). In plant research, the knowledge of microarray data is useful to compare expression profiles of huge miRNA data generated from *in-silico* tools to draw various conclusions, which have added new insights and are quite promising to cope up with agriculture issues.

6.2.5 *In-Situ* Hybridization

This method renders the visualization and subcellular localisation of tagged ncRNA molecules in cells through hybridization with a labelled antisense RNA probe. Those probes can be designed through an *in-vitro* transcription process (Javelle et al. 2011). It helps to get relative expression patterns among the miRNA and their target in a spatio-temporal manner. Methodology includes probe designing, which is specific to the desired sequence and depending upon the type of visualization which could be radiolabelled or fluorescent based. If hybridization gives fluorescence then it is known as FISH (fluorescence in-situ hybridisation). This is further modified as multiplexed miRNA FISH, where multiple differentially regulated miRNAs in specific tissues can be visualised (Renwick et al. 2014). Then comes sample preparation through fixatives like paraformaldehyde (PFA). Normally 4% PFA, 15% DMSO, and 0.1% Tween 20 in phosphate-buffered saline (PBS) are used as fixative for whole mount (Gautam et al. 2019). After fixation, the non-specific sites are blocked by incubating samples with a pre-hybridization solution followed by hybridization with probes. Post-hybridization washes remove unbound probes and the sample is blocked again. The final step includes the detection method based on the labelled probe. Limitations of this approach include the concentration of probes and non-specific binding, which might result in weaker or false signals (Javelle et al. 2011). Adjustments in hybridization temperature and washing temperature can improve the signal intensity. Also, if the signal background is high in the sample, the reduced concentration of probes works better in hybridization. In a study, the whole mount in situ method was used against miR166 and *PHABULOSA*

(PHB) in *Arabidopsis*. They used NBT/BCIP as a substrate that gives colour in alkaline conditions (pH, 9.5) (Gautam et al. 2019). By this method, one can localize the candidate miRNAs and their target transcripts and correlate the results with the expression data.

6.2.6 RNA Ligation Mediated-Rapid Identification of cDNA Ends - RLM-RACE

RLM-RACE is a common approach for the detection of RNA cleavage intermediates (Ueno et al. 2020). A linker is joined at the end of the free phosphate of UTR of an uncapped mRNA generated through various nucleolytic processes. If the linker ligates at the 5′ end then it is called 5′-RLM-RACE whereas if it attaches towards the 3′ end then it is said to be 3′-RLM-RACE (Adkar-Purushothama et al. 2017). 5′ RLM-RACE is not appropriate if the cleavage occurs near the 3′end of the target. The ligation product is then reverse-transcribed using a forward primer against the adapter and a gene-specific reverse primer. The methodology of RLM-RACE starts with the adapter ligation step in which total RNA is treated with phosphatase so that 5′ phosphate can be removed from the cleaved RNA fragments. A synthetic RNA adapter having 5′ phosphate is ligated to the 5′ end of the target mRNA via T4 RNA ligase (Hafner et al. 2008). The product of ligation is then reverse transcribed using gene-specific primers. So that only the desired copy of cDNA would be generated. After this, the first round of nested PCR has a forward primer specific to the adaptor sequences and a gene-specific reverse primer is used to amplify the synthesized cDNA which is followed by a second round of PCR through nested primers (Adkar-Purushothama et al. 2017). The amplified product is run on the gel for visualization which is further cloned and sequenced to determine the 5′ end of the targets. However, this technique has some disadvantages like incomplete outcome sequences which may lead to the missing out of important sequences which have regulatory roles. The second reason could be the non-specific ligation due to the partially digested samples and secondary structures (Song et al. 2017). Overall, this approach is widely used to characterize RNA ends and has great potential if the limitations are minimised.

6.2.7 RNA Degradome Sequencing

This technique is like RLM-RACE but it can map genome-wide target transcripts which makes it high throughput and it is also known as PARE (parallel analysis of RNA ends) or GMUCT (genome-wide mapping of uncapped transcripts) (Ding et al. 2012; He et al. 2022). The very first step is the enrichment where poly (A) tail containing RNAs are retained before cDNA preparation so that rRNAs can be removed. This decreases the huge noise because of a high percentage of rRNAs (Weinberg et al. 2017). Sometimes, the ligation of adapters becomes a challenge due to the low efficiency of ligase or when there are secondary structures between adapters and the end of the transcript (Song et al. 2017). In degradome, sequence motif analysis is used to get the sequence features near degradation intermediates because these motifs affect RNA stability (Hou et al. 2014; Ueno et al. 2020). Through this method, many tissue-specific degradome libraries are prepared which are proven to be very useful in target validation. *In-silico* databases and their upcoming pipelines have a major contribution to making the efficient use of large data generated from degradome sequencing (Ding et al. 2012). One study in maize identified, 260 targets corresponding to 60 miRNAs belonging to 25 families using the same technique. They further point out the involvement of a negative feedback loop in the RNAi pathway by taking the example of transcripts miR159c and miR169e cleaved by mature forms of miR159 and miR169 (He et al. 2022). Another study, by Schon et al. integrates the RNA 5′ end data with NGS and named nanoPARE. It has the potential to identify transcription start sites with the resolution of a single nucleotide (Schon et al. 2018). Although this technique is useful in understanding RNA degradation and ncRNA target identification, more efforts are required to analyze degradome data.

6.2.8 Luciferase and GFP Reporter Assays

In dual luciferase assay, the firefly LUC gene is fused with the predicted target sequences whose expression is used to report the miRNA interaction with the target sequence. Renilla luciferase cassette is used as an internal control for standardization (Liu et al. 2014; Liu and Axtell 2015; Moyle et al. 2017). The light emitted by the LUC cassette has target sequences detected by a luminometer. Sometimes, the plant extracts hamper LUC expression which can be overcome by certain dilution methods (Moyle and Birch 2013). Other parameters such as the optimal ratio of the pre-miR construct, LUC-target construct and post-agroinfiltration incubation period are crucial to get the maximum repression. Moyle's group recommended that 10:1 is the optimal ratio and 3 days post-agroinfiltration is the incubation period for the highest repression of the target. Due to the emergence of a modified vector construct, it is possible to clone multiple miRNA and target sequences that will be helpful in the rapid screening of huge predicted interactions (Moyle et al. 2017). Another commonly used reporter is the green fluorescence protein (GFP) gene which can be utilised to check different promoter activities. Not only this, but the GFP reporter assay also helps to visualize the degree of repression of the miRNA-target. Co-infiltration of the target-reporter construct with the pre-miRNA construct is also useful for better insights into miRNA-target interactions (Franco-Zorrilla and Solano 2017). Researchers also replaced the synthetic promoter with the endogenous promoter that mimics the natural condition of the plant system (Fig. 6.2a). After cloning the desired promoter sequences in the MCS, a construct is mobilized into agrobacterium followed by agroinfiltration in plants (Varchenko et al. 2020). Thus, we can say that this method is quite promising for detecting miRNA-directed translational inhibition, promoter activities, and the localization of the gene of interest which cannot be possible through earlier approaches.

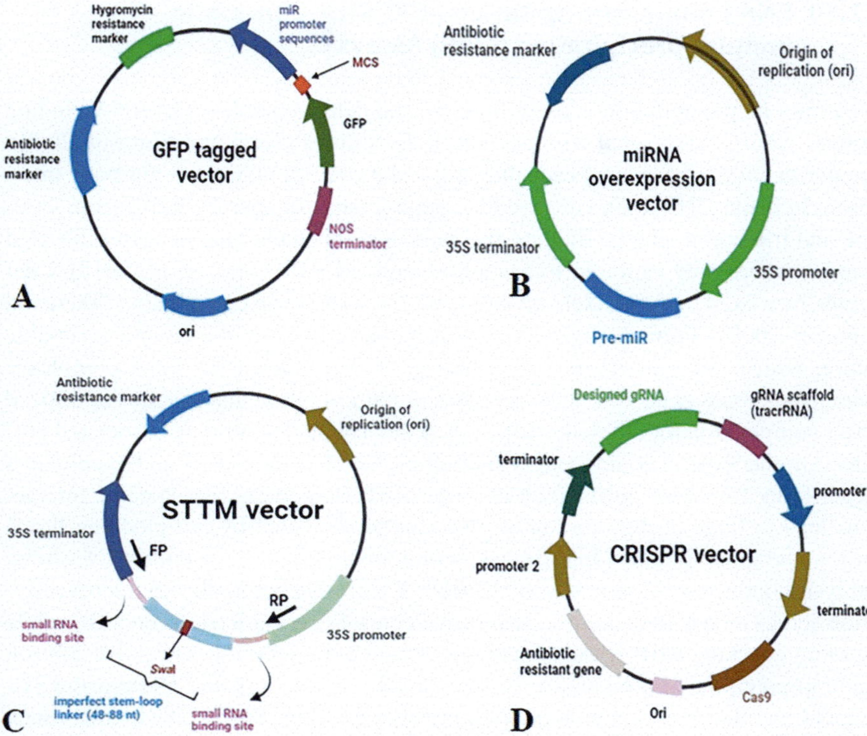

Fig. 6.2 Common cloning approaches available for ncRNA characterization in plants. (**a**) 35S promoter replaced with miRNA promoter/endogenous promoter which drives GFP expression, (**b**) precursor miRNA cloned in multiple cloning site (MCS) under the control of 35S promoter, (**c**) STTM construct having 48-88 nt long imperfect stem loop flanked with small ncRNA binding sites (having 3 nt mismatches), (**d**) Guide RNA specific to target miRNA designed and cloned in CRISPR vector having Cas9

6.2.9 miRNA-mRNA Pull-Down Assay

Experimental methods for examining and describing the interactions between RNA molecules and proteins are called RNA pull-down assays. These tests offer a means of identifying miRNA-binding mRNAs. An RNA pull-down assay used biotinylated miRNA as bait to extract interacting miRNA-mRNA complex from biological materials, such as cell lysates. Biotinylation can be done via *in-vitro* transcription in which biotin-labeled nucleotides get incorporated in RNA. Tagged RNAs are transfected into suitable cell lines and then total RNA is extracted. After extraction, the RNA sample is immobilized onto streptavidin-coated beads to separate the biotin-tagged miRNAs from the total RNA pool. miRNA and mRNA target complex confirmed through microarray and deep sequencing (Dash et al. 2018). There are some challenges such as background noise due to non-specific targets. Incubation time, blocking, and washing are some crucial steps that can give better results if optimized correctly.

6.2.10 RNA Immunoprecipitation (RIP)/RNA-Chromatin Immunoprecipitation (ChIP) Assays

RNA immunoprecipitation is used to purify and detect complexes of RNA binding proteins (RBPs) associated with specific RNA strands (Terzi and Simpson 2009). The purification is based on beads that utilize the concept of antigen-antibody interaction. In plants, RIP assays are used to capture native or specific RBPs from wild-type and transgenic plants. Studies also validated this technique in agrion-filtrated samples where they captured transiently expressed RBP (Shen et al. 2015). Leaf samples are fixed by formaldehyde through *in vivo* chemical crosslinking (Terzi and Simpson 2009). Formaldehyde is a widely used fixative due to its reversible nature, which helps in reverse crosslinking of the bound RBP-RNA complexes (Niranjanakumari et al. 2002). Then comes protein isolation that requires optimized buffer composition for efficient extraction of nuclear and membrane-associated proteins. Another crucial component is antibodies, which should be of high quality and high affinity for better purification of target RBP. Sometimes non-specific interactions lead to huge background noise that can be mitigated by pre-blocking beads with non-cognate proteins like Bovine serum albumin (BSA), optimized washing, and lysis conditions (Marmisollé et al. 2018). Cell lysate before the immunoprecipitation serves as the input sample or positive control (having RBP of interest) while the sample lacking RBP or use of non-specific antibody serves as a negative control. Immuno-precipitated proteins can be recovered from the organic phase using Tri-Reagent and their validation is done by western blotting. Further, techniques such as qPCR and northern blot are used for the detection of different sizes and types of small ncRNAs and their precursors which are precipitated with proteins (Marmisollé et al. 2018). This technique helps us to get insights into conserved RNA mechanisms and novel interactions between RNA machinery intermediates and exogenous proteins like viral suppressors of RNA silencing (VSRs).

6.3 Overexpression and Knockdown Studies of miRNA in Plants

miRNAs play important roles in plant phenotype, physiology, growth, and development (Kamthan et al. 2015). Cloning and sequencing of small ncRNAs come under experimental approaches which can become quite challenging. The expression of miRNAs is low and spatio-temporal i.e. most of them are expressed in response to stimulus and can coexist with their cleaved target mRNAs (Unver et al. 2009).To explore their potential in plants, their expression could be checked by creating overexpression and mutant cassettes followed by plant transformation. In overexpression studies, the pre-miRNA, rather than the mature sequence, is usually required for fusion under a constitutive or inducible promoter (Fig. 6.2b). This ensures that the pre-miRNA is correctly processed and localised in the cell. Also, the pre-miRNA sequence is much conserved and important for expression, processing, and mimicking the natural biogenesis pathway.

The other aspect is to bring the miRNA level either too low (knockdown) or can be completely removed (knockout). For stable knockdown lines, one of the most widely used approach, short tandem target mimic (STTM) is advantageous. As the name suggests, these STTMs act as target mimics so that the cleavage of target mRNA can be minimized. The typical length of STTM is around 21–24 nt just like of miRNA. The structure has tandem repeats, resembling the target, and a small bulge in the center that traps small RNA by stacking at the 10th and 11th nt positions. Between two small ncRNA binding sites (tandem repeats) there is a 48–88 nt long weak secondary structure known as spacer (Tang et al. 2012). Below tandem repeats some flanking sequences help in joining with the promoter and terminator so that the transcript production remains continuous. Other than STTM, several techniques can be used for loss of function studies such as amiRNAs targetting pri−/pre-miRNAs (Carbonell et al. 2015). TALENS for miRNA sequence editing (Bi et al. 2020) or CRISPR/Cas9 for incorporation of indels at miR genes (based) (Damodharan et al. 2018), AMOs (anti-miRNA oligonucleotides) to pair with mature miRNAs and block their activity (He et al. 2016), artificial TMs (target mimics) and molecular sponges (SPs) to chelate specific mature miRNAs (Todesco et al. 2010; Reichel et al. 2015) or an entire family of mature miRNAs (Yan et al. 2012). Hence, multiple types of decoys are beneficial to get better results (Chen et al. 2021). STTM approach is quite successful especially, in the case of agronomic traits of rice. Transgenics of STTM398, STTM156, STTM159 (Zhang et al. 2017), STTM166 (Zhang et al. 2018), and STTM167 (Peng et al. 2018) showed alteration in seed size, grain weight, and panicle length. STTM159d, 160, 166,171,172,398,441 alter plant height, STTM156 and STTM166 showed variation in tiller number and tiller angle (Chen et al. 2021). Some mimicking techniques utilized the cleavage-resistant target genes so that their overexpression upregulateds the miRNA-resistant target and creates loss-of-function state (Li and Millar 2013) (Fig. 6.1, Table 6.2).

There are various cloning vectors available for STTM cloning. pOT2 is one such vector having a 35S promoter, a Poly-Cis DNA fragment, and a 35S terminator. The Poly-Cis is replaced by STTM structure with the help of PCR having primers 3′ complementary (binds with vector) and a 5′ tail part (binds with half of STTM sequence). 5′ end of each primer there is a *Swa*I restriction site and the 5′ tail of each primer was amplified with the vector promoter, terminator, antibiotic-resistant marker, and *ori* (Figure 6.2c). After that, the PCR product is subjected to *Swa*I digestion and self-ligation resulting in pOT2-STTM generation (Tang et al. 2012).

Overexpression and knockdown lines are genetically modified hence these are under strict GMO regulations. The need for transgene-free editing led to the implementation of another approach, which is Cluster regularly interspaced short palindromic repeats (CRISPR). It has also proven to be more robust, more stable, and can by-pass the regulations (Chen and Gao 2020). CRISPR/Cas mediated induced mutations within ncRNAs genome sequences are helpful in functional characterization and provide new ideas to increase yield and stress adaptability in plants (Bravo-Vázquez et al. 2024). Advancement of gene editing triggers discoveries of multiple Cas enzymes such as Cas10 for induction of short indels and long deletions, Cas12a for transcriptional regulation (Puchta et al. 2022), Cas12b for

Table 6.2 Examples of gain of function and loss of function studies on plant miRNAs

Functional validation methods	Target	Plant	Outcome	References
	Gain-of-function			
Overexpression of miRNA	miR156	Rice	Delayed flowering, reduced panicle size	Xie et al. (2006)
	miR156b	Soybean	High phosphorus efficiency, higher yield	Lu et al. (2023)
	miR408	Arabidopsis Tobacco Rice	Improved photosynthesis and grain yield	Pan et al. (2018)
	miR168a	Arabidopsis	Drought tolerance and hypersensitive to ABA	Li et al. (2012)
	miR162a	Rice	Resistance to Brown plant hopper (BPH)	Shen et al. (2021)
	Gma-miR394a	Arabidopsis	Drought tolerance	Ni et al. (2012)
	miR268	Rice	Inhibit seedling growth under cadmium stress	Ding et al. (2017)
	miR319	Rice	Promotes cold tolerance	Yang et al. (2013)
	miR397	Rice	Enhance panicle branching, grain size and grain yield	Zhang et al. (2013)
		Banana	Improved biomass	Patel et al. (2019)
	miR169o	Rice	Increased nitrogen efficiency but reduced resistance against bacterial blight	Yu et al. (2018)
Overexpression of artificial miRNA	amiR528	Rice	Albino phenotype, elongated upper internode and spontaneous lesions	Warthmann et al. (2008)

(continued)

Table 6.2 (continued)

Functional validation methods	Target	Plant	Outcome	References
	Loss-of-function			
TALENs	miR390	Rice	Shoot less phenotype	Bi et al. (2020)
CRISPR-based	Pre-miR160a	Arabidopsis	Severe leaf serration	Bi et al. (2020)
	miR168a	Rice	Dwarf, shorter spikelet and seeds	Zhou et al. (2022)
	miR482b and miR482c	Tomato	Resistance against *Phytophthora infestans*	Hong et al. (2021)
	miR482 (10 isoforms)	Cotton	Evaluation of disease response against *Verticillium dahlia*	Zhu et al. (2022)
	miR408 and miR528	Rice	Regulate abiotic stresses	Zhou et al. (2017)
	miR396e and miR396f	Rice	Improves grain yield under nitrogen deficiency	Zhang et al. (2020)
	miR818b and miR171f	Rice	Drought tolerance	Chung et al. (2020)
	miR160b-5p and miR390-5p	Potato	Involved in cell death and auxin signaling	Lukan et al. (2022)
AMOs	miR160a-d, miR164abf, miR167d-h, miR171g, miR390	Rice	Defective root phenotype in seedlings and affect auxin signaling	He et al. (2016)
SPs	miR159,	Arabidopsis	Leaf curling and reduced growth	Reichel et al. (2015)
	miR165/166		Floral clumps and sterile	
TMs	MIM396	Arabidopsis	Improved plant defences against fungi	Soto-Suárez et al. (2017)

(continued)

Table 6.2 (continued)

Functional validation methods	Target	Plant	Outcome	References
STTM	STTM165/166	Arabidopsis	Severe developmental defects	Yan et al. (2012)
	STTM398	Rice	Shorter plant height, reduced grain length and width	Zhang et al. (2017)
	STTM172		Defective culms with impaired panicle development	
	STTM156		Fewer tillers, thicker culms and roots, reduced shoot in seedlings	
	STTM159		Reduced plant height, grain length and width	
	STTM159, STTM319, and STTM165/166	Arabidopsis	Altered leafdevelopment, small size and no. of siliques, smaller seeds	Peng et al. (2018)
	STTM319/159		Completely sterile	
	STTM159 (35S)		Smaller seed grains	
	STTM167 (Gt13a) STTM1432(Gt13a)		Increase seed size	
	STTM166, STTM172	Rice	Rolled leaves	
	STTM166,	Maize	Longer maturity time	
		Tomato	Twisted leaves with abnormal phyllotaxy	
	STTM396	Tomato	Taller with bigger leaves and fruits	
	STTM393	Poplar	Improved plant growth and biomass	Chu et al. (2021)
Overexpression of miRNA-resistant target genes	miR172-resistant AtTOE3	Arabidopsis	Indeterminate flowers with several stamens and carpals	Jung et al. (2014)
	miR156-resistant AtSPL3		Early flowering	Kim et al. (2012)
	miR164-resistant SlNAM2	Tomato	Establishment of flower boundaries	Hendelman et al. (2013)
	miR166-resistant OsHB4	Rice	Drought resistance, vascular development	Zhang et al. (2018)
	miR319-resistant OsTCP21		Virus resistance or mild symptoms against *Rice ragged stunt virus*	Zhang et al. (2016)

longer deletions than Cas9, Cas13a for mRNA, lncRNA and viral RNA knockdown (Ohyanagi et al. 2022), Cas14a for reverse transcription and demethylation free miRNA detection (Yang et al. 2021), Cas14R for miR156a expression pattern. Researchers also utilize a dual-sgRNA/Cas9 system that can mutate two regions in the target sequence (Pauwels et al. 2018). Another novel approach is CRISPR interference (CRISPRi) which uses deadCas9 to target the promoter or transcription start site (TSS) and inhibits the transcription factors binding. This technique is beneficial for selective repression of miRNAs having high sequence homology (Figure 6.2d) (Drobna-Śledzińska et al. 2022).

6.4 Summary and Future Perspectives

ncRNAs are differentially expressed and processed in a spatio-temporal manner to regulate the numerous growth and developmental processes of plants. The availability of several *in-silico* tools is useful to predict the ncRNAs and their desired functions. To evaluate their functions in biology, it is important to have techniques that allow their isolation, quantification, expression analysis, generation of transgenic lines, and so on. Classical methods such as RT-qPCR and northern blot are quantification methods but are limited to some extent. Other than quantification, modern techniques like in-situ hybridization, RLM-RACE, degradome sequencing, NGS, reporter assays, RNA pull down assays, etc. allow validation of target transcripts, subcellular localization and interaction networks. The newer methods have high throughput, and give high confirmation, which overcomes many limitations of classical approaches. However, the issues with sensitivity and specificity are still there. To better understand the function of any gene, we must see the outcomes either in their presence or absence. Thus, the transgenic approach is very useful to characterize ncRNAs within the living system. A large pool of research studies showcases their overexpression and knockdown studies with outcomes in different plant species. For overexpression purposes, usually the 35S promoter is used while for knockdown/knockout, there are target mimics, molecular sponges, STTM and nowadays CRISPR are used. Plants generated through these approaches showed pleiotropic effects that become a challenge in the dissection of a specific trait. Also, there are many isoforms in a single miRNA family so if one becomes non-functional then the other will take over. Not only is this but the detailed research on novel regulatory RNAs such as circRNAs and lncRNAs equally important to connect the gaps in RNAi biology. To address all these challenges the functional methods and their consistent improvements are very much important.

References

Adkar-Purushothama CR, Bru P, Perreault J-P (2017) 3′ RNA ligase mediated rapid amplification of cDNA ends for validating viroid induced cleavage at the 3′ extremity of the host mRNA. J Virol Methods 250:29–33

Babak T, Zhang W, Morris Q, Blencowe BJ, Hughes TR (2004) Probing microRNAs with microarrays: tissue specificity and functional inference. RNA 10(11):1813–1819

Barad O, Meiri E, Avniel A, Aharonov R, Barzilai A, Bentwich I, Einav U, Gilad S, Hurban P, Karov Y (2004) MicroRNA expression detected by oligonucleotide microarrays: system establishment and expression profiling in human tissues. Genome Res 14(12):2486–2494

Bi H, Fei Q, Li R, Liu B, Xia R, Char SN, Meyers BC, Yang B (2020) Disruption of miRNA sequences by TALENs and CRISPR/Cas9 induces varied lengths of miRNA production. Plant Biotechnol J 18(7):1526–1536

Bravo-Vázquez LA, Méndez-García A, Chamu-García V, Rodríguez AL, Bandyopadhyay A, Paul S (2024) The applications of CRISPR/Cas-mediated microRNA and lncRNA editing in plant biology: shaping the future of plant non-coding RNA research. Planta 259(2):1–21

Bustin S (2002) Quantification of mRNA using real-time reverse transcription PCR (RT-PCR): trends and problems. J Mol Endocrinol 29(1):23–39

Carbonell A, Fahlgren N, Mitchell S, Cox KL Jr, Reilly KC, Mockler TC, Carrington JC (2015) Highly specific gene silencing in a monocot species by artificial micro RNA s derived from chimeric mi RNA precursors. TPJ 82(6):1061–1075

Castoldi M, Schmidt S, Benes V, Noerholm M, Kulozik AE, Hentze MW, Muckenthaler MU (2006) A sensitive array for microRNA expression profiling (miChip) based on locked nucleic acids (LNA). RNA 12(5):913–920

Chen K, Gao C (2020) Genome-edited crops: how to move them from laboratory to market. Front Agric Sci Eng 7:181–187

Chen C, Ridzon DA, Broomer AJ, Zhou Z, Lee DH, Nguyen JT, Barbisin M, Xu NL, Mahuvakar VR, Andersen MR (2005) Real-time quantification of microRNAs by stem–loop RT–PCR. Nucleic Acids Res 33(20):e179–e179

Chen J, Teotia S, Lan T, Tang G (2021) MicroRNA techniques: valuable tools for agronomic trait analyses and breeding in Rice. Front Plant Sci 12:744357

Chu L, He X, Shu W, Wang L, Tang F (2021) Knockdown of miR393 promotes the growth and biomass production in poplar. Front Plant Sci 12:714907

Chung PJ, Chung H, Oh N, Choi J, Bang SW, Jung SE, Jung H, Shim JS, Kim J-K (2020) Efficiency of recombinant CRISPR/rCas9-mediated miRNA gene editing in rice. IJMS 21(24):9606

Damodharan S, Corem S, Gupta SK, Arazi T (2018) Tuning of Sl ARF 10A dosage by sly-miR160a is critical for auxin-mediated compound leaf and flower development. TPJ 96(4):855–868

Dash S, Balasubramaniam M, Dash C, Pandhare J (2018) Biotin-based pulldown assay to validate mRNA targets of cellular miRNAs. JoVE 136:e57786

Dash PK, Gupta P, Pradhan SK, Shasany AK, Rai R (2022) Analysis of homologous regions of small RNAs MIR397 and MIR408 reveals the conservation of microsynteny among rice crop-wild relatives. Cells 11(21):3461

Ding J, Zhou S, Guan J (2012) Finding microRNA targets in plants: current status and perspectives. GPB 10(5):264–275

Ding Y, Wang Y, Jiang Z, Wang F, Jiang Q, Sun J, Chen Z, Zhu C (2017) MicroRNA268 overexpression affects rice seedling growth under cadmium stress. J Agric Food Chem 65(29):5860–5867

Dong H, Lei J, Ding L, Wen Y, Ju H, Zhang X (2013) MicroRNA: function, detection, and bioanalysis. Chem Rev 113(8):6207–6233

Drobna-Śledzińska M, Maćkowska-Maślak N, Jaksik R, Dąbek P, Witt M, Dawidowska M (2022) CRISPRi for specific inhibition of miRNA clusters and miRNAs with high sequence homology. Sci Rep 12(1):6297

Fahmideh L, Kord H, Shiri Y (2016) Importance of microarray technology and its applications. J Curr Res Sci 4(3):25

Fischer S, Handrick R, Aschrafi A, Otte K (2015) Unveiling the principle of microRNA-mediated redundancy in cellular pathway regulation. RNA Biol 12(3):238–247

Franco-Zorrilla JM, Solano R (2017) Identification of plant transcription factor target sequences. BBA-Gene Regul Mech 1860(1):21–30

Gautam V, Singh A, Verma S, Singh S, Chatterjee S, Sarkar AK (2019) Whole mount in situ localization of miRNAs and target mRNA transcripts in plants. 3 Biotech 9:1–7

German MA, Pillay M, Jeong D-H, Hetawal A, Luo S, Janardhanan P, Kannan V, Rymarquis LA, Nobuta K, German R (2008) Global identification of microRNA–target RNA pairs by parallel analysis of RNA ends. Nat Biotechnol 26(8):941–946

Hafner M, Landgraf P, Ludwig J, Rice A, Ojo T, Lin C, Holoch D, Lim C, Tuschl T (2008) Identification of microRNAs and other small regulatory RNAs using cDNA library sequencing. Methods 44(1):3–12

He L, Xie M, Huang J, Zhang T, Shi S, Tang T (2016) Efficient and specific inhibition of plant microRNA function by anti-microRNA oligonucleotides (AMOs) in vitro and in vivo. Plant Cell Rep 35:933–945

He J, Xu C, You C, Mo B, Chen X, Gao L, Liu L (2022) Parallel analysis of RNA ends reveals global microRNA-mediated target RNA cleavage in maize. TPJ 112(1):268–283

Hendelman A, Stav R, Zemach H, Arazi T (2013) The tomato NAC transcription factor SlNAM2 is involved in flower-boundary morphogenesis. J Exp Bot 64(18):5497–5507

Hong Y, Meng J, He X, Zhang Y, Liu Y, Zhang C, Qi H, Luan Y (2021) Editing miR482b and miR482c simultaneously by CRISPR/Cas9 enhanced tomato resistance to Phytophthora infestans. Phytopathology 111(6):1008–1016

Hou C-Y, Wu M-T, Lu S-H, Hsing Y-I, Chen H-M (2014) Beyond cleaved small RNA targets: unraveling the complexity of plant RNA degradome data. BMC Genomics 15:1–16

Javelle M, Klein-Cosson C, Vernoud V, Boltz V, Maher C, Timmermans M, Depege-Fargeix N, Rogowsky PM (2011) Genome-wide characterization of the HD-ZIP IV transcription factor family in maize: preferential expression in the epidermis. Plant Physiol 157(2):790–803

Jeck WR, Sorrentino JA, Wang K, Slevin MK, Burd CE, Liu J, Marzluff WF, Sharpless NE (2013) Circular RNAs are abundant, conserved, and associated with ALU repeats. RNA 19(2):141–157

Jung J-H, Lee S, Yun J, Lee M, Park C-M (2014) The miR172 target TOE3 represses AGAMOUS expression during Arabidopsis floral patterning. Plant Sci 215:29–38

Kamthan A, Chaudhuri A, Kamthan M, Datta A (2015) Small RNAs in plants: recent development and application for crop improvement. Front Plant Sci 6:124569

Kang K, Peng X, Luo J, Gou D (2012) Identification of circulating miRNA biomarkers based on global quantitative real-time PCR profiling. J Anim Sci Biotechnol 3:1–9

Kim JJ, Lee JH, Kim W, Jung HS, Huijser P, Ahn JH (2012) The microRNA156-SQUAMOSA PROMOTER BINDING PROTEIN-LIKE3 module regulates ambient temperature-responsive flowering via FLOWERING LOCUS T in Arabidopsis. Plant Physiol 159(1):461–478

Kramer MF (2011) Stem-loop RT-qPCR for miRNAs. CPMB 95(1):15.10. 11-15.10. 15

Kristensen LS, Andersen MS, Stagsted LV, Ebbesen KK, Hansen TB, Kjems J (2019) The biogenesis, biology and characterization of circular RNAs. Nat Rev Genet 20(11):675–691

Kulcheski FR, Christoff AP, Margis R (2016) Circular RNAs are miRNA sponges and can be used as a new class of biomarker. J Biotech 238:42–51

Lagos-Quintana M, Rauhut R, Lendeckel W, Tuschl T (2001) Identification of novel genes coding for small expressed RNAs. Science 294(5543):853–858

Lau NC, Lim LP, Weinstein EG, Bartel DP (2001) An abundant class of tiny RNAs with probable regulatory roles in Caenorhabditis elegans. Science 294(5543):858–862

Lee RC, Ambros V (2001) An extensive class of small RNAs in Caenorhabditis elegans. Science 294(5543):862–864

Lee Y, Jeon K, Lee J-T, Kim S, Kim VN (2002) MicroRNA maturation: stepwise processing and subcellular localization. EMBO J 21(17):4663–4670

Li J, Millar AA (2013) Expression of a microRNA-resistant target transgene misrepresents the functional significance of the endogenous microRNA: target gene relationship. Mol Plant 6(2):577–580

Li W, Ruan K (2009) MicroRNA detection by microarray. ABC 394:1117–1124

Li W, Cui X, Meng Z, Huang X, Xie Q, Wu H, Jin H, Zhang D, Liang W (2012) Transcriptional regulation of Arabidopsis MIR168a and argonaute1 homeostasis in abscisic acid and abiotic stress responses. Plant Physiol 158(3):1279–1292

Liang R-Q, Li W, Li Y, Tan C-Y, Li J-X, Jin Y-X, Ruan K-C (2005) An oligonucleotide microarray for microRNA expression analysis based on labeling RNA with quantum dot and nanogold probe. Nucleic Acids Res 33(2):e17–e17

Liu Q, Axtell MJ (2015) Quantitating plant microRNA-mediated target repression using a dual-luciferase transient expression system. Plant Funct Genom 1284:287–303

Liu Q, Wang F, Axtell MJ (2014) Analysis of complementarity requirements for plant microRNA targeting using a Nicotiana benthamiana quantitative transient assay. Plant Cell 26(2):741–753

Livak KJ, Schmittgen TD (2001) Analysis of relative gene expression data using real-time quantitative PCR and the 2− ΔΔCT method. Methods 25(4):402–408

Lu G, Tian Z, Hao Y, Xu M, Lin Y, Wei J, Zhao Y (2023) Overexpression of soybean microRNA156b enhanced tolerance to phosphorus deficiency and seed yield in Arabidopsis. Sci Rep 13(1):652

Lukan T, Veillet F, Křižnik M, Coll A, Mahkovec Povalej T, Pogačar K, Stare K, Chauvin L, Chauvin J-E, Gruden K (2022) CRISPR/Cas9-mediated fine-tuning of miRNA expression in tetraploid potato. Hortic Res 9:uhac147

Marmisollé FE, Garcia ML, Reyes C (2018) RNA-binding protein immunoprecipitation as a tool to investigate plant miRNA processing interference by regulatory proteins of diverse origin. Plant Methods 14:1–9

Memczak S, Jens M, Elefsinioti A, Torti F, Krueger J, Rybak A, Maier L, Mackowiak SD, Gregersen LH, Munschauer M (2013) Circular RNAs are a large class of animal RNAs with regulatory potency. Nature 495(7441):333–338

Miska EA, Alvarez-Saavedra E, Townsend M, Yoshii A, Šestan N, Rakic P, Constantine-Paton M, Horvitz HR (2004) Microarray analysis of microRNA expression in the developing mammalian brain. Genome Biol 5:1–13

Moyle RL, Birch RG (2013) Sugarcane loading stem gene promoters drive transgene expression preferentially in the stem. Plant Mol Biol 82:51–58

Moyle RL, Carvalhais LC, Pretorius L-S, Nowak E, Subramaniam G, Dalton-Morgan J, Schenk PM (2017) An optimized transient dual luciferase assay for quantifying microRNA directed repression of targeted sequences. Front Plant Sci 8:1631

Nagalakshmi U, Wang Z, Waern K, Shou C, Raha D, Gerstein M, Snyder M (2008) The transcriptional landscape of the yeast genome defined by RNA sequencing. Science 320(5881):1344–1349

Nelson PT, Baldwin DA, Scearce LM, Oberholtzer JC, Tobias JW, Mourelatos Z (2004) Microarray-based, high-throughput gene expression profiling of microRNAs. Nat Methods 1(2):155–161

Ni Z, Hu Z, Jiang Q, Zhang H (2012) Overexpression of gma-MIR394a confers tolerance to drought in transgenic Arabidopsis thaliana. BBRC 427(2):330–335

Niranjanakumari S, Lasda E, Brazas R, Garcia-Blanco MA (2002) Reversible cross-linking combined with immunoprecipitation to study RNA–protein interactions in vivo. Methods 26(2):182–190

Ohyanagi H, Yano K, Yamamoto E, Kitazumi A (2022) Plant omics: advances in big data biology. CABI, London

Owusu Adjei M, Zhou X, Mao M, Rafique F, Ma J (2021) MicroRNAs roles in plants secondary metabolism. Plant Signaling Behave 16(7):1915590

Pan J, Huang D, Guo Z, Kuang Z, Zhang H, Xie X, Ma Z, Gao S, Lerdau MT, Chu C (2018) Overexpression of microRNA408 enhances photosynthesis, growth, and seed yield in diverse plants. J Integr Plant Biol 60(4):323–340

Panda AC, Gorospe M (2018) Detection and analysis of circular RNAs by RT-PCR. Bio-Protocol 8(6):e2775

Pandey PR, Munk R, Kundu G, De S, Abdelmohsen K, Gorospe M (2020) Methods for analysis of circular RNAs. Wiley Interdiscip Rev RNA 11(1):e1566
Patel P, Yadav K, Srivastava AK, Suprasanna P, Ganapathi TR (2019) Overexpression of native Musa-miR397 enhances plant biomass without compromising abiotic stress tolerance in banana. Sci Rep 9(1):16434
Pauwels L, De Clercq R, Goossens J, Iñigo S, Williams C, Ron M, Britt A, Goossens A (2018) A dual sgRNA approach for functional genomics in Arabidopsis thaliana. G3 8(8):2603–2615
Peng T, Qiao M, Liu H, Teotia S, Zhang Z, Zhao Y, Wang B, Zhao D, Shi L, Zhang C (2018) A resource for inactivation of microRNAs using short tandem target mimic technology in model and crop plants. Mol Plant 11(11):1400–1417
Ponchel F, Toomes C, Bransfield K, Leong FT, Douglas SH, Field SL, Bell SM, Combaret V, Puisieux A, Mighell AJ (2003) Real-time PCR based on SYBR-green I fluorescence: an alternative to the TaqMan assay for a relative quantification of gene rearrangements, gene amplifications and micro gene deletions. BMC Biotechnol 3:1–13
Puchta H, Jiang J, Wang K, Zhao Y (2022) Updates on gene editing and its applications. Plant Physiol 188(4):1725–1730
Rao X, Huang X, Zhou Z, Lin X (2013) An improvement of the 2^ (−delta delta CT) method for quantitative real-time polymerase chain reaction data analysis. Biostat Bioinform Biomath 3(3):71
Reichel M, Li Y, Li J, Millar AA (2015) Inhibiting plant micro RNA activity: molecular SPONGE s, target MIMIC s and STTM s all display variable efficacies against target micro RNA s. Plant Biotechnol J 13(7):915–926
Renwick N, Cekan P, Bognanni C, Tuschl T (2014) Multiplexed miRNA fluorescence in situ hybridization for formalin-fixed paraffin-embedded tissues. Methods Mol Biol 1211:171–187
Repele A, Manu (2019) Robust normalization of luciferase reporter data. Methods Protocols 2(3):62
Ro S, Yan W (2010) Detection and quantitative analysis of small RNAs by PCR. RNA Ther 629:293–303
Robinson SJ, Cram DJ, Lewis CT, Parkin IA (2004) Maximizing the efficacy of SAGE analysis identifies novel transcripts in Arabidopsis. Plant Physiol 136(2):3223–3233
Schmittgen TD, Jiang J, Liu Q, Yang L (2004) A high-throughput method to monitor the expression of microRNA precursors. Nucleic Acids Res 32(4):e43–e43
Schon MA, Kellner MJ, Plotnikova A, Hofmann F, Nodine MD (2018) NanoPARE: parallel analysis of RNA 5′ ends from low-input RNA. Genome Res 28(12):1931–1942
Schwab R, Ossowski S, Riester M, Warthmann N, Weigel D (2006) Highly specific gene silencing by artificial microRNAs in Arabidopsis. Plant Cell 18(5):1121–1133
Shen W-X, Au PCK, Smith NA, Dennis ES, Wang M-B (2015) Satellite RNAs interfere with the function of viral RNA silencing suppressors. Front Plant Sci 6:134782
Shen W, Cao S, Liu J, Zhang W, Chen J, Li J-F (2021) Overexpression of an Osa-miR162a derivative in rice confers cross-kingdom RNA interference-mediated brown planthopper resistance without perturbing host development. Int J Mol Sci 22(23):12652
Shingara J, Keiger K, Shelton J, Laosinchai-Wolf W, Powers P, Conrad R, Brown D, Labourier E (2005) An optimized isolation and labeling platform for accurate microRNA expression profiling. RNA 11(9):1461–1470
Siddika T, Heinemann IU (2021) Bringing MicroRNAs to light: methods for MicroRNA quantification and visualization in live cells. Front Bioeng Biotechnol 8:619583
Song H, Zhang J, Li D, Cooper AM, Silver K, Li T, Liu X, Ma E, Zhu KY, Zhang J (2017) A double-stranded RNA degrading enzyme reduces the efficiency of oral RNA interference in migratory locust. Insect Biochem Mol Biol 86:68–80
Soto-Suárez M, Baldrich P, Weigel D, Rubio-Somoza I, San Segundo B (2017) The Arabidopsis miR396 mediates pathogen-associated molecular pattern-triggered immune responses against fungal pathogens. Sci Rep 7(1):44898

Suzuki H, Zuo Y, Wang J, Zhang MQ, Malhotra A, Mayeda A (2006) Characterization of RNase R-digested cellular RNA source that consists of lariat and circular RNAs from pre-mRNA splicing. Nucleic Acids Res 34(8):e63–e63
Tang G, Yan J, Gu Y, Qiao M, Fan R, Mao Y, Tang X (2012) Construction of short tandem target mimic (STTM) to block the functions of plant and animal microRNAs. Methods 58(2):118–125
Terzi LC, Simpson GG (2009) Arabidopsis RNA immunoprecipitation. Plant J 59(1):163–168
Thomson JM, Parker J, Perou CM, Hammond SM (2004) A custom microarray platform for analysis of microRNA gene expression. Nat Methods 1(1):47–53
Todesco M, Rubio-Somoza I, Paz-Ares J, Weigel D (2010) A collection of target mimics for comprehensive analysis of microRNA function in Arabidopsis thaliana. PLoS Genet 6(7):e1001031
Ueno D, Mukuta T, Yamasaki S, Mikami M, Demura T, Matsui T, Sawada K, Katsumoto Y, Okitsu N, Kato K (2020) Different plant species have common sequence features related to mRNA degradation intermediates. Plant Cell Physiol 61(1):53–63
Unver T, Namuth-Covert DM, Budak H (2009) Review of current methodological approaches for characterizing microRNAs in plants. Int J Plant Genomics 2009:262463
Válóczi A, Hornyik C, Varga N, Burgyán J, Kauppinen S, Havelda Z (2004) Sensitive and specific detection of microRNAs by northern blot analysis using LNA-modified oligonucleotide probes. Nucleic Acids Res 32(22):e175–e175
Várallyay E, Burgyán J, Havelda Z (2008) MicroRNA detection by northern blotting using locked nucleic acid probes. Nat Protocols 3(2):190–196
Varchenko O, Kuchuk M, Parii M, Symonenko YV (2020) Comparison of gfp gene expression levels after agrobacterium-mediated transient transformation of Nicotiana rustica L. by constructs with different promoter sequences. Cytol Genet 54:531–538
Warthmann N, Chen H, Ossowski S, Weigel D, Hervé P (2008) Highly specific gene silencing by artificial miRNAs in rice. PLoS One 3(3):e1829
Weinberg Z, Lünse CE, Corbino KA, Ames TD, Nelson JW, Roth A, Perkins KR, Sherlock ME, Breaker RR (2017) Detection of 224 candidate structured RNAs by comparative analysis of specific subsets of intergenic regions. Nucleic Acids Res 45(18):10811–10823
Xie K, Wu C, Xiong L (2006) Genomic organization, differential expression, and interaction of SQUAMOSA promoter-binding-like transcription factors and microRNA156 in rice. Plant Physiol 142(1):280–293
Yan J, Gu Y, Jia X, Kang W, Pan S, Tang X, Chen X, Tang G (2012) Effective small RNA destruction by the expression of a short tandem target mimic in Arabidopsis. Plant Cell 24(2):415–427
Yan X, Ma L, Yang M (2020) Identification and characterization of long non-coding RNA (lncRNA) in the developing seeds of Jatropha curcas. Sci Rep 10(1):10395
Yang C, Li D, Mao D, Liu X, Ji C, Li X, Zhao X, Cheng Z, Chen C, Zhu L (2013) Overexpression of micro RNA 319 impacts leaf morphogenesis and leads to enhanced cold tolerance in rice (O ryza sativa L.). Plant Cell Environ 36(12):2207–2218
Yang H, Chen J, Yang S, Zhang T, Xia X, Zhang K, Deng S, He G, Gao H, He Q (2021) CRISPR/Cas14a-based isothermal amplification for profiling plant microRNAs. Anal Chem 93(37):12602–12608
Yazaki J, Gregory BD, Ecker JR (2007) Mapping the genome landscape using tiling array technology. Curr Opin plant biol 10(5):534–542
Yu C, Chen Y, Cao Y, Chen H, Wang J, Bi Y-M, Tian F, Yang F, Rothstein SJ, Zhou X (2018) Overexpression of miR169o, an overlapping microRNA in response to both nitrogen limitation and bacterial infection, promotes nitrogen use efficiency and susceptibility to bacterial blight in rice. Plant Cell Physiol 59(6):1234–1247
Yuan J, Zhang Y, Dong J, Sun Y, Lim BL, Liu D, Lu ZJ (2016) Systematic characterization of novel lncRNAs responding to phosphate starvation in Arabidopsis thaliana. BMC Genomics 17:1–16
Zhang Y-C, Yu Y, Wang C-Y, Li Z-Y, Liu Q, Xu J, Liao J-Y, Wang X-J, Qu L-H, Chen F (2013) Overexpression of microRNA OsmiR397 improves rice yield by increasing grain size and promoting panicle branching. Nat Biotechnol 31(9):848

Zhang C, Ding Z, Wu K, Yang L, Li Y, Yang Z, Shi S, Liu X, Zhao S, Yang Z (2016) Suppression of jasmonic acid-mediated defense by viral-inducible microRNA319 facilitates virus infection in rice. Mol Plant 9(9):1302–1314

Zhang H, Zhang J, Yan J, Gou F, Mao Y, Tang G, Botella JR, Zhu J-K (2017) Short tandem target mimic rice lines uncover functions of miRNAs in regulating important agronomic traits. Proc Nati Acad Sci 114(20):5277–5282

Zhang J, Zhang H, Srivastava AK, Pan Y, Bai J, Fang J, Shi H, Zhu J-K (2018) Knockdown of rice microRNA166 confers drought resistance by causing leaf rolling and altering stem xylem development. Plant Physiol 176(3):2082–2094

Zhang J, Zhou Z, Bai J, Tao X, Wang L, Zhang H, Zhu J-K (2020) Disruption of MIR396e and MIR396f improves rice yield under nitrogen-deficient conditions. Natl Sci Rev 7(1):102–112

Zhou J, Deng K, Cheng Y, Zhong Z, Zhang T, Qi Y, Zhang Y (2017) CRISPR-Cas9 based genome editing reveals new insights into microRNA function and regulation in rice. Front Plant Sci 8:288343

Zhou J, Zhang R, Jia X, Tang X, Guo Y, Yang H, Zheng X, Qian Q, Qi Y, Zhang Y (2022) CRISPR-Cas9 mediated OsMIR168a knockout reveals its pleiotropy in rice. Plant Biotechnol J 20(2):310–322

Zhu QH, Jin S, Yuan Y, Liu Q, Zhang X, Wilson I (2022) CRISPR/Cas9-mediated saturated mutagenesis of the cotton MIR482 family for dissecting the functionality of individual members in disease response. Plant Direct 6(6):e410

Exploring RNAi Mechanisms: Computational Approaches and Meta-Analysis in OMICS Studies

7

Shachi Gahoi and Budhayash Gautam

Abstract

RNA interference (RNAi) is a gene silencing technology that involves sequence-specific silencing of target genes driven by non-coding RNAs (ncRNAs). The roles of RNAi in crop improvement have been illustrated in plant biomass regulation, enhancement of crop yield and productivity, development of seedless fruits, shelf-life enhancement, secondary metabolite regulation, nutritional improvements, enhanced biotic (bacteria, fungi, viruses, nematodes, insects) and abiotic stress tolerance (drought, salinity, cold, etc.). Advances in high-throughput techniques have paved the way for a new generation of various omics approaches such as genomics, transcriptomics, proteomics and metabolomics, which played an important role in crop science. The increasing use of NGS and high-throughput gene expression and quantification technologies over the last two decades has increased the published data stored in public repositories which triggered an explosion of research studies available through public databases. This stored information in public databases offers an invaluable resource for meta-analysis to reuse these data and generate new scientific findings. Consequently, there has been an enormous advancement in the number of systematic reviews and meta-analyses over the past decade. This chapter highlights the importance of next-generation sequencing technology, various bioinformatics methods and meta-analysis in understanding RNAi phenomena and their role in crop improvement. Additionally, this chapter also elaborates on the application of gene expression/meta-analysis in animals and plants and the workflow used in a common meta-analysis experiment, highlighting the key points that the researcher has to consider while conducting the analysis.

S. Gahoi · B. Gautam (✉)
Department of Computational Biology and Bioinformatics, Jacob Institute of Biotechnology and Bioengineering, Sam Higginbottom University of Agriculture, Technology and Sciences, Prayagraj, Uttar Pradesh, India
e-mail: budhayash.gautam@shuats.edu.in

K. Goswami et al. (eds.), *Non-coding RNAs for Crop Improvement*,
https://doi.org/10.1007/978-981-97-7213-1_7

Keywords

Bioinformatics · Crop improvement · Next-generation sequencing · Meta-analysis · RNA interference · Small RNAs

7.1 Introduction

The present world population size is 7.8 billion and it is projected to reach 9.9 billion by 2050, which represents an increase of more than 25% from 2020 (as per 2020 World Population Data Sheet). To feed the world population by 2050, a significant increase of about 70% in food production will be needed (FAO 2009), and it is a big task as there is already a huge gap between population size and food production. Moreover, extreme climatic conditions negatively influence agricultural productivity globally and are the major causes of abiotic and biotic stresses resulting in more than 50% crop yield loss annually worldwide (Singhal et al. 2017). Therefore, we have to acquire numerous technologies, especially plant breeding and biotechnology, including RNA interference (RNAi)-based strategies to meet the estimated demand of food production and for the improvement of crop yield and quality by protecting crop plants against biotic (pathogens and pests) and abiotic (salinity, drought, heat, light, cold etc.) stresses (Rajam 2020).

RNAi is a universal phenomenon of gene silencing in most eukaryotes, including plants and are evolutionary conserved across eukaryotic systems, and are involved in regulation of genes associated with plant developmental processes and stress responses and in defense against viruses, transgenes and transposons (Kamthan et al. 2015; Mamta and Rajam 2018). Moreover, RNA silencing has demonstrated to be novel and a probable reverse genetics tool for functional genomics to unravel the function of genes through genome-wide screening in distinct eukaryotic organisms, including plants, drosophila, and *c. elegans* (Fire 1999; McGinnis 2010). RNAi has various advantages over traditional transgenic methods as genetically modified RNAi plants do not contain transgene protein, though the risk assessment of these plants should be investigated to exclude any off-target effects (Ramon et al. 2014).

RNAi involves the processing of long dsRNAs into small RNA (sRNA) molecules of ~21–24 nucleotide (nt) length. sRNAs have been shown to play important regulatory functions in development, genome stability, response to biotic and abiotic stressors, and transposon control (Bucher et al. 2012). Recent findings suggested that plant sRNAs are involved in plant defense against viruses, nematodes (Niu et al. 2012; Papolu et al. 2013), insect pests (Zhang et al. 2015a) and fungi (Tinoco et al. 2010). The utilization of RNA-mediated gene silencing could provide a promising solution for plant protection against various pests and diseases. Therefore, identification and analysis of sRNAs can be extremely important in discovering the regulatory mechanisms of numerous biological processes. In plants, sRNAs are mainly of two types, viz., small interfering RNAs (siRNAs) and microRNAs (miRNAs) (Borges and Martienssen 2015; Mamta and Rajam 2018). siRNAs and miRNAs are about 20–25 and 20–24 nt, respectively, and they combine with argonaute (AGO) proteins to form RNA-induced silencing complexes (RISC) which can cleave target mRNA or suppress the translation to regulate biological processes (Carthew and Sontheimer 2009; Jagtap et al. 2011).

Based on the high degree of sequencing depth and coverage, NGS technologies have recently appeared as a cost-effective means of generating enormous sequence data in a very short time and revolutionized the study of sRNAs in plants on a genome-wide or transcriptome-wide scale. As NGS generates large amounts of data, several bioinformatics tools have been developed to help biologists mine meaningful information from these huge quantities of data. Rapid development of NGS techniques and bioinformatics tools has led to the discovery of many novel and conserved miRNA families related to grain development in cereal crops such as bread wheat, maize, rice, and barley (Curaba et al. 2012; Lan et al. 2012).

7.2 OMICS Technologies in RNAi Analysis

7.2.1 Transcriptomics

It plays a crucial role in RNAi analysis by providing a dynamic and comprehensive snapshot of gene expression changes induced by the targeted silencing of specific genes (Younis et al. 2014). Techniques such as RNA-seq and microarrays can systematically profile the transcriptome, revealing alternative splicing events, alterations in mRNA abundance and the expression of ncRNAs (Han et al. 2015). This extensive information allows for the identification of genes directly affected by RNAi, unraveling complex regulatory networks and pathways. Transcriptomics not only presents as a potent tool for verifying the efficacy of gene silencing but also facilitates the discovery of off-target effects and provides valuable insights into the comprehensive cellular consequences of RNAi, providing a deeper understanding of gene function and regulatory mechanisms.

7.2.2 Genomics

It plays a pivotal role in RNAi study by facilitating the identification, selection, and optimization of siRNAs for targeted gene silencing. High-throughput genomic technologies, such as NGS and bioinformatics tools investigate entire genomes, identify potential target genes, and predict off-target effects (Huang et al. 2022). Genomic data guide the rational design of highly specific and efficient siRNAs, enhancing the precision of RNAi experiments. Additionally, genomics contributes to the understanding of gene function and regulatory networks, providing a foundation for the strategic application of RNAi in functional genomics studies and therapeutic interventions (Lu et al. 2005). The integration of genomics with other omics technologies further enriches the comprehensive analysis of RNAi effects on cellular processes.

7.2.3 Proteomics

Proteomics plays crucial role in RNAi analysis by elucidating the downstream effects of gene silencing at the protein level (Chen et al. 2009). Through techniques like mass spectrometry and protein microarrays, proteomics enables the identification and quantification of proteins affected by RNAi-mediated knockdown or overexpression. This information is crucial for understanding post-transcriptional regulation, uncovering changes in protein expression, and revealing alterations in signaling pathways (Halbeisen et al. 2008). Proteomics also aids in identifying potential off-target effects and provides a holistic view of the molecular landscape influenced by RNAi. Integrating proteomic analyses into RNAi studies enhances our comprehension of the functional consequences of gene silencing and contributes to the broader understanding of cellular responses to targeted genetic manipulation (Shanker et al. 2014).

7.2.4 Metabolomics

Metabolomics plays a significant role in RNAi study by providing insights into the metabolic consequences of gene silencing (Johnson et al. 2016) By profiling the small molecules involved in cellular processes, metabolomics reveals alterations in metabolic pathways. This information is crucial for understanding the broader cellular consequences of gene silencing, identifying potential biomarkers, and uncovering metabolic signatures associated with specific RNAi interventions. Metabolomic data enrich the comprehensive analysis of RNAi effects, shedding light on the intricate connections between gene expression changes and cellular metabolism (Johnson et al. 2016). Integrating metabolomics into RNAi studies enhances our understanding of the functional implications at the cellular and physiological contexts.

7.3 Role of NGS and Bioinformatics in sRNA Analysis

7.3.1 Importance of NGS Technology in sRNA Analysis

The recent advent of NGS also known as high-throughput sequencing generates enormous sequence data in a very short time and cost-effective manner. NGS has several advantages over Sanger sequencing and microarray techniques. Firstly, it uses massively parallel sequencing in place of Sanger sequencing or microarray-based approaches and offers various potential advantages of scalability, dynamic range and flexibility of input library (Sims et al. 2011). Secondly, novel sRNAs discovered from NGS have a better signal to noise ratio than microarrays and do not experience saturation effects. Additionally, NGS has a higher throughput than qPCR or northern blotting and also allows for the detection of single base differences (Liu et al. 2020). Thus, NGS has become an important technology for sRNA detection and profiling and several research have utilized NGS to reveal the regulatory pathways of plant sRNAs (Liu et al. 2016).

7.3.2 Importance of miRNA in Crop Protection

One of the well-studied sRNA classes is miRNA (Jones-Rhoades et al. 2006). miRNAs are 20-24 nt long single-stranded RNA molecules and are being given ample attention in recent years because of their role in gene expression, growth, and numerous developmental processes including host-pathogen interaction, tumorigenesis and many more (Riahi Rad et al. 2021). Next-generation sequencing of mRNA in specific tissues and at different developmental stages in combination with a better alignment algorithm, have accelerated the process of predicting novel miRNAs in a multiple plant species. With the use of high-throughput sequencing and computational analysis, many miRNAs have been identified in various plant species such as wheat, rice, maize, sorghum, cotton and sunflower under different stress conditions, suggesting the potential role of miRNAs in regulating stress response in plants (Khraiwesh et al. 2012; Kumar et al. 2014). Many miRNAs are also found to be involved in insect endosymbiont cross-talks (Hussain 2012). Moreover, microRNAs emerged as a novel target in the area of genetic engineering and have been used to develop stress-tolerant and high-yield crop varieties (Hussain 2012; Zhang and Wang 2015).

7.3.3 Role of Bioinformatics in sRNA Study

With advances in NGS technology and computational methods, various bioinformatics resources continue to emerge and update rapidly and regularly, including workflow for *in silico* sRNA analysis, databases, and tools dedicated to sRNA identification and functional annotation which assist biologists in pinpointing the precise role of miRNA and siRNA and their targets (Monga and Kumar 2019). Additionally, based on the increasing quantity of miRNA sequence data and with the availability of annotated miRNA sequence information, many ab initio-based algorithms for miRNA prediction have been developed independent of genome sequence knowledge. The developed miRNA prediction tools are mainly based on four approaches: (1) homology-based, (2) machine learning-based, (3) target-based, and (4) scoring-based approaches (Yu et al. 2020).

Currently, numerous miRNAs have been successfully predicted and identified in multiple animals and plants using a variety of bioinformatics software and tools (Yu et al. 2020). Most of the miRNA analysis tools are available online, enabling the researchers to access and use them for further study (Shukla et al. 2017). However, many of these tools are developed for the study of miRNAs in animals rather than plants (Akhtar et al. 2016). The details of bioinformatics tools and databases developed for miRNA and siRNA prediction in plants are given in Table 7.1.

+

Table 7.1 List of tools and databases for sRNA (miRNA, siRNA) prediction and identification in plants

Software package	Description	Link	Reference
miRDeep-P2	Analyzing the microRNA transcriptome in plants	Https://sourceforge.Net/projects/mirdp2/	Kuang et al. (2019)
UEA sRNA workbench	A suite of tools for analysing and visualizing next-generation sequencing microRNA and small RNA datasets	https://github.com/sRNAworkbenchuea/UEA_sRNA_Workbench	Stocks et al. (2018)
PmiRDiscVali	An integrated pipeline for plant microRNA discovery and validation	https://github.com/unincrna/pmirdv	Yu et al. (2019)
miRPlant	An integrated tool for identification of plant MiRNA from RNA sequencing data	https://sourceforge.net/projects/mirplant/	An et al. (2014)
CAP-miRSeq	A comprehensive analysis pipeline for microRNA sequencing data	http://bioinformaticstools.mayo.edu/research/cap-mirseq/	Sun et al. (2014)
microRPM		http://microrpm.itps.ncku.edu.tw/	Tseng et al. (2018)
miRCat2	Accurate prediction of plant and animal microRNAs from next-generation sequencing datasets	http://srna-workbench.cmp.uea.ac.uk/mircat2	Paicu et al. (2017)
sRNAnalyzer	A flexible and customizable small RNA sequencing data analysis pipeline	http://srnanalyzer.systemsbiology.net/	Wu et al. (2017)
miRDis	A web tool for endogenous and exogenous microRNA discovery based on deep-sequencing data analysis	http://sbbi.unl.edu/miRDis/index.php	Zhang et al. (2018)
miRNA digger	A comprehensive pipeline for genome-wide novel miRNA mining	http://bis.zju.edu.cn/miRNA_Digger/	Yu et al. (2016)
Mir-PREFeR v0.24	microRNA PREdiction from small RNA-seq data	https://github.com/hangelwen/miR-PREFeR	Lei and Sun (2014)
Mirinho	An efficient and general plant and animal pre-miRNA predictor for genomic and deep-sequencing data	http://mirinho.gforge.inria.fr/	Higashi et al. (2015)
MIRPIPE v1.2	Open source RNA-seq bioinformatics analysis docker for identification of miRNAs and piRNAs	https://github.com/loosolab/mirpipe	Kuenne et al. (2014)
BioVLAB-MMIA-NGSc	microRNA–mRNA integrated analysis using high-throughput sequencing data	http://epigenomics.snu.ac.kr/biovlab_mmia_ngs/	Chae et al. (2015)
MTide		http://bis.zju.edu.cn/MTide/	Zhang et al. (2015b)

7.4 Meta-Analysis in RNAi Studies

7.4.1 Importance of Meta-Analysis in sRNA Study

The increasing use of NGS and high-throughput gene expression quantification technologies over the past two decades has significantly increased the research studies and therefore increased the published data in public repositories which triggered an explosion of research work and analysis available through public databases. This stored information in public databases offers an invaluable resource for reuse to generate new scientific findings. In this context, great interest has been focused on meta-analysis to integrate and jointly analyze different scientific studies. Additionally, computational resources such as web servers, software and tools facilitating large-scale RNA-seq and sRNA sequence expression meta-analysis along with integrative and comparative interpretation have started to accumulate. Thus the NGS technology, computational systems biology and advanced statistical analysis such as meta-analysis have provided novel opportunities to overcome biological complexity.

Meta-analysis approaches are a set of statistical methods that integrate multiple independent studies to attain a single common and significant result. In the context of gene expression, datasets from various studies can be analyzed in a meta-analysis to identify common and reliable biomarkers. Meta-analysis incorporates samples from different studies which increases sample size, statistical power and robustness of the results. Recently, the research work involving miRNA differential gene expression meta-analysis has increased drastically, with most studies conducted on human samples (Toro-Domínguez et al. 2019; Toro-Domínguez et al. 2021). In humans, meta-analysis of miRNA expression has been widely used in identification of biomarkers related to different types of cancers, diabetes, heart diseases, autoimmune diseases, malignancies, myopia, alzheimer's disease, parkinson's disease, rheumatoid arthritis, sclerosis, steatohepatitis, and other diseases (Toro-Domínguez et al. 2019). The identified markers can be used to diagnose, prognosis and for better treatment of that particular disease (Condrat et al. 2020).

Meta-analysis of miRNA expression has been used extensively in animal systems and many potential biomarkers have been developed dissecting complex human diseases. However, miRNA/siRNA expression meta-analysis studies in major crops are limited, despite the increasing number of gene expression studies. This is possible due to many issues, including, but not restricted to, high levels of inter-study differences, unavailability of metadata, and limited technical information for conducting a meta-analysis. In the recent past, many transcriptome meta-analysis studies have been performed to identify important gene families in multiple plant species viz. salt tolerant genes in rice (Amirbakhtiar et al. 2021), drought-responsive genes in wheat (Shojaee et al. 2022), pathogen stress-responsive genes in Arabidopsis (Biniaz et al. 2022), biotic and abiotic stress-responsive genes in tomato (Ashrafi-Dehkordi et al. 2018), pathogen responsive genes in maize (Wang et al. 2022), and many more. The identified common genes and pathways from meta-analysis during biotic and abiotic stresses may provide potential candidate targets for the genetic manipulation of crop plants and for the development of stress-resistance breeding programs.

Also, bioinformatics resources enabling meta-analysis of miRNA expression across various plant species are still evolving. In order to gain new insights into the role of sRNA in the improvement of crop yield and protection, a comprehensive meta-analysis of sRNA sequence data should be performed. Though, data-driven meta-analysis of raw, original omics data emerging from individual studies is computationally intensive, could produce more robust and reliable results. Moreover, with the growing accessibility of enormous raw data repositories such as the National Center for Biotechnology Information, Sequence Read Archive (Leinonen et al. 2010) and Gene Expression Omnibus (Barrett 2013), such novel computational meta-analyses are now possible.

7.4.2 Meta-Analysis Workflow in miRNA Expression Data

In this segment, we described crucial considerations and key steps in miRNA expression meta-analysis Fig. 7.1.

7.4.2.1 Hypothesis, Systematic Review, and Data Retrieval

Based on the aim of the study, eligibility criteria must be defined to search datasets in public repositories using keywords and further comprehensive review must be performed to find out relevant studies and datasets to be included in meta-analysis (Begum et al. 2012). Also, various parameters must be considered for the selection of datasets since these might influence the outcomes. Additionally, it must consider the required level of heterogeneity for the inclusion of datasets in our meta-study as selecting low or high inter-study heterogeneity can provide distinct purposes. Higher level of inter-study heterogeneity decreases statistical power, but increases the generalized ability of the results (Bailey 1987). Moreover, the isolated effect of a meta-study could be minimized with the inclusion of a larger number of studies, prioritizing the general purpose of the results.

7.4.2.2 Data Pre-Processing

To minimize technical heterogeneity, data pre-processing steps should be standardized. Normalization is an important pre-processing step to minimize non-biological variations (Zyprych-Walczak et al. 2015) and the normalization techniques; Trimmed mean of M-values (TMM), Upper quartile (UQ) and Relative log expression (RLE) are standard normalization algorithms and can be applied on raw miRNA sequence sets (Tam et al. 2015). Also, quality control analysis must be performed to detect outliers, technical problems and inconsistent measurements in the samples. Detection and removal of outliers is a very important step as outliers introduce bias into following analysis leading to under- or over-estimated resulting values.

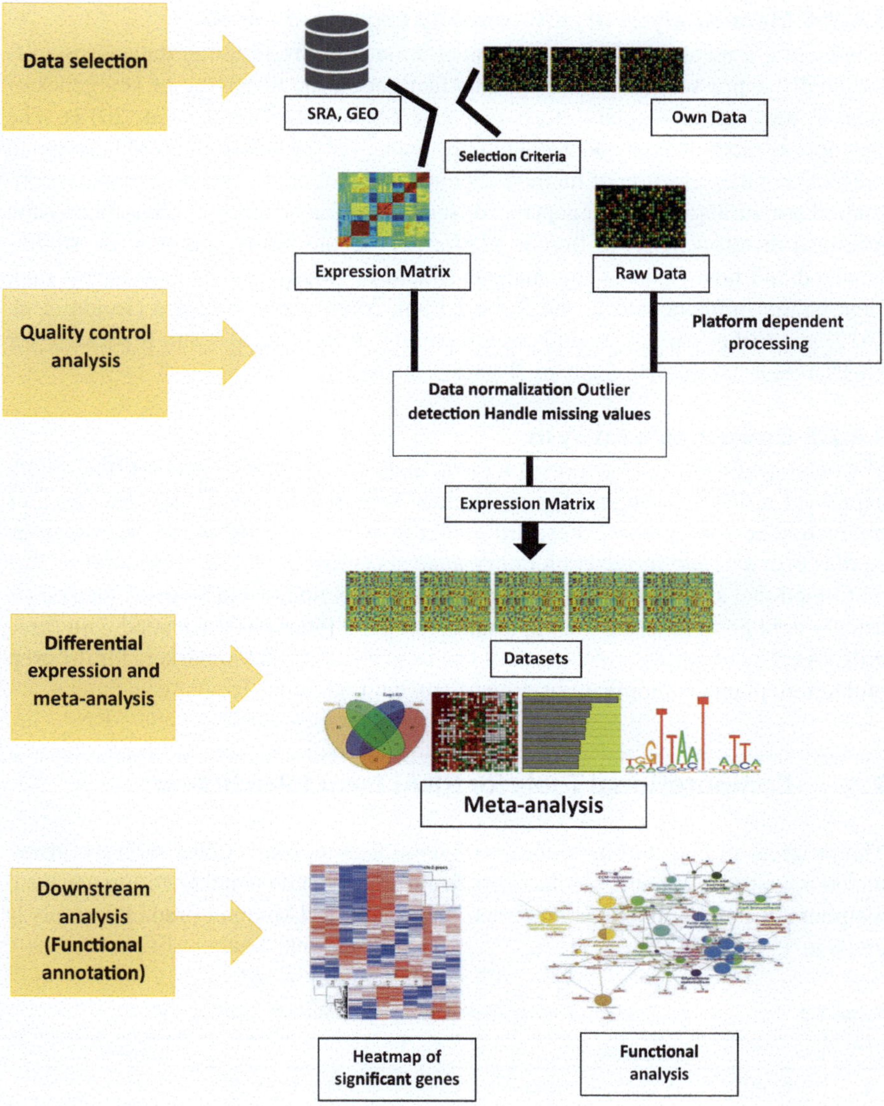

Fig. 7.1 Meta-analysis workflow

7.4.2.3 Differential Expression Analysis

After the data cleaning steps, differential expression analysis must be conducted for each miRNA sequence set and the miRNAs with Benjamini-Hochberg adjusted *p*-value <0.05 and absolute log2Fold Change (FC) > 1 and could be considered as differentially expressed (Chen et al. 2014).

7.4.2.4 Meta-Analysis of Differentially Expressed miRNA

Conducting a meta-analysis permits integration of transcriptomic data across differentially expressed genes (DEG), thus facilitating the discovery of key genes of general importance to plant stress-response (Ashrafi-Dehkordi et al. 2018). The appropriate method must be selected for the meta-analysis of differentially expressed miRNA and the selection of method depends on many features which must be considered are: biological heterogeneity (different study conditions or phenotypes) and technical heterogeneity (different platforms) of the study, number of studies involved and how rigorous the analysis should be. Many studies used robust rank aggregation implemented in the Robust Rank Aggregation package (Kolde et al. 2012) and Fisher's p-value combining method via the Meta Volcano R package for miRNA expression meta-analysis (Prada et al. 2020).

7.4.2.5 Downstream Analysis

The outcome of a miRNA expression meta-analysis comprises a list of differentially expressed miRNA through different studies. Visualization tools like heat maps or interaction network plots of these differentially expressed miRNAs in the numerous studies provide help in the exploratory analysis (Pirrò et al. 2021). Moreover, further functional annotation using Gene Ontology enrichment and pathway analysis is usually applied to describe the important biological functions associated with these miRNAs. The identified miRNAs could be used as potential biomarkers for the crop protection, plant development or disease prevention.

7.5 Computational Tools for RNAi Target Prediction

The progress in high-throughput experimental techniques, coupled with the growing computational capabilities, has significantly advanced the field of computational methods for predicting RNAi targets. List of a few RNAi target prediction tools is given in Table 7.2.

Table 7.2 List of tools/databases for the prediction of plant miRNA/siRNAs

S.N.	Tool	Description	Reference
1.	P-TarPmiR	P-TarPmiR method is used to predict targets for all miRNA within the soybean genome	Ajila et al. (2023)
2.	psRNATarget	A plant small RNA target analysis server	Dai et al. (2018)
4.	PlantPAN 3.0	Integrates information on miRNA-target interactions in plants. Provides functional annotations and pathway analysis. Allows users to search for miRNA targets in multiple plant species	Chow et al. (2016)
5.	Targetfinder	Suitable for high-throughput analyses	Bo and Wang (2005)
8.	siRNA-finder	Software for RNAi-target design and off-target prediction	Lück et al. (2019)
9.	sPARTA-plant	Predicts miRNA-target sites and evaluates siRNA specificity	Wang et al. (2017)

7.6 Network Analysis of RNAi Pathways

7.6.1 Construction of Gene Regulatory Networks (GRNs) Affected by RNAi

Following RNAi, GRNs can be constructed by integrating data from high-throughput technologies, such as transcriptomics and proteomics, to capture changes in gene expression and protein abundance (Qian and Huang 2020). Bioinformatics tools are then employed to analyze these datasets, identifying direct and indirect regulatory relationships between genes. Network inference algorithms, including correlation-based methods, mutual information approaches, and Bayesian frameworks, play a key role in predicting regulatory connections based on the observed changes in molecular profiles (Banf and Rhee 2017).

Validation of predicted regulatory interactions often involves experimental techniques such as chromatin immunoprecipitation (ChIP) (Das et al. 2004), quantitative PCR (qPCR), and functional assays (Smith and Osborn 2009). Integration with existing biological knowledge databases and literature mining further refines the accuracy of the constructed GRNs. The resulting gene regulatory networks provide insights into the hierarchical organization of regulatory elements, key molecular players, and feedback loops affected by RNAi. These networks contribute to a deeper understanding of the functional consequences of gene silencing, helping to unravel intricate cellular processes and signaling cascades. Additionally, GRNs offer a valuable resource for identifying potential therapeutic targets, biomarkers, and regulatory hubs within biological systems impacted by RNAi interventions.

7.6.2 Identification of Key Nodes and Pathways in RNAi-Mediated Gene Regulation

The identification of key nodes and pathways in RNAi-mediated gene regulation involves pinpointing critical genes and signaling pathways that play crucial roles in the cellular response to gene silencing. This process is essential for unraveling the functional consequences of RNAi and understanding the broader impact on biological systems (Agrawal et al. 2003). Analysing high-throughput data from transcriptomic, proteomics, and other omics technologies is a fundamental step in identifying key nodes. Computational methods, such as network analysis and pathway enrichment analysis, help to pinpoint genes that serve as central regulators or hubs within gene regulatory networks affected by RNAi. These central nodes often have a substantial influence on the overall network dynamics. Pathway analysis tools assess the enrichment of differentially expressed genes within specific biological pathways. The identification of key nodes and pathways is integral to discerning the underlying mechanisms and orchestrating principles of RNAi-mediated gene regulation (Chang et al. 2012). It not only advances our understanding of gene function and cellular processes but also provides potential targets for therapeutic interventions and offers insights into the development of RNAi-based treatments for various diseases.

7.6.3 Visualization Techniques for Representing RNAi Networks

The advent of high-throughput RNAi screening has transformed gene discovery and therapeutic development (Chen et al. 2007). Nevertheless, the visualization of intricate RNAi interaction networks poses a challenge owing to the immense volume of generated data. Various visualization techniques, ranging from force-directed layouts and graph classification algorithms to newer methods like heatmaps and adjacency matrices, have emerged to depict RNAi networks effectively (Koutrouli et al. 2020). These strategies play a crucial role in pinpointing essential elements within RNAi networks and streamlining the analysis of gene regulatory pathways.

7.7 Functional Annotation and Enrichment Analysis

Functional annotation and enrichment analysis such as Gene Ontology enrichment and metabolic pathway analysis play a pivotal role in translating large-scale RNAi data into meaningful biological insights, in understanding the impact of gene silencing on cellular processes and pathways (Zhao et al. 2016). These analyses are crucial for uncovering the functional consequences of RNAi experiments and identifying potential therapeutic targets.

7.8 Integration of Multi-OMICS Data

The integration of multi-omics data in RNAi analysis is a powerful strategy that involves combining information from various high-throughput technologies, such as genomics, transcriptomics, proteomics, and metabolomics, to gain a comprehensive understanding of the molecular landscape influenced by RNAi (Mahmood et al. 2022). This integrative approach unravels complex regulatory networks, identify key players, and elucidate the functional consequences of gene silencing. Several methods available for integrating multi-omics data, such as;

7.8.1 Data Fusion

Using this method, datasets from several omics technologies are combined into an integrated dataset (Baldwin et al. 2020). Multi-omics data can be combined and analyzed using a variety of statistical techniques, including canonical correlation analysis (CCA), independent component analysis (ICA), and principal component analysis (PCA) (Karhunen et al. 2013).

7.8.2 Network-Based Integration

Network-based techniques focus on integrating multi-omics data to create biological networks, such as gene regulatory networks or protein-protein interaction networks (Zhou et al. 2020).

7.8.3 Machine Learning Approaches

Multi-omics data can be used to train machine learning algorithms, such as support vector machines (SVM), random forests, neural networks, and deep learning models, to predict phenotypic outcomes, classify samples, and find patterns or correlations between various omics layers (Kang et al. 2022; Li et al. 2022).

7.8.4 Pathway Analysis

In order to discover dysregulated pathways linked to disease or biological processes, pathway analysis entails mapping omics data onto biological pathways or functional modules (Dubovenko et al. 2017). Finding new routes and regulatory mechanisms is made possible by the robustness and accuracy of pathway analysis enhanced by the integration of multi-omics data (Maghsoudi et al. 2022).

7.9 Conclusion and Future Prospects

The phenomenon of RNAi widely found in eukaryotes has been developed as a promising technology for crop health protection. The biology of sRNA is complex and produces various computational challenges. The prediction of sRNA based on sequencing data is either inaccurate because of false positives or requires dedicated tools altogether. Also, less attention has been paid to plants than to animals. Currently, most of the development of tools are focused on miRNA and this bias is most likely because, in comparison with other sRNA categories, miRNA have well-defined structural features and miRNAs can be easily validated experimentally, which helps in assessing computational algorithms for miRNA detection and prediction. Proper software or tools should coevolve with experimental strategies for acquiring sRNA data. Novel insights gained through data mining in publically available rich resources of sRNA NGS data and meta-analysis of multiple studies will help in designing biotechnology-based approaches for crop improvement to enhance plant yield and nutritional value. Meta-analysis of miRNA expression has been used extensively in animal systems and many potential biomarkers have been developed for diagnosis, prognosis and treatment of multiple human diseases. However, to the best of our knowledge, there are no literature reviews so far on miRNA/siRNA expression meta-analysis in plants and in order to gain new insights into the role of sRNA in the improvement of crop yield and protection, a comprehensive meta-analysis of sRNA sequence data should be performed to identify common and reliable biomarkers which could be used for further validation for crop improvement programs.

References

Agrawal N, Dasaradhi P, Mohmmed A, Malhotra P, Bhatnagar RK, Mukherjee SK (2003) RNA interference: biology, mechanism, and applications. Microbiol Mol Biol Rev 67(4):657–685

Ajila V, Colley L, Ste-Croix DT, Nissan N, Golshani A, Cober ER, Mimee B, Samanfar B, Green JR (2023) P-TarPmiR accurately predicts plant-specific miRNA targets. Sci Rep 13(1):332

Akhtar MM, Micolucci L, Islam MS, Olivieri F, Procopio AD (2016) Bioinformatic tools for microRNA dissection. Nucleic Acids Res 44(1):24–44

Amirbakhtiar N, Ismaili A, Ghaffari M-R, Mirdar Mansuri R, Sanjari S, Shobbar Z-S (2021) Transcriptome analysis of bread wheat leaves in response to salt stress. PLoS One 16(7):e0254189

An J, Lai J, Sajjanhar A, Lehman ML, Nelson CC (2014) miRPlant: an integrated tool for identification of plant miRNA from RNA sequencing data. BMC Bioinformatics 15:1–4

Ashrafi-Dehkordi E, Alemzadeh A, Tanaka N, Razi H (2018) Meta-analysis of transcriptomic responses to biotic and abiotic stress in tomato. Peer J 6:e4631

Bailey KR (1987) Inter-study differences: how should they influence the interpretation and analysis of results? Stat Med 6(3):351–358

Baldwin E, Han J, Luo W, Zhou J, An L, Liu J, Zhang HH, Li H (2020) On fusion methods for knowledge discovery from multi-omics datasets. Comput Struct Biotechnol J 18:509–517

Banf M, Rhee SY (2017) Computational inference of gene regulatory networks: approaches, limitations and opportunities. Biochim Biophys Acta (BBA) Gene Regul Mech 1860(1):41–52

Barrett T (2013) Gene expression omnibus (GEO). In: The NCBI handbook [internet], 2nd edn. National Center for Biotechnology Information (US), Bethesda

Begum F, Ghosh D, Tseng GC, Feingold E (2012) Comprehensive literature review and statistical considerations for GWAS meta-analysis. Nucleic Acids Res 40(9):3777–3784

Biniaz Y, Tahmasebi A, Afsharifar A, Tahmasebi A, Poczai P (2022) Meta-analysis of common and differential transcriptomic responses to biotic and abiotic stresses in Arabidopsis thaliana. Plan Theory 11(4):502

Bo X, Wang S (2005) TargetFinder: a software for antisense oligonucleotide target site selection based on MAST and secondary structures of target mRNA. Bioinformatics 21(8):1401–1402

Borges F, Martienssen RA (2015) The expanding world of small RNAs in plants. Nat Rev Mol Cell Biol 16(12):727–741

Bucher E, Reinders J, Mirouze M (2012) Epigenetic control of transposon transcription and mobility in Arabidopsis. Curr Opin Plant Biol 15(5):503–510

Carthew RW, Sontheimer EJ (2009) Origins and mechanisms of miRNAs and siRNAs. Cell 136(4):642–655

Chae H, Rhee S, Nephew KP, Kim S (2015) BioVLAB-MMIA-NGS: microRNA–mRNA integrated analysis using high-throughput sequencing data. Bioinformatics 31(2):265–267

Chang S-S, Zhang Z, Liu Y (2012) RNA interference pathways in fungi: mechanisms and functions. Ann Rev Microbiol 66:305–323

Chen M, Du Q, Zhang H-Y, Wang X, Liang Z (2007) High-throughput screening using siRNA (RNAi) libraries. Expert Rev Mol Diagn 7(3):281–291

Chen Z, Brown R, Guo B, Menkir A, Cleveland T (2009) Identifying aflatoxin resistance-related proteins/genes through proteomics and RNAi gene silencing. Peanut Sci 36(1):35–41

Chen Y, McCarthy D, Robinson M, and Smyth GK (2014) edgeR: differential expression analysis of digital gene expression data User's guide. Bioconductor User's Guide

Chow C-N, Zheng H-Q, Wu N-Y, Chien C-H, Huang H-D, Lee T-Y, Chiang-Hsieh Y-F, Hou P-F, Yang T-Y, Chang W-C (2016) PlantPAN 2.0: an update of plant promoter analysis navigator for reconstructing transcriptional regulatory networks in plants. Nucleic Acids Res 44(D1):D1154–D1160

Condrat CE, Thompson DC, Barbu MG, Bugnar OL, Boboc A, Cretoiu D, Suciu N, Cretoiu SM, Voinea SC (2020) miRNAs as biomarkers in disease: latest findings regarding their role in diagnosis and prognosis. Cells 9(2):276

Curaba J, Spriggs A, Taylor J, Li Z, Helliwell C (2012) miRNA regulation in the early development of barley seed. BMC Plant Biol 12:1–16

Dai X, Zhuang Z, Zhao PX (2018) psRNATarget: a plant small RNA target analysis server (2017 release). Nucleic Acids Res 46(W1):W49–W54

Das PM, Ramachandran K, vanWert J, Singal R (2004) Chromatin immunoprecipitation assay. BioTechniques 37(6):961–969

Dubovenko A, Nikolsky Y, Rakhmatulin E, Nikolskaya T (2017) Functional analysis of OMICs data and small molecule compounds in an integrated "knowledge-based" platform. Biol Net Path Anal 1613:101–124

Fire A (1999) RNA-triggered gene silencing. Trends Genet 15(9):358–363

Halbeisen RE, Galgano A, Scherrer T, Gerber A (2008) Post-transcriptional gene regulation: from genome-wide studies to principles. Cell Mol Life Sci 65:798–813

Han Y, Gao S, Muegge K, Zhang W, Zhou B (2015) Advanced applications of RNA sequencing and challenges. Bioinformatics Biol Insights 9:BBI. S28991

Higashi S, Fournier C, Gautier C, Gaspin C, Sagot M-F (2015) M irinho: An efficient and general plant and animal pre-miRNA predictor for genomic and deep sequencing data. BMC Bioinform 16:1–14

Huang Y, Shang M, Liu T, Wang K (2022) High-throughput methods for genome editing: the more the better. Plant Physiol 188(4):1731–1745

Hussain MU (2012) Micro-RNAs (miRNAs): genomic organisation, biogenesis and mode of action. Cell Tissue Res 349(2):405–413

Jagtap UB, Gurav RG, Bapat VA (2011) Role of RNA interference in plant improvement. Naturwissenschaften 98:473–492

Johnson CH, Ivanisevic J, Siuzdak G (2016) Metabolomics: beyond biomarkers and towards mechanisms. Nat Rev Mol Cell Biol 17(7):451–459

Jones-Rhoades MW, Bartel DP, Bartel B (2006) MicroRNAs and their regulatory roles in plants. Annu Rev Plant Biol 57:19–53

Kamthan A, Chaudhuri A, Kamthan M, Datta A (2015) Small RNAs in plants: recent development and application for crop improvement. Front Plant Sci 6:124569

Kang M, Ko E, Mersha TB (2022) A roadmap for multi-omics data integration using deep learning. Brief Bioinform 23(1):bbab454

Karhunen J, Hao T, Ylipaavalniemi J (2013) Finding dependent and independent components from related data sets: a generalized canonical correlation analysis based method. Neurocomputing 113:153–167

Khraiwesh B, Zhu J-K, Zhu J (2012) Role of miRNAs and siRNAs in biotic and abiotic stress responses of plants. Biochim Biophys Acta (BBA) Gene Regul Mech 1819(2):137–148

Kolde R, Laur S, Adler P, Vilo J (2012) Robust rank aggregation for gene list integration and meta-analysis. Bioinformatics 28(4):573–580

Koutrouli M, Karatzas E, Paez-Espino D, Pavlopoulos GA (2020) A guide to conquer the biological network era using graph theory. Front Bioeng Biotechnol 8:34

Kuang Z, Wang Y, Li L, Yang X (2019) miRDeep-P2: accurate and fast analysis of the microRNA transcriptome in plants. Bioinformatics 35(14):2521–2522

Kuenne C, Preussner J, Herzog M, Braun T, Looso M (2014) MIRPIPE: quantification of microRNAs in niche model organisms. Bioinformatics 30(23):3412–3413

Kumar D, Singh D, Kanodia P, Prabhu KV, Kumar M, Mukhopadhyay K (2014) Discovery of novel leaf rust responsive microRNAs in wheat and prediction of their target genes. J Nucleic Acids 2014:1

Lan Y, Su N, Shen Y, Zhang R, Wu F, Cheng Z, Wang J, Zhang X, Guo X, Lei C (2012) Identification of novel MiRNAs and MiRNA expression profiling during grain development in indica rice. BMC Genomics 13:1–10
Lei J, Sun Y (2014) miR-PREFeR: an accurate, fast and easy-to-use plant miRNA prediction tool using small RNA-Seq data. Bioinformatics 30(19):2837–2839
Leinonen R, Sugawara H, Shumway M, Collaboration, I. N. S. D (2010) The sequence read archive. Nucleic Acids Res 39(suppl_1):D19–D21
Li R, Li L, Xu Y, Yang J (2022) Machine learning meets omics: applications and perspectives. Brief Bioinform 23(1):bbab460
Liu B, Liu C-M, Li D, Li Y, Ting H-F, Yiu S-M, Luo R, Lam T-W (2016) BASE: a practical de novo assembler for large genomes using long NGS reads. BMC Genomics 17:561–569
Liu K, Tong H, Li T, Wang X, Chen Y (2020) Research progress in molecular biology related quantitative methods of MicroRNA. Am J Transl Res 12(7):3198
Lu PY, Xie F, Woodle MC (2005) In vivo application of RNA interference: from functional genomics to therapeutics. Adv Genet 54:115–142
Lück S, Kreszies T, Kuhlmann M, Douchkov D (2019) siRNA-finder (si-fi) software for RNAi-target design and off-target prediction. Front Plant Sci 10:450758
Maghsoudi Z, Nguyen H, Tavakkoli A, Nguyen T (2022) A comprehensive survey of the approaches for pathway analysis using multi-omics data integration. Brief Bioinform 23(6):bbac435
Mahmood U, Li X, Fan Y, Chang W, Niu Y, Li J, Qu C, Lu K (2022) Multi-omics revolution to promote plant breeding efficiency. Front Plant Sci 13:1062952
Mamta B, Rajam MV (2018) RNA interference: a promising approach for crop improvement. In: Biotechnologies of crop improvement, transgenic approaches, vol 2. Springer, Cham, pp 41–65
McGinnis KM (2010) RNAi for functional genomics in plants. Brief Funct Genom 9(2):111–117
Monga I, Kumar M (2019) Computational resources for prediction and analysis of functional miRNA and their targetome. Methods Mol Biol 1912:215–250
Niu J, Jian H, Xu J, Chen C, Guo Q, Liu Q, Guo Y (2012) RNAi silencing of the Meloidogyne incognita Rpn7 gene reduces nematode parasitic success. Eur J Plant Pathol 134:131–144
Paicu C, Mohorianu I, Stocks M, Xu P, Coince A, Billmeier M, Dalmay T, Moulton V, Moxon S (2017) miRCat2: accurate prediction of plant and animal microRNAs from next-generation sequencing datasets. Bioinformatics 33(16):2446–2454
Papolu PK, Gantasala NP, Kamaraju D, Banakar P, Sreevathsa R, Rao U (2013) Utility of host delivered RNAi of two FMRF amide like peptides, flp-14 and flp-18, for the management of root knot nematode, Meloidogyne incognita. PLoS One 8(11):e80603
Pirrò S, Matic I, Colizzi V, Galgani A (2021) The microRNA analysis portal is a next-generation tool for exploring and analyzing miRNA-focused data in the literature. Sci Rep 11(1):9007
Prada C, Lima D, Nakaya H (2020) MetaVolcanoR: gene expression meta-analysis visualization tool. R Package Version 1:2020
Qian Y, Huang SSC (2020) Improving plant gene regulatory network inference by integrative analysis of multi-omics and high resolution data sets. Curr Opin Syst Biol 22:8–15
Rajam MV (2020) RNA silencing technology: a boon for crop improvement. J Biosci 45(1):118
Ramon M, Devos Y, Lanzoni A, Liu Y, Gomes A, Gennaro A, Waigmann E (2014) RNAi-based GM plants: food for thought for risk assessors. Plant Biotechnol J 12(9):1271–1273
Riahi Rad Z, Riahi Rad Z, Goudarzi H, Goudarzi M, Mahmoudi M, Yasbolaghi Sharahi J, Hashemi A (2021) MicroRNAs in the interaction between host–bacterial pathogens: a new perspective. J Cell Physiol 236(9):6249–6270
Shanker AK, Maheswari M, Yadav S, Desai S, Bhanu D, Attal NB, Venkateswarlu B (2014) Drought stress responses in crops. Funct Integr Genomics 14:11–22
Shojaee S, Ravash R, Shiran B, Ebrahimie E (2022) Meta-analysis highlights the key drought responsive genes in genes: PEPC and TaSAG7 are hubs response networks. J Genet Eng Biotechnol 20(1):127
Shukla V, Varghese VK, Kabekkodu SP, Mallya S, Satyamoorthy K (2017) A compilation of web-based research tools for miRNA analysis. Brief Funct Genom 16(5):249–273

Sims D, Mendes-Pereira AM, Frankum J, Burgess D, Cerone M-A, Lombardelli C, Mitsopoulos C, Hakas J, Murugaesu N, Isacke CM (2011) High-throughput RNA interference screening using pooled shRNA libraries and next generation sequencing. Genome Biol 12:1–13
Singhal RK, Sodani R, Chauhan J, Sharma MK, Yashu BR (2017) Physiological adaptation and tolerance mechanism of rice (Oryza sativa L.) in multiple abiotic stresses. Int J Pure App Biosci 5(3):459–466
Smith CJ, Osborn AM (2009) Advantages and limitations of quantitative PCR (Q-PCR)-based approaches in microbial ecology. FEMS Microbiol Ecol 67(1):6–20
Stocks MB, Mohorianu I, Beckers M, Paicu C, Moxon S, Thody J, Dalmay T, Moulton V (2018) The UEA sRNA workbench (version 4.4): a comprehensive suite of tools for analyzing miRNAs and sRNAs. Bioinformatics 34(19):3382–3384
Sun Z, Evans J, Bhagwate A, Middha S, Bockol M, Yan H, Kocher J-P (2014) CAP-miRSeq: a comprehensive analysis pipeline for microRNA sequencing data. BMC Genomics 15:1–10
Tam S, Tsao M-S, McPherson JD (2015) Optimization of miRNA-seq data preprocessing. Brief Bioinform 16(6):950–963
Tinoco MLP, Dias BB, Dall'Astta RC, Pamphile JA, Aragão FJ (2010) In vivo trans-specific gene silencing in fungal cells by in planta expression of a double-stranded RNA. BMC Biol 8:1–11
Toro-Domínguez D, Martorell-Marugán J, López-Domínguez R, García-Moreno A, González-Rumayor V, Alarcón-Riquelme ME, Carmona-Sáez P (2019) ImaGEO: integrative gene expression meta-analysis from GEO database. Bioinformatics 35(5):880–882
Toro-Domínguez D, Villatoro-Garciá JA, Martorell-Marugán J, Román-Montoya Y, Alarcón-Riquelme ME, Carmona-Saéz P (2021) A survey of gene expression meta-analysis: methods and applications. Brief Bioinform 22(2):1694–1705
Tseng K-C, Chiang-Hsieh Y-F, Pai H, Chow C-N, Lee S-C, Zheng H-Q, Kuo P-L, Li G-Z, Hung Y-C, Lin N-S (2018) microRPM: a microRNA prediction model based only on plant small RNA sequencing data. Bioinformatics 34(7):1108–1115
Wang G, Giannakis GB, Chen J, Akçakaya M (2017) SPARTA: sparse phase retrieval via truncated amplitude flow. In: 2017 IEEE international conference on acoustics, speech and signal processing (ICASSP). IEEE, New York
Wang Y, Li T, Yang Q (2022) Comparative transcriptome meta-analysis reveals a set of genes involved in the responses to multiple pathogens in maize. Front Plant Sci 13:971371
Wu X, Kim T-K, Baxter D, Scherler K, Gordon A, Fong O, Etheridge A, Galas DJ, Wang K (2017) sRNAnalyzer—a flexible and customizable small RNA sequencing data analysis pipeline. Nucleic Acids Res 45(21):12140–12151
Younis A, Siddique MI, Kim C-K, Lim K-B (2014) RNA interference (RNAi) induced gene silencing: a promising approach of hi-tech plant breeding. Int J Biol Sci 10(10):1150
Yu L, Shao C, Ye X, Meng Y, Zhou Y, Chen M (2016) miRNA digger: a comprehensive pipeline for genome-wide novel miRNA mining. Sci Rep 6(1):18901
Yu D, Wan Y, Ito H, Ma X, Xie T, Wang T, Shao C, Meng Y (2019) PmiRDiscVali: an integrated pipeline for plant microRNA discovery and validation. BMC Genomics 20:1–7
Yu T, Xu N, Haque N, Gao C, Huang W, Huang Z (2020) Popular computational tools used for miRNA prediction and their future development prospects. Interdiscip Sci 12:395–413
Zhang B, Wang Q (2015) MicroRNA-based biotechnology for plant improvement. J Cell Physiol 230(1):1–15
Zhang J, Khan SA, Hasse C, Ruf S, Heckel DG, Bock R (2015a) Full crop protection from an insect pest by expression of long double-stranded RNAs in plastids. Science 347(6225):991–994
Zhang Z, Jiang L, Wang J, Gu P, Chen M (2015b) MTide: an integrated tool for the identification of miRNA–target interaction in plants. Bioinformatics 31(2):290–291
Zhang H, Vieira Resende e Silva B, Cui J (2018) miRDis: a web tool for endogenous and exogenous microRNA discovery based on deep-sequencing data analysis. Brief Bioinform 19(3):415–424

Zhao L, Yang M, Shen Q, Liu X, Shi Z, Wang S, Tang B (2016) Functional characterization of three trehalase genes regulating the chitin metabolism pathway in rice brown planthopper using RNA interference. Sci Rep 6(1):27841

Zhou G, Li S, Xia J (2020) Network-based approaches for multi-omics integration. Methods Mol Biol 2104:469–487

Zyprych-Walczak J, Szabelska A, Handschuh L, Górczak K, Klamecka K, Figlerowicz M, Siatkowski I (2015) The impact of normalization methods on RNA-Seq data analysis. Biomed Res Int 2015:2015

RNA interference (RNAi)-Based Techniques for Management of Plant Viruses

8

Abisola Banji, Titilayo Ajose, Joshua O. Matthew, Sekinat O. Azeez, and Olawale Arogundade

Abstract

RNA interference (RNAi) is a highly conserved intracellular regulatory phenomenon in plants to regulate gene expression by either suppressing transcription or degrading the mRNA. The well-documented harmful effects of chemicals like pesticides on human and animal health have driven researchers to investigate RNAi as a promising biotechnological approach. This strategy aims to tackle the significant problem of crop damage caused by biotic stressors, which are major obstacles to achieving global food security. The mechanism of RNAi consists of three major components: the inducer (dsRNA), the messenger RNA, and the set of proteins that initiate the degradative process, which are comparable in ancestry across most organisms. Transgene-induced RNAi has demonstrated efficacy in gene suppression across numerous plant species, presenting significant potential as a tool for functional genomics in various taxa. The capacity to modify the

A. Banji
Department of Crop Protection and Environmental Biology, University of Ibadan, Ibadan, Oyo State, Nigeria

T. Ajose · J. O. Matthew
Fruits Research Programme, National Horticultural Research Institute, Jericho Reservation Area, Idi-Ishin, Ibadan, Oyo State, Nigeria

S. O. Azeez
Department of Botany, Obafemi Awolowo University, Ile-Ife, Osun State, Nigeria

O. Arogundade (✉)
Department of Crop Protection and Environmental Biology, University of Ibadan, Ibadan, Oyo State, Nigeria

Fruits Research Programme, National Horticultural Research Institute, Jericho Reservation Area, Idi-Ishin, Ibadan, Oyo State, Nigeria

K. Goswami et al. (eds.), *Non-coding RNAs for Crop Improvement*,
https://doi.org/10.1007/978-981-97-7213-1_8

expression of a specific transcript enables the precise association of an individual gene with its consequent phenotypic effects, which are absent in non-transgenic plants. Since its development, RNAi has become a valuable tool for gene silencing in various domains of biological research. In recent years, the application of this mechanism in plant protection has significantly increased. Utilizing this approach in plants for antiviral defense is highly effective, as it suppresses viral transcripts in plants which enhances the plants' innate or engineered resistance to viruses. When a virus attacks, RNA viruses produce double-stranded (ds) RNA molecules through activation or as replication intermediates, which trigger the host's RNAi defense mechanism. This leads to the specific degradation of viral RNAs, thereby inhibiting viral expression. This chapter delves into the detailed applications and mechanisms of RNAi in managing viral diseases in plants.

Keywords

Plants · Plant viruses · PTGS · RNA silencing · TGS

8.1 Introduction

RNAi is a conserved pathway that regulates gene expression in a sequence-specific manner, aided by specific host molecules and proteins. This prevents the target mRNA from being translated and thus the synthesis of its corresponding protein (Martineau and Pyrah 2007). In nature, this mechanism is used to regulate target genes as well as to induce resistance against viruses (Kumar et al. 2017). The process is triggered by the introduction of double-stranded (ds) RNA, naturally present in the cell, into the cytosol. This dsRNA serves as a substrate for the DICER enzyme, which cleaves it into small RNA fragments. These fragments then make association with RISC, which targets and degrades homologous mRNA sequences. This specific degradation of target mRNA leads to a reduction in protein synthesis (Elbashir et al. 2001a, b). The discovery of RNAi (Napoli et al. 1990) triggered an explosion of research in the field of genomics, and by 2002, it had earned the title of "Technology of the Year" (Couzin 2002).

RNAi is an innate mechanism used by eukaryotes to combat pathogenic organisms such as fungi, bacteria, and viruses, and regulate gene expression during normal developmental processes. Many organisms, from bacteria to humans, have different forms of gene silencing that are interconnected in the process of achieving the accumulation and degradation of specific RNA transcripts (Hamilton and Baulcombe 1999). This phenomenon, known as co-suppression in plants (Napoli et al. 1990), quelling in fungi (Fagard et al. 2000), and RNA interference in nematodes (Guo and Kemphues 1995), is a common gene silencing phenomenon of PTGS. RNAi is now routinely employed to determine gene function across multiple biological disciplines. It is also being exploited for medical interventions in disease pathogenesis (Singh et al. 2017). To achieve this, dsRNA is now designed for silencing targeted RNA, which is introduced into the cell, to facilitate the activation of RNAi machinery. In plants, it is a powerful means of silencing target gene of invading virus, when the appropriate inducer molecule is existing in the virus-invaded cell (Wang et al. 2011)). Some virus-susceptible plants display natural recovery, which is attributed to late activation RNA silencing, while some others lack the ability to regain normalcy after virus infection, irrespective of the buildup of siRNAs, which could be as a result of potent silencing effectors encoded by these viruses (Harvey et al. 2011; Carbonell et al. 2012; Carbonell and Carrington 2015).

8.2 Unraveling the RNAi Phenomenon

The first report on RNAi in plants was made by Rich Jorgenson and his associates (Napoli et al. 1990) as an unexpected result of efforts to intensify pigment genes in petunia for deeper purple coloration (Martineau and Pyrah 2007). An experiment aimed at overexpressing the chalcone synthase (CHS) gene, which is crucial for anthocyanin pigment production, involved the insertion of a CHS transgene linked to a strong promoter sequence. The study observed silencing of both the native and introduced CHS genes. As a result, the flowers exhibited variegated or white patterns rather than the anticipated increase in pigment intensity (Martineau and Pyrah 2007). The understanding of the underlying mechanisms was rapidly advanced in the years that followed these discoveries, due to experiments conducted in several model systems, which showed that the observed inhibition was a result of the joint action of sense and antisense strand precursor of dsRNA. In *C. elegans*, the establishment of the anterior-posterior polarity of embryo was controlled by par-1 gene (Guo and Kemphues 1995). At the time, a prevalent method for investigating gene function in nematodes involved injecting substantial quantities of antisense single-stranded RNA (ssRNA). The aim was for ssRNA to hybridize with its complementary mRNA, thereby inhibiting translation and inducing an efficient loss-of-function of the mRNA (Nellen and Lichtenstein 1993).

Fire et al. (1998) found that introducing either an exogenous antisense strand or sense RNA strand could weakly induce gene silencing, but only dsRNA efficiently led to the loss of the target mRNA. This observation led to the hypothesis that dsRNA is a potent trigger for RNAi-mediated gene silencing, being much more effective than ssRNA. So, RNAi was coined to describe the process where exogenous RNA interferes with the function of an endogenous gene. Before this study, there were reports of gene silencing by unidentified substrates in various organisms, including insects, plants, and fungi. Instances of gene silencing due to plant or fungal genome transformation with a virus, resulting in the silencing of homologous genes, had been well presented.

To understand the molecular mechanism underlying RNAi, Zamore et al. (2000) used *D. melanogaster* embryo lysates and demonstrated that RNAi relies on long dsRNA strands, which are cleaved into siRNA fragments of 22 nucleotides. These siRNA fragments then specifically cleave the mRNA, indicating the necessity of dsRNA for the reaction. Later, Elbashir et al. (2001a) showed that synthetic, short interfering duplexes of 21-nucleotide siRNA with 3′ overhangs could efficiently induce RNA cleavage in *D. melanogaster* embryo lysates. However, substituting one or both siRNA strands with deoxy or methyl oligonucleotides inhibited the reaction. Similarly, Elbashir et al. (2001b) demonstrated that siRNA could directly mediate RNAi in various cultured mammalian cell lines, with 21-nucleotide siRNA duplexes effectively suppressing the expression of both endogenous and heterologous genes in these cells. However, since siRNA does not integrate into the genome, the RNAi response is transient. Advancing this research, Hannon (2002) employed RNA Pol. III promoter-driven (e.g., U6 or H1) expression of short hairpin RNAs (shRNAs) to induce stable gene suppression in mammalian cells.

8.3 Molecular Interplay in RNA Silencing

RNA silencing can occur at different stages viz., transcriptional gene silencing (TGS) and posttranscriptional gene silencing (PTGS) (Gura 2000). TGS involves altering promoter and enhancer efficiencies, gene methylation status, and deleting parts of genes using homologous recombination, TALENs, and CRISPR/Cas9 systems (Kim and Kim 2014). The PTGS techniques rely on mRNA degradation using various technologies such as antisense RNA, ribozymes, DNAzymes, miRNAs, and RNAi (Scherer and Rossi 2003). RNAi is the most essential tool for targeted gene silencing among these techniques. The RNAi mechanism starts with the processing of sRNAs by an external trigger, which can be dsRNA transgenes, pathogens like viruses, bacteria, fungi, or transposons. RNAi occurs primarily within the cell's cytoplasm and is triggered by the introduction of a dsRNA into the cytoplasm (Martineau and Pyrah 2007). The process is thought to be divided into two phases, based on the discoveries and studies of key components involved. These include an initiation phase in which long dsRNAs are cleaved into discrete sRNA fragments of 21–25 nt, and an effector phase in which small RNAs incorporate with the multi-nuclease complex called RISC to activate cutting and degradation of a homologous target mRNA (Xu et al. 2011). In plants, five endogenous pathways involved in RNA silencing have been identified, including siRNAs, miRNAs, tncRNAs, lncRNAs, and PiwiRNAs, which differ in their biogenesis but share some resemblance in their mode of action (Hamilton and Baulcombe 1999; Pasquinelli et al. 2000; Bartel 2009; Aalto and Pasquinelli 2012). However, siRNAs and miRNAs have received the most attention and research, with the role of siRNAs and their biogenesis being better understood, while that of miRNAs is currently a major focus of many RNAi studies and experiments.

The RNAi assembly phase involves binding the RNA nucleases to a large dsRNA and cleaving it into distinctly separated 21–25-nucleotide RNA fragments (siRNA). The endonuclease Dicer (an RNAse III family enzyme (DCL), followed by the RISC activated to mediate this mechanism (Chiu and Rana 2002). Higher plants, insects and protozoa have multiple DCL genes; some being required mainly for the generation of small RNAs. Long dsRNA molecules are cut into 21–24-nt siRNAs or miRNAs once they enter plant cells by highly conserved ATP-dependent Dicer-like endoribonuclease enzymes, depending on the plant developmental stage and structure of each molecule and its substrate composition (Liu et al. 2009; Fukudome and

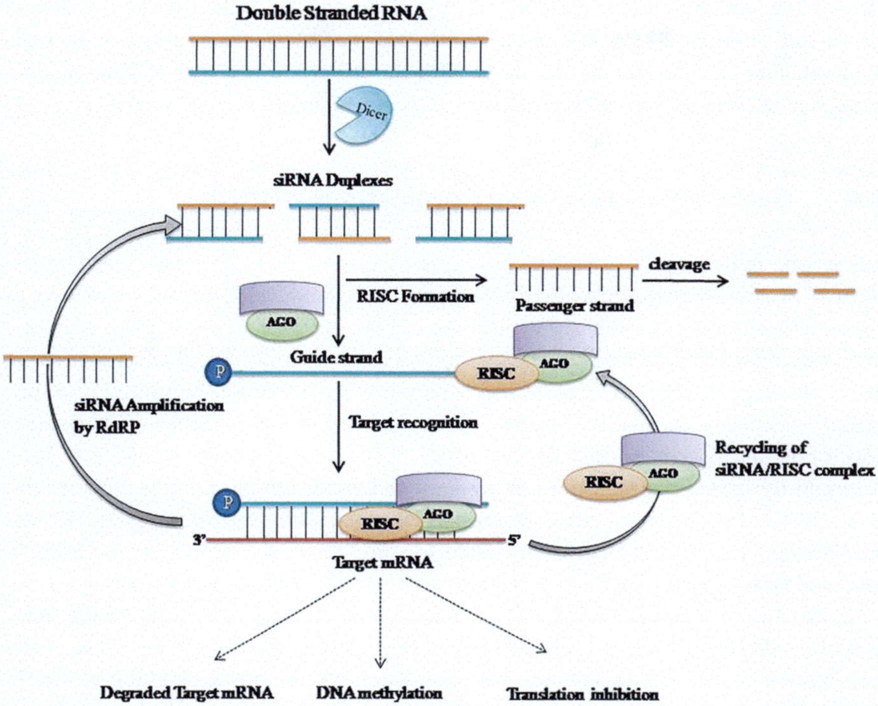

Fig. 8.1 Pictorial representation of the RNAi mechanism

Fukuhara 2017). AGO proteins sort both miRNA and siRNA, mostly based on different mechanisms such as thermodynamic stability, 5′-nucleotide identity; and one of the two small RNA strands is loaded and incorporated into the RISC (Fig. 8.1) (Kim 2008; Takeda et al. 2008).

RISC then uses a helicase to unwind the double-stranded siRNA before binding to the free antisense strand. With the help of AGO2 protein, this RISC complex distinguishes between the two siRNA strands as guide or passenger (Liu et al. 2004; Song et al. 2004; Kumar et al. 2017). Passenger or strands (which have the same sequence as the target gene) are removed and probably degraded, whereas the guide strands are incorporated into the RISC. These are used as sequence-specific complementary guides to cleave target mRNAs by the AGO protein. Activated RISC can repeatedly cut mRNA, thereby taking part in mRNA degradation and as such, inhibiting protein synthesis (Kumar et al. 2017). The introduction of long-stranded dsRNA into the cytoplasm of plants and invertebrates successfully manipulated RNAi (Martineau and Pyrah 2007).

The last stage of siRNA molecule amplification completes the RNAi process. The next generation of siRNAs is well established to be derived from RdRP using target mRNA as template by existing siRNAs. The second generation of siRNAs induces secondary RNAi, also known as transitive RNAi. In plants and *C. elegans*, transitive RNAi causes systemic genetic interference. Interestingly, because both

Drosophila and mammals lack RdRP, transitive and systemic RNAi are absent (Denli and Hannon 2003). The process of RNAi-mediated gene regulation depends on sequence-specific complementarity with the mRNA (Abbary and Rezk 2018). Sequence specificity provides a quality control mechanism (Voinnet 2005).

8.4 General Functions and Applications of RNAi

Eukaryotic cells employ RNAi for multiple functions, including antiviral defense, cellular protection, heterochromatin silencing, and the developmental regulation of endogenous gene expression (Martineau and Pyrah 2007). It not only plays a crucial role in cellular regulation but also constitutes a major technological advancement in the medical sector. However, technical constraints have frequently limited its application to subsets of model systems, particularly in in-vivo studies of vertebrate development (Martineau and Pyrah 2007). The mechanism of RNAi has been explored by functional genomics using gain or loss of function studies (Vaucheret et al. 2001). Furthermore, through proper promoter and construct design, RNAi-mediated gene silencing enabled scientists to study the genes entirely or partially, in time and space (Senthil-Kumar and Mysore 2011; Abbary and Rezk 2018).

In functional genomics research, a prevalent method involves inhibiting the function of a subset of genes and subsequently analyzing the resulting phenotypes in the knockdown lines to elucidate the roles of the targeted genes (McGinnis 2010). Often, identifying a phenotype requires the presence of specific marker gene alleles, necessitating several generations of genetic crosses to test a particular mutant allele. Utilizing transgene-induced RNAi can streamline this process, as a single copy of the transgene is typically sufficient to induce gene silencing, thereby simplifying many of the assays needed to assess phenotypes in functional genomics studies (Abbary and Rezk 2018).

Patil et al. (2011) demonstrated the creation of genetically modified melon expressing a transcript that forms an RNA hairpin structure capable of initiating specific silencing of Cm-eIF4E. They observed a decrease in the expression of Cm-eIF4E in the transgenic plants, while the expression of its isoform Cm-eIF (iso)4E remained unaffected. These melon plants were subsequently inoculated with economically significant viruses that infect melons, namely CVYV, MNSV, MWMV, and ZYMV. The results indicated resistance to these viruses, suggesting that Cm-eIF4E functions as a susceptibility factor necessary for these infections.

By utilizing small ncRNA sequences capable of selectively silencing gene expression, RNAi presents a novel potential tool for agricultural improvement in plants (Abbary and Rezk 2018). This ability to suppress specific gene expression facilitates a deeper understanding of gene functions and their associated phenotypic impacts (Abbary and Rezk 2018). The advent of RNAi has transformed plant breeding by enabling precise exploration of gene expression patterns and providing a toolkit for both quantitative and qualitative regulation of gene expression (Hirai and Kodama 2008). Additionally, RNAi has been instrumental in managing pathogenicity in plant parasites (Runo et al. 2011), enhancing resistance to various biotic and abiotic stresses

(Wani et al. 2010; Jagtap et al. 2011), and successfully developing male sterile rice plants (Kumar et al. 2017). Recent advancements include the development and regulatory approval of RNAi-based non-browning varieties of apple and potato, engineered to reduce natural carcinogens and increase fruit consumption (Kumar et al. 2017). Building on the foundational work by Napoli et al. (1990), researchers have explored applications such as modifying flowering fragrance (Kessler et al. 2008) and regulating flowering time (Baurle and Dean 2006) using this technology.

RNAi is not restricted to elucidating gene function but also holds promise for treating viral diseases in humans and animals (Whitehurst et al. 2007; Smith et al. 2010; Singh et al. 2016). In efforts to develop viral-resistant host cells or combat viral infections, RNAi can be leveraged against essential viral and human genes involved in viral replication (Smith et al. 2010). RNAi strategies can also target oncogenes implicated in cancer progression (Chaika et al. 2012; Shukla et al. 2015) impeding tumor growth (Swanton et al. 2007).

In the pharmaceutical sector, RNAi got substantial attention for its potential in drug discovery and validation (Lavery and King 2003; Lu et al. 2003). Demonstrating therapeutic potential, RNAi has shown clinical efficacy in models of neurodegenerative diseases (Xia et al. 2004) and there are some successful stories for age-related macular degeneration (Alnylam Pharmaceutical, Acuity Pharma) and the mRNA vaccine for COVID-19. Additionally, RNAi holds promise as an antiviral therapy, with examples of siRNA targeting viruses in animal models, including Herpes simplex virus-2, respiratory syncytial virus, parainfluenza virus, SARS, and influenza A (Cristofaro and Ramratnam 2006; Palliser et al. 2006).

8.5 Applications of RNAi in Plant Virus Resistance

Since the 1920s, researchers have noted that plants can acquire resistance against a severe virus by first being infected with a milder strain of the same (Agrawal et al. 2003). The precise mechanism underlying this phenomenon, known as cross-protection, remained unclear for some time. It was postulated that this could be attributed, at least in part, to PTGS induced by the mild strain, which targets the genetic material of the virulent virus (Lindbo and Dougherty 1992).

One example illustrating this concept is viral recovery, a form of virus-induced gene silencing. For instance, when *Brassica napus* plants were inoculated with Cauliflower mosaic virus, visible lesions appeared at the injection site within 5–7 days, followed by systemic infection symptoms becoming evident 10–14 days post-inoculation. Symptoms peaked around 30–40 days after inoculation and gradually decreased thereafter, with newly formed leaves showing no symptoms by day 50 (Al-Kaff et al. 1998; Agrawal et al. 2003). Substantial evidence supports the hypothesis that PTGS evolved as a defense mechanism against viral sequences in plants. This is underscored by findings showing that plant viruses produce proteins that suppress PTGS (Anandalakshmi et al. 1998; Brigneti et al. 1998; Voinnet et al. 1999), thereby protecting viral RNA genomes from degradation by the host plant's machinery. Various viral RNA silencing suppressors have been identified (Voinnet et al. 1999).

RNAi investigations in agriculture are primarily directed by virologists aiming to enhance pathogen-derived resistance (PDR) in crops (Abhary et al. 2006). Patil et al. (2011) illustrated the direct inhibitory role of dsRNA in viral infections. Certain host genes identified as crucial for viral replication, such as proteins facilitating cell-to-cell or long-distance movement, are targeted by RNAi to develop virus resistance (Abbary and Rezk 2018). RNAi has effectively been employed against diverse viruses including *Turnip mosaic virus* (Potyvirus) (Jan et al. 1999), *Citrus tristeza virus* (closterovirus; Olivares-Fuster et al. 2003), and *Cucumber green mottle mosaic virus.*

Kamachi et al. (2007) provided initial evidence of successful RNAi-based control against an Ipomovirus, offering cross-protection against distinct virus strains causing *Cassava brown streak disease* (CBSD). RNAi strategies are also applied to manage geminivirus-induced diseases in various plants, suppressing viral expression of begomoviruses causing diseases like Cotton Leaf Curl Disease (CLCuD) and Mungbean yellow mosaic virus-Vigna (MYMV-Vig) (Pooggin et al. 2003), African cassava mosaic virus (ACMV) (Walsh et al. 2019), as well as in tobacco (Pooggin and Hohn 2004), effectively reducing viral load in both plants and their insect vectors.

Besides the highly effectiveness for controlling numerous viral diseases (Tenllado et al. 2004; Waterhouse and Fusaro 2006), concerns over GM crops spurred research into dsRNA and their application against biotic stresses (Zhang et al. 2022). However, the susceptibility of a virus to RNAi depends on its type and life cycle within the host, which determines whether it can be targeted effectively by the RNAi machinery or if it can evade it.

8.6 Evidence of Endogenous RNAi Pathways in Plants and Their Role in Antiviral Defense

Utilizing *Arabidopsis* and tobacco as experimental models, scientists have extensively studied the genetic prerequisites governing the production and function siRNAs. In *Arabidopsis*, four distinct DCLs are identified as crucial for generating viral siRNAs. The efficacy of these enzymes in antiviral defense varies depending on the viral type (Blevins et al. 2006). Specifically, DCL4 and DCL2 predominantly target RNA viruses replicating in the cytoplasm, yielding siRNAs (Blevins et al. 2006; Bouché et al. 2006; Fusaro et al. 2011). In contrast, DNA viruses, which commence replication in the nucleus, are targeted by all four DCLs (DCL1–4), producing siRNAs (Blevins et al. 2006; Blevins et al. 2011; Aregger et al. 2012). Plant antiviral defenses also rely on multiple AGOs, such as AGO1, AGO2, and AGO4 (Carbonell and Carrington 2015). AGO2 is particularly pivotal due to its ability to bind viral siRNAs (Wang et al. 2011; Harvey et al. 2011) and facilitate RISC formation, which includes slicer activity (Carbonell et al. 2012). AGO1, which associates with 21–22 nucleotide 5′U-RNAs, exerts both direct and indirect antiviral effects through interactions with viral siRNAs (Garcia-Ruiz et al. 2010; Harvey et al. 2011; Wang et al.

2011; Carbonell et al. 2012; Carbonell and Carrington 2015), as well as with endogenous miRNAs of the host plant. Conversely, AGO4, known to bind 24 nucleotide 5′A-RNAs, is primarily implicated in defense against DNA viruses, although its role with 24-nucleotide viral siRNAs remains less substantiated (Blevins et al. 2011; Pooggin 2013).

Even if a virus is initially localized, virus-derived siRNAs have the capacity to induce systemic gene silencing across the entire plant (Voinnet and Baulcombe 1997; Jones et al. 1999). Himber et al. (2003) proposed a model suggesting that primary siRNAs originating from the viral genome, processed by DCL4, can move approximately 10–15 cells within leaves. However, their spread may be constrained by symplastic barriers, limiting their transmission to other tissues (Dunoyer et al. 2005; Kobayashi and Zambryski 2007). Yoo et al. (2004) reported the presence of viral siRNAs in the phloem sap of pumpkins, along with some endogenous miRNAs. Nevertheless, it remains unclear whether these findings reflect short-distance export from companion cells or are indicative of long-distance effects (Ossowski et al. 2008).

8.7 Development of RNAi-Based Strategies and RNAi Delivery Methods

In plants, one of the primary challenges in RNAi research concerns the effective delivery of molecules that activate the RNAi pathway (Shabir et al. 2010). Several strategies have been developed to express dsRNAs in plants (Lemgo et al. 2013). Among these, a pioneering and highly effective approach involves constructing hairpin-shaped RNA molecules. These constructs are formed by linking two segments of a target viral gene sequence in an inverted repeat orientation, with a spacer or intron between them (Frizzi and Huang 2010; Ilardi and Di Nicola-Negri 2011; Duan et al. 2012). A particularly potent variation employs a hairpin structure where the terminal loop is initially created by a short intron. This method has demonstrated success rates as high as 90%, although variability exists across studies (Wesley et al. 2001; Kerschen et al. 2004). It has been widely adopted for various plant species.

Agro-infiltration is the most employed technique to deliver RNAi products, which involves introducing transgenes carried by agrobacterium strains into the leaf cells to induce RNA silencing (Hilly and Liu 2007). Typically, this method is used to initiate widespread silencing throughout the plant or to study the effects of suppressor genes (Shabir et al. 2010; Wani et al. 2010; Jagtap et al. 2011). Microbombardment, another effective approach, employs biolistic pressure to deliver synthetic siRNAs directly into plant tissues (Shabir et al. 2010). This technique involves bombarding cells with particles coated with dsRNA, siRNA, or DNA-encoding hairpin constructs, thereby activating the plant's endogenous RNAi pathway. The RNAi-induced silencing effect can manifest within a day of bombardment and persist for 3–4 days thereafter (Shabir et al. 2010).

Additionally, VIGS operates at the post-transcriptional level, utilizing genetically modified viruses to trigger RNAi. Various RNA and DNA viruses have been adapted successfully as carriers for RNAi (Timmermans et al. 1994; Pogue et al. 2002). RNA viruses have generally shown greater efficacy compared to DNA viruses, which face challenges due to their larger size in terms of uptake and movement within plant tissues (Kjemtrup et al. 1998). However, not all RNA virus-derived vectors are suitable for gene silencing, as some possess strong anti-silencing proteins that directly disrupt host RNAi mechanisms (Angell and Baulcombe 1999; MacFarlane and Popovich 2000; Pogue et al. 2002; Palmer and Rybicki 2001; Mallory et al. 2022).

Spray-induced gene silencing (SIGS) represents an innovative technique for administering gene products into plants to combat viral infections. This method has demonstrated efficacy in managing pathogen infections in both monocots and dicots (Qi et al. 2019). Other various approaches for delivering dsRNA externally include infiltration, irrigation, foliar spray, and the use of adjuvants in formulations (Dalakouras et al. 2020). Among these, infiltration is considered a straightforward method for introducing transgenes into experimental plants.

Generally, multiple studies have demonstrated that plant cells efficiently uptake dsRNAs, which are subsequently processed into siRNAs, thereby prolonging resistance against certain plant diseases (Koch et al. 2016; Konakalla et al. 2016; Kaldis et al. 2018). Application of dsRNAs, specifically targeting the nuclear inclusion b (NIb) protein coding region (BCMVNIb-dsRNA) or the coat protein (CP) coding region (BCMVCP-dsRNA), via topical methods conferred protection to tobacco and cowpea plants from BCMV upon mechanical inoculation (Mitter et al. 2017). Mitter et al. (2017) further investigated the efficacy of BioClay containing BCMV-specific dsRNAs in safeguarding cowpea and *N. benthamiana* plants against BCMV transmission by aphids. Additionally, Numata et al. (2014) observed that direct infiltration of dsRNA targeting the YFP transgene into *Arabidopsis* leaves did not induce RNAi. However, employing a carrier complex composed of a lysine copolymer alongside dsRNA significantly improved delivery efficacy, leading to reduced accumulation of the targeted genes.

8.8 Target Selection and Specificity

Initially, efforts to improve siRNA-mediated gene silencing in both plants and animals focused primarily on enhancing effectiveness. However, there has been increasing recognition of specificity challenges (Ossowski et al. 2008). To validate the impact of siRNAs on target genes, it is common practice to use two or three independent siRNAs or hairpins that generate siRNAs complementary to the target gene. Another approach involves enhancing siRNA effectiveness by introducing transgenes with silent mutations in the target gene, rendering them resistant to siRNA-induced silencing (Lens et al. 2003). This approach is straightforward in mammalian cell cultures where single siRNA duplexes are typically used, and RNA-dependent RNA polymerases, which could produce secondary siRNAs, are absent (Ossowski et al. 2008).

In plants, potential off-target effects of hpRNAi may arise from specific characteristics of hairpin transcripts. For conferring resistance against both RNA and DNA viruses, the selection of target sequences within the viral genome is crucial (Poogin 2018). In rice, Sasaya et al. (2014) demonstrated that transgenic RNAi resistance against segmented RNA viruses required targeting a vulnerable viral gene. The reasons why some viral target genes or genome regions are resistant to transgenic siRNAs have not been thoroughly investigated. One possible explanation is that GC-rich target sequences may form secondary structures that hinder access by the RISC, whereas AU-poor hairpin arms may produce low levels of 5′A- and 5′U-siRNAs, reducing the formation of active RISCs with AGO proteins (Poogin 2018). Additionally, viral genome regions that produce fewer siRNAs in infected plants might be more suitable targets compared to viral siRNA hotspot regions that could serve as decoys (Blevins et al. 2011; Rajeswaran et al. 2014).

Due to the sequence specificity of RNAi, virus resistance typically applies only to viral strains or closely related viruses sharing over 90% sequence identity. Broad-spectrum resistance could be achieved by transgenically expressing hairpin RNAs with chimeric arms comprising sequences from multiple viruses, or by using constructs containing several separate intron-hairpin RNAs targeting different viruses (Bucher et al. 2006; Liu et al. 2007; Zhang et al. 2011; Arif et al. 2012; Chung et al. 2013). The persistence of RNAi-mediated resistance may depend on the ability of target viruses to overcome RNAi; viruses with potent silencing suppressors can sometimes defeat RNAi-mediated resistance (Hassani-Mehraban et al. 2009). Moreover, a non-target virus might suppress RNAi resistance against a target virus. Additionally, a target virus could evade RNAi by mutating or replacing the RNAi target sequence through recombination with distantly related viruses, as observed in RNAi-transgenic tomato plants under field conditions (Fuentes et al. 2016).

8.9 Success Stories and Case Studies

In various plant-virus interactions, RNAi technology has induced immune responses against diverse viruses. The efficacy of this method in creating virus-resistant plants was initially demonstrated with Potato Virus Y (PVY) in potatoes, utilizing vectors that express both sense and antisense transcripts of the helper-component proteinase (HC-Pro) gene (Waterhouse et al. 1998). Potyvirus P1/HC-Pro suppressors were found to block RNA silencing downstream of dsRNA processing, likely by impeding unwinding of duplex siRNAs, their incorporation into the RISC, or both (Chapman et al. 2004). In studies on geminiviruses, RNAi-mediated silencing was achieved using a transient protoplast assay, where co-transfer with siRNAs targeting the replication (Rep)-coding sequence of African cassava mosaic virus (ACMV) resulted in a drastic reduction of Rep transcripts by 99% and viral DNA by 66% (Vanitharani et al. 2003). Successful engineering of virus resistance has been achieved through small RNA-based strategies such as artificial microRNAs (amiRNAs) and hpRNA targeting economically significant plant viruses. Examples include Cassava mosaic virus (CMV) (Ntui et al. 2015), Citrus tristeza virus (CTV)

(Soler et al. 2012), Maize streak virus (MSV) (Shepherd et al. 2007), Maize dwarf virus (MDV) (Zhang et al. 2013), Soyabean mosaic virus (SMV) (Gao et al. 2015), Potato leaf roll virus (PLRV) (Zhao et al. 2020), Banana bract mosaic virus (BBrMV) and Banana bunchy top virus (BBTV) (Shekhawat et al. 2012), Cucumber mosaic virus (CMV) (Duan et al. 2012), Potato virus X (PVX) (Ai et al. 2011), Tomato leaf curl burewala virus (TLCBV) and Tomato leaf curl Delhi virus (TLCDV), Wheat streak mosaic virus (WSMV) (Sivamani et al. 2002), Barley yellow dwarf virus (BYDV) (Wang et al. 2006), Potato virus Y (PVY) (Ai et al. 2011), Tomato yellow leaf curl virus (TYLCV) (Yang et al. 2004), Sugarcane mosaic virus (SCMV) (Tabassum et al. 2013), and Rice tungro bacilliform virus (RTBV) (Tyagi et al. 2008).

8.10 Conclusion

Recent advances in RNAi technology have demonstrated significant potential in both reverse genetics and therapeutic applications. However, the intricate complexities associated with RNAi mechanisms remain largely unexplored. RNAi is increasingly acknowledged as a transformative tool for functional genomics across various organisms, enabling precise full genomic screens to pinpoint genes that govern specific traits. Ongoing research aims to further elucidate the expansive capabilities of RNAi in enhancing human welfare and bolstering food security.

In agriculture, the topical application of dsRNA molecules has emerged as a promising strategy for protecting plants against viral infections, presenting an alternative to transgenic approaches. This method circumvents the complexities associated with plant transformation, screening, and biosafety concerns. Studies have demonstrated that exogenous dsRNA designed to target viral sequences, whether produced in vitro or in vivo, can effectively confer resistance against several plant viruses, including those from the Tobamovirus, Potyvirus, Alfamovirus, and Potexvirus families.

Despite its potential, RNAi technology has limitations. Notably, its effectiveness is highly dependent on sequence specificity, making viruses with significant nucleotide diversity, typically exceeding 10%, less susceptible to RNAi-mediated suppression. Addressing this challenge often requires constructing RNAi molecules from multiple viral fragments to achieve broad-spectrum resistance. However, integrating numerous gene fragments into a single construct can pose challenges during cloning, transformation, and subsequent expression within transgenic plants, potentially affecting their stability and ability to form functional RNAi structures. Viral genomes exhibit high variability, allowing them to rapidly adapt to host and environmental conditions. It is imperative to adopt a comprehensive approach that includes additional precautionary measures beyond genetic resistance strategies (Singh et al. 2016).

8.11 Future Prospects

The application of RNAi through foliar spray has generated considerable interest due to its potential as a targeted biopesticide suitable for commercial use. SIGS offers a non-transgenic approach to control plant pests and pathogens (Puyam and Kaur 2020; Sang and Kim 2020). This method involves administering siRNAs and dsRNAs to affected plants via foliar spray, inducing RNAi upon ingestion by the targeted organisms. SIGS allows for precise management of harmful pathogens while potentially mitigating adverse impacts on the ecosystem associated with chemical pesticides (Puyam and Kaur 2020). It has demonstrated effectiveness against various pathogens and pests. However, a notable drawback is the relative instability of ssRNAs and dsRNAs, particularly when exposed to environmental elements (Sang and Kim 2020; Koeppe et al. 2023).

Ongoing research focuses on enhancing the stability and longevity of dsRNA to meet environmental and industry standards. Strategies such as polymeric, lipid-based, and inorganic nanoparticles are being explored to improve the survival and uptake of exogenous RNA. Synthetic biology advancements offer opportunities to optimize traits by influencing metabolic and immune pathways through siRNA delivery. The future holds promising avenues for further development and applications of RNA interference in enhancing crop yields, managing pests, and advancing human healthcare therapies (Koeppe et al. 2023).

References

Aalto AP, Pasquinelli AE (2012) Small non-coding RNAs mount a silent revolution in gene expression. Curr Opin Cell Biol 24:333–340. https://doi.org/10.1016/j.ceb.(2012)03.006

Abbary M, Rezk A (2018) In: Al-Khayri JM et al (eds) Advances in plant breeding strategies: breeding, biotechnology and molecular tools. Springer International Publishing, Cham, pp 397–427. https://doi.org/10.1007/978-3-319-22521-0_14

Abhary MK, Anfoka GH, Nakhla MK, Maxwell DP (2006) Post-transcriptional gene silencing in controlling viruses of the tomato yellow leaf curl virus complex. Arch Virol 151(12):2349–2363. https://doi.org/10.1007/s00705-006-0819-7

Agrawal PVN, Dasaradhi AM, Pawan M, Raj KB, Sunil KM (2003) RNA interference: biology, mechanism, and applications. Microbiol Mol Biol Rev 67:657–685. https://doi.org/10.1128/MMBR.67.4

Ai T, Zhang L, Gao Z, Zhu CX, Guo X (2011) Highly efficient virus resistance mediated by artificial microRNAs that target the suppressor of PVX and PVY in plants. Plant Biol 13:304–316. https://doi.org/10.1111/j.1438-8677.2010.00374.x

Al-Kaff NS, Covey SN, Kreike MM, Page AM, Pinder R, Dale P (1998) Transcriptional and post-transcriptional plant gene silencing in response to a pathogen. Science 279(5359):2113–2115. https://doi.org/10.1126/science.279.5359.2113

Anandalakshmi R, Pruss GJ, Ge X, Marathe R, Mallory AC, Smith TH, Vance VB (1998) A viral suppressor of gene silencing in plants. Proc Natl Acad Sci 95:13079–13084

Angell SM, Baulcombe DC (1999) Technical advance: potato virus X amplicon mediated silencing of nuclear genes. Plant J 20:357–362

Aregger M, Borah BK, Seguin J, Rajeswaran R, Gubaeva EG, Zvereva AS, Windels D, Vazquez F, Blevins T, Farinelli L, Pooggin MM (2012) Primary and secondary siRNAs in geminivirus-induced gene silencing. PLoS Pathog 8:e1002941

Arif M, Azhar U, Arshad M, Zafar Y, Mansoor S, Asad S (2012) Engineering broad-spectrum resistance against RNA viruses in potato. Transgenic Res 21:303–311

Bartel DP (2009) MicroRNAs: target recognition and regulatory functions. Cell 136:215–233. https://doi.org/10.1016/j.cell.(2009)01.002

Baurle I, Dean C (2006) The timing of developmental transitions in plants. Cell 125:655–664

Blevins T, Rajeswaran R, Shivaprasad PV, Beknazariants D, Si-Ammour A, Park HS, Vazquez F, Robertson D, Meins F Jr, Hohn T, Pooggin MM (2006) Four plant dicers mediate viral small RNA biogenesis and DNA virus induced silencing. Nucleic Acids Res 34:6233–6246

Blevins T et al (2011) Massive production of small RNAs from a non-coding region of Cauliflower mosaic virus in plant defense and viral counter-defense. Nucleic Acids Res 39:5003–5014

Bouché N, Lauressergues D, Gasciolli V, Vaucheret H (2006) An antagonistic function for *Arabidopsis* DCL2 in development and a new function for DCL4 in generating viral siRNAs. EMBO J 25:3347–3356

Brigneti G, Voinnet O, Li WX, Ji LH, Ding SW, Baulcombe BC (1998) Viral pathogenicity determinants are suppressors of transgene silencing in *Nicotiana benthamiana*. EMBO J 17:6739–6746

Bucher E, Lohuis D, van Poppel PM, Geerts-Dimitriadou C, Goldbach R, Prins M (2006) Multiple virus resistance at a high frequency using a single transgene construct. J Gen Virol 87:3697–3701

Carbonell A, Carrington JC (2015) Antiviral roles of plant ARGONAUTES. Curr Opin Plant Biol 27:111–117

Carbonell A, Fahlgren N, Garcia-Ruiz H, Gilbert KB, Montgomery TA, Nguyen T, Cuperus JT, Carrington JC (2012) Functional analysis of three *Arabidopsis* ARGONAUTES using slicer defective mutants. Plant Cell 24:3613–3629

Chaika NV, Gebregiworgis T, Lewallen ME, Purohit V, Radhakrishnan P, Liu X, Zhang B, Mehla K, Brown RB, Caffrey T, Yu F, Johnson KR, Powers R, Hollingsworth MA, Singh PK (2012) MUC1 mucin stabilizes and activates hypoxia-inducible factor 1 alpha to regulate metabolism in pancreatic cancer. Proc Natl Acad Sci 109:13787–13792. https://doi.org/10.1073/pnas.1203339109

Chapman EJ, Prokhnevsky AI, Gopinath K, Dolja VV, Carrington JC (2004) Viral RNA silencing suppressors inhibit the micro-RNA pathway at an interphase step. Genes Dev 18:1179–1186

Chiu Y, Rana TM (2002) RNAi in human cells: basic structural and functional features of small interfering RNA. Mol Cell 10:549–561

Chung BN, Yoon JY, Palukaitis P (2013) Engineered resistance in potato against potato leafroll virus, potato virus a and potato virus Y. Virus Genes 47:86–92

Couzin J (2002) Small RNAs make big splash. Science 298:2296–2297

Cristofaro P, Ramratnam B (2006) RNAi tackles a sexually transmitted disease. Nat Biotechnol 24(1):48–49. https://doi.org/10.1038/nbt0106-48

Dalakouras A, Wassenegger M, Dadami E, Ganopoulos I, Pappas ML, Papadopoulou K (2020) Genetically modified organism-free RNA interference: exogenous application of RNA molecules in plants1[open]. Plant Physiol 182:38–50. https://doi.org/10.1104/pp.19.00570

Denli AM, Hannon GJ (2003) RNAi: an ever-growing puzzle. Trends Biochem Sci 28:196–201. https://doi.org/10.1016/S0968-0004(03)00058-6

Duan CG, Wang CH, Guo HS (2012) Application of RNA silencing to plant disease resistance. Silence 3:5. https://doi.org/10.1186/1758-907X-3-5

Dunoyer P, Himber C, Voinnet O (2005) DICER-LIKE 4 is required for RNA interference and produces the 21-nucleotide small interfering RNA component of the plant cell-to-cell silencing signal. Nat Genet 37:1356–1360

Elbashir SM, Lendeckel W, Tuschl T (2001a) RNA interference is mediated by 21-and 22-nucleotide RNAs. Genes Dev 15:188–200. https://doi.org/10.1101/gad.862301

Elbashir SM, Harborth J, Lendeckel W, Yalcin A, Weber K, Tuschl T (2001b) Duplexes of 21-nucleotide RNAs mediate RNA interference in cultured mammalian cells. Nature 411:494–498. https://doi.org/10.1038/35078107

Fagard M, Boutet S, Morel JB, Bellini C, Vaucheret H (2000) AGO1, QDE-2, and RDE-1 are related proteins required for post-transcriptional gene silencing in plants, quelling in fungi, and RNA interference in animals. Proc Natl Acad Sci 97:11650–11654

Fire A, Xu S, Montgomery MK, Kostas SA, Driver SE, Mello CC (1998) Potent and specific genetic interference by double-stranded RNA in *Caenorhabditis elegans*. Nature 391:806–811

Frizzi A, Huang S (2010) Tapping RNA silencing pathways for plant biotechnology. Plant Biotechnol J 8:655–677

Fuentes A, Carlos N, Ruiz Y, Callard D, Sánchez Y, Ochagavía ME, Seguin J, MalpicaLópez N, Hohn T, Lecca MR (2016) Field trial and molecular characterization of RNAi-transgenic tomato plants that exhibit resistance to tomato yellow leaf curl Geminivirus. Mol Plant-Microbe Interact 29:197–209

Fukudome A, Fukuhara T (2017) Plant dicer-like proteins: double-stranded RNA-cleaving enzymes for small RNA biogenesis. J Plant Res 130:33–44

Fusaro AF, Matthew L, Smith NA, Curtin SJ, Dedic-Hagan J, Ellacott GA, Watson JM, Blevins T, Rajeswaran R, Aregger M, Borah BK, Schepetilnikov M, Baerlocher L, Farinelli L, Meins F Jr, Hohn T, Pooggin MM (2011) Massive production of small RNAs from a non-coding region of cauliflower mosaic virus in plant defense and viral counter-defense. Nucleic Acids Res 39:5003–5014

Gao L, Ding X, Li K, Liao W, Zhong Y, Ren R, Liu Z, Adhimoolam K, Zhi H (2015) Genetics a: characterization of soybean mosaic virus resistance derived from inverted repeat-SMV-HC-pro genes in multiple soybean cultivars. Theor Appl Genet 128:1489–1505

Garcia-Ruiz H, Takeda A, Chapman EJ, Sullivan CM, Fahlgren N, Brempelis KJ, Carrington JC (2010) *Arabidopsis* RNA-dependent RNA polymerases and dicer-like proteins in antiviral defense and small interfering RNA biogenesis during turnip mosaic virus infection. Plant Cell 22:481–496

Guo S, Kemphues KJ (1995) Par-1, a gene required for establishing polarity in C. elegans embryos, encodes a putative Ser/Thr kinase that is asymmetrically distributed. Cell 81:611–620

Gura TA (2000) Silence that speaks volumes. Nature 404:804–808. https://doi.org/10.1038/35009245

Hamilton A, Baulcombe D (1999) A species of small antisense RNA in posttranscriptional gene silencing in plants. Science 286:950–952

Hannon GJ (2002) RNA interference. Nature 418:244–251. https://doi.org/10.1038/418244a

Harvey JJ, Lewsey MG, Patel K, Westwood J, Heimstädt S, Carr JP, Baulcombe DC (2011) An antiviral defense role of AGO2 in plants. PLoS One 6:e14639

Hassani-Mehraban A, Brenkman AB, Van den Broek NJ, Goldbach R, Kormelink R (2009) RNAi-mediated transgenic Tospovirus resistance broken by intraspecies silencing suppressor protein complementation. Mol Plant-Microbe Interact 22:1250–1257

Hilly JM, Liu Z (2007) An overview of small RNAs. In: Bassett CL (ed) Regulation of gene expression in plants. Springer-Verlag, Berlin, pp 123–147

Himber C, Dunoyer P, Moissiard G, Ritzenthaler C, Voinnet O (2003) Transitivity-dependent and -independent cell-to-cell movement of RNA silencing. EMBO 22:4523–4533

Hirai S, Kodama H (2008) RNAi vectors for manipulation of gene expression in higher plants. Open Plant Sci J 2:21–30

Ilardi V, Di Nicola-Negri E (2011) Genetically engineered resistance to plum pox virus infection in herbaceous and stone fruit hosts. GM Crops 2:24–33

Jagtap UB, Gurav RG, Bapat VA (2011) Role of RNA interference in plant improvement. Naturwissenschaften 98:473–492

Jan F-J, Pang S-Z, Fagoaga C, Gonsalves D (1999) Turnip mosaic potyvirus resistance in Nicotiana benthamiana derived by post-transcriptional gene silencing. Transgenic Res 8:203–213

Jones L, Hamilton AJ, Voinnet O, Thomas CL, Maule AJ, Baulcombe DC (1999) RNA–DNA interactions and DNA methylation in post-transcriptional gene silencing. Plant Cell 11:2291–2301

Kaldis A, Berbati M, Melita O, Reppa C, Holeva M, Otten P (2018) Exogenously applied dsRNA molecules deriving from the zucchini yellow mosaic virus (ZYMV) genome move systemically and protect cucurbits against ZYMV. Mol Plant Pathol 19:883–895. https://doi.org/10.1111/mpp.12572

Kamachi S, Mochizuki A, Nishiguchi M, Tabei Y (2007) (2007) Transgenic Nicotiana benthamiana plants resistant to cucumber green mottle mosaic virus based on RNA silencing. Plant Cell Rep 26:1283–1288. https://doi.org/10.1007/s00299-007-0358-z

Kerschen A, Napoli CA, Jorgensen RA, Muller AE (2004) Effectiveness of RNA interference in transgenic plants. FEBS Lett 566:223–228

Kessler D, Gase K, Baldwin IT (2008) Field experiments with transformed plants reveal the sense of floral scents. Science 321(5893):1200–1202

Kim VN (2008) Sorting out small RNAs. Cell 133(1):25–26. https://doi.org/10.1016/j.cell.2008.03.015

Kim H, Kim JSA (2014) Guide to genome engineering with programmable nucleases. Nat Rev Genet 15:321–334. https://doi.org/10.1038/nrg3686

Kjemtrup S, Sampson KS, Peele CG, Nguyen LV, Conkling MA (1998) Gene silencing from plant DNA carried by a Geminivirus. Plant J 14:91–100

Kobayashi K, Zambryski P (2007) RNA silencing and its cell-tocell spread during *Arabidopsis* embryogenesis. Plant J 50:597–604

Koch A, Biedenkopf D, Furch A, Weber L, Rossbach O, Abdellatef E et al (2016) An RNAi-based control of Fusariumgraminearum infections through spraying of long dsRNAs involves a plant passage and is controlled by the fungal silencing machinery. PLoS Pathog 12:e1005901. https://doi.org/10.1371/journal.ppat.1005901

Koeppe S, Kawchuk L, Kalischuk M (2023) RNA interference past and future applications in plants. Int J Mol Sci 24:9755. https://doi.org/10.3390/ijms24119755

Konakalla NC, Kaldis A, Berbati M, Masarapu H, Voloudakis AE (2016) Exogenous application of double-stranded RNA molecules from TMV p126 and CP genes confers resistance against TMV in tobacco. Planta 244:961–969. https://doi.org/10.1007/s00425-016-2567-6

Kumar P, Mundiyara R, Jitarwal RC (2017) RNA interference: new approach of gene silencing in plants. Int J Pure App Biosci 5(3):421–425. https://doi.org/10.18782/2320-7051.5066

Lavery KS, King TH (2003) Antisense and RNAi: powerful tools in drug target discovery and validation. Curr Opin Drug Discov Dev 4:561–569

Lemgo GNY, Sabbadini S, Pandolfini T, Mezzetti B (2013) Biosafety considerations of RNAi-mediated virus resistance in fruit-tree cultivars and in rootstock. Transgenic Res 22:1073–1088. https://doi.org/10.1007/s11248-013-9728-1

Lens SM, Wolthuis RM, Klompmaker R, Kauw J, Agami R, Brummelkamp T, Kops G, Medema RH (2003) Surviving is required for a sustained spindle checkpoint arrest in response to lack of tension. EMBO J 22:2934–2947

Lindbo JA, Dougherty WG (1992) Untranslatable transcripts of tobacco etch virus coat protein gene sequence can interfere with tobacco etch virus replication in transgenic plants and protoplasts. Virology 189:725–733

Liu J, Carmell MA, Rivas FV, Marsden CG, Thomson JM, Song J, Hammond SM, Joshua-Tor L, Hannon GJ (2004) Argonaute 2 is the catalytic engine of mammalian RNAi. Science 305:1437–1441

Liu Z, Scorza R, Hily J-M, Scott SW, James D (2007) Engineering resistance to multiple Prunus viruses through expression of chimeric hairpins. J Am Soc Hort Sci 132(3):407–414. https://doi.org/10.21273/JASHS.132.3.407

Liu Q, Feng Y, Zhu Z (2009) Dicer-like (DCL) proteins in plants. Func Integr Genom 9:277–286

Lu PY, Xie FY, Woodle MC (2003) siRNA-mediated anti-tumorigenesis for drug target validation and therapeutics. Curr Opin Mol Therap 5:225–234

MacFarlane SA, Popovich AH (2000) Efficient expression of foreign proteins in roots from Tobravirus vectors. Virology 267:29–35

Mallory AC, Parks G, Endres MW, Baulcombe D, Bowman LH (2022) The amplicon-plus system for high-level expression of transgenes in plants. Nat Biotechnol 20:622–625

Martineau M, Pyrah IT (2007) Review of the application of RNA interference Technology in the Pharmaceutical Industry. Toxicol Pathol 35:327–336. https://doi.org/10.1080/01926230701197107

McGinnis KM (2010) RNAi for functional genomics in plants. Brief Funct Genomics 9(2):111–117. https://doi.org/10.1093/bfgp/elp052

Mitter N, Worrall EA, Robinson KE, Xu ZP, Carroll BJ (2017) Induction of virus resistance by exogenous application of double-stranded RNA. Curr Opin Virol 26:49–55

Napoli C, Lemieux C, Jorgensen R (1990) Introduction of a chimeric chalcone synthase gene into petunia results in reversible co-suppression of homologous genes in transgenic plants. Plant Cell 2:279–289. https://doi.org/10.1105/tpc.2.4.279

Nellen W, Lichtenstein C (1993) What makes an mRNA anti-sensitive? Trends Biochem Sci 18:419–423

Ntui VO, Kong K, Khan RS, Igawa T, Janavi GJ, Rabindran R, Nakamura I, Mii M (2015) Resistance to Sri Lankan cassava mosaic virus (SLCMV) in genetically engineered cassava cv. KU50 through RNA silencing. PLoS One 10:e0120551

Numata K, Ohtani M, Yoshizumi T, Demura T, Kodama Y (2014) Local gene silencing in plants via synthetic ds RNA and carrier peptide. Plant Biotechnol J 12:1027–1034

Olivares-Fuster O, Fleming GH, Albiach-Marti MR, Gowda S, Dawson WO, Crosser JW (2003) Citrus tristeza virus (CTV) resistance in transgenic citrus based on virus challenge of protoplasts. In-Vitro Cell Dev Biol Plant 39:567–572

Ossowski S, Schwab R, Weigel D (2008) Gene silencing in plants using artificial microRNAs and other small RNAs. Plant J 53:674–690

Palliser D, Chowdhury D, Wang QY, Lee SJ, Bronson RT, Knipe DM, Lieberman J (2006) An siRNA-based microbicide protects mice from lethal herpes simplex virus 2 infection. Nature 439:89–94

Palmer KE, Rybicki EP (2001) Investigation of the potential of maize streak virus to act as an infectious gene vector in maize plants. Arch Virol 146:1089–1104. https://doi.org/10.1007/s007050170108

Pasquinelli AE, Reinhart BJ, Slack F, Martindale MQ, Kuroda MI, Maller B et al (2000) Conservation of the sequence and temporal expression of let-7 heterochronic regulatory RNA. Nature 408:86–89. https://doi.org/10.1038/35040556

Patil BL, Ogwok E, Wagaba H, Mohammed IU, Yadav JS, Bagewadi B, Taylor NJ, Kreuze JF, Maruthi MN, Alicai T, Fauquet CM (2011) RNAi-mediated resistance to diverse isolates belonging to two virus species involved in cassava brown streak disease. Mol Plant Pathol 12(1):31–41. https://doi.org/10.1111/J.1364-3703.(2010)00650

Pogue GP, Lindbo JA, Garger SJ, Fitzmaurice WP (2002) Making an ally from an enemy: plant virology and the new agriculture. Annu Rev Phytopathol 40:45–74. https://doi.org/10.1146/annurev.phyto.40.021102.150133

Pooggin MM (2013) How can plant DNA viruses evade siRNA-directed DNA methylation and silencing? Int J Mol Sci 14:15233–15259

Pooggin M, Shivaprasad PV, Veluthambi K, Hohn T (2003) RNAi targeting of DNA virus in plants. NatBiotechnol 21:131–132

Pooggin MM, Hohn T (2004) Fighting geminiviruses by RNAi and vice versa. Plant Molecular Biology, Springer, Netherlands 55:149–152

Pooggin MM (2018) Small RNA-Omics for Plant Virus Identification, Virome Reconstruction, and Antiviral Defense Characterization. Front Microbiol 9:2779. https://doi.org/10.3389/fmicb.2018.02779

Puyam A, Kaur K (2020) Exploiting RNA interference mechanism in plants for disease resistance. In: UlHaq I, Ijaz S (eds) Plant disease management strategies for sustainable agriculture through traditional and modern approaches. Sustainability in plant and crop protection, vol 13. Springer, Cham, pp 217–236

Qi T, Guo J, Peng H, Liu P, Kang Z, Guo J (2019) Host-induced gene silencing: a powerful strategy to control diseases of wheat and barley. Int J Mol Sci 20(1):206. https://doi.org/10.3390/ijms20010206

Rajeswaran R, Golyaev V, Seguin J, Zvereva AS, Farinelli L, Pooggin MM (2014) Interactions of Rice tungro bacilliform pararetrovirus and its protein P4 with plant RNA-silencing machinery. Mol Plant-Microbe Interact 27:1370–1378

Runo S, Alakonya A, Machuka J, Sinha N (2011) RNA interference as a resistance mechanism against crop parasites in Africa: a 'Trojan horse' approach. Pest Manag Sci 67:129–136. https://doi.org/10.1002/Ps.2052

Sang H, Kim JI (2020) Advanced strategies to control plant pathogenic fungi by host-induced gene silencing (HIGS) and spray-induced gene silencing (SIGS). Plant Biotechnol Rep 14:1–8

Sasaya T, Nakazono-Nagaoka E, Saika H, Aoki H, Hiraguri A, Netsu O, Uehara-Ichiki T, Onuki M, Toki S, Saito K, Yatou O (2014) Transgenic strategies to confer resistance against viruses in rice plants. Front Microbiol 4:409

Scherer LJ, Rossi JJ (2003) Approaches for the sequence-specific knockdown of mRNA. Nat Biotechnol 21:1457–1465. https://doi.org/10.1038/nbt915

Senthil-Kumar M, Mysore KS (2011) New dimensions for VIGS in plant functional genomics. Trends Plant Sci 16(12):656–665

Shabir HW, Gulzar SS, Singh NB (2010) Biotechnology and plant disease control-role of RNA interference. Am J Plant Sci 1:55–68

Shekhawat UK, Ganapathi TR, Hadapad AB (2012) Transgenic banana plants expressing small interfering RNAs targeted against viral replication initiation gene display high-level resistance to banana bunchy top virus infection. J Gen Virol 93:1804–1813

Shepherd DN, Mangwende T, Martin DP, Bezuidenhout M, Thomson JA, Rybicki EP (2007) Inhibition of maize streak virus (MSV) replication by transient and transgenic expression of MSV replication-associated protein mutants. J Gen Virol 88:325–336

Shukla SK, Gunda V, Abrego J, Haridas D, Mishra A, Souchek J, Chaika NV, Yu F, Sasson AR, Lazenby AJ, Batra SK, Singh PK (2015) MUC16-mediated activation of mTOR and c-Myc reprograms pancreatic cancer metabolism. Oncotarget 6:19118–19131

Singh D, Chaudhary S, Kumar J, Sirohi P, Mehla K, Sirohi A, Kumar S, Chand P, Singh P (2016) RNA interference technology-applications and limitations. Intech Open Science, London, pp 21–36. https://doi.org/10.5772/61760

Singh IK, Singh S, Mogilicherla K, Shukla JN, Palli SR (2017) Comparative analysis of double-stranded RNA degradation and processing in insects. Sci Rep 7:17059. https://doi.org/10.1038/s41598-017-17134-2

Sivamani E, Brey CW, Talbert LE, Young MA, Dyer WE, Kaniewski WK, Qu R (2002) Resistance to wheat streak mosaic virus in transgenic wheat engineered with the viral coat protein gene. Transgenic Res 11:31–41

Smith JA, White EA, Sowa ME, Powell ML, Ottinger M, Harper JW, Howley PM (2010) Genome-wide siRNA screen identifies SMCX, EP400, and Brd4 as E2-dependent regulators of human papillomavirus oncogene expression. Proc Natl Acad Sci 107:3752–3757. https://doi.org/10.1073/pnas.0914818107

Soler N, Plomer M, Fagoaga C, Moreno P, Navarro L, Flores R, Peña L (2012) Transformation of Mexican lime with an intron hairpin construct expressing untranslatable versions of the genes coding for the three silencing suppressors of citrus tristeza virus confers complete resistance to the virus. Plant Biotechnol J 10:597–608

Song J, Smith SK, Hannon GJ, Joshua-Tor L (2004) Crystal structure of Argonaute and its implication for RISC slicer activity. Science 305:1434–1437

Swanton C, Marani M, Pardo O, Warne PH, Kelly G, Sahai E, Elustondo F, Chang J, Temple J, Ahmed AA, Brenton JD, Downward J, Nicke B (2007) Regulators of mitotic arrest and ceramide metabolism are determinants of sensitivity to paclitaxel and other chemotherapeutic drugs. Cancer Cell 11:498–512. https://doi.org/10.1016/j.ccr.(2007)04.011

Tabassum B, Sher Z, Tariq M, Khan A, Shahid N, Bilal M, Ramzan M, Shehzad IM, Nasir IA, Husnain T (2013) Overview of acquired virus resistance in transgenic plants. Exp Agri Hortic 2:12–28

Takeda A, Iwasaki S, Watanabe T, Utsumi M, Watanabe Y (2008) The mechanism selecting the guide strand from small RNA duplexes is different among Argonaute proteins. Plant Cell Physiol 49:493–500

Tenllado F, Llave C, Díaz-Ruíz JR (2004) RNA interference as a new biotechnological tool for the control of virus diseases in plants. Virus Res 102:85–96

Timmermans M, Das O, Messing J (1994) Geminivirus and their uses as extrachromosomal replicons. Annu Rev Plant Physiol 45:79–112

Tyagi H, Rajasubramaniam S, Rajam MV, Dasgupta I (2008) RNA interference in rice against Rice tungro bacilliform virus results in its decreased accumulation in inoculated rice plants. Transgenic Res 17:897–904

Vanitharani R, Chellappan P, Fauquet CM (2003) Short interfering RNA-mediated interference of gene expression and viral DNA accumulation in cultured plant cells. Proc Natl Acad Sci USA 100:9632–9636. https://doi.org/10.1073/pnas.1733874100

Vaucheret H, Béclin C, Fagard M (2001) Post-transcriptional gene silencing in plants. J Cell Sci 114:3083–3091

Voinnet O (2005) Induction and suppression of RNA silencing: insights from viral infections. Nat Rev Genet 6:206–220

Voinnet O, Baulcombe DC (1997) Systemic signalling in gene silencing. Nature 389:389–553

Voinnet O, Pinto YM, Baulcombe DC (1999) Suppressor of gene silencing: a general strategy used by diverse DNA and RNA viruses of plants. Proc Natl Acad Sci 96:14147–14152

Walsh HA, Vanderschuren H, Taylor S, Rey MEC (2019) RNA silencing of South African cassava mosaic virus in transgenic cassava expressing AC1/AC4 hp- RNA induces tolerance. Biotechnol Rep (Amst) 24:e00383. https://doi.org/10.1016/j.btre.2019.e00383

Wang MB, Brosnan C, Carroll BJ, Waterhouse PM (2006) RNA interference-inducing hairpin RNAs in plants act through the viral defense pathway. EMBO Rep 7:1168–1175

Wang XB, Jovel J, Udomporn P, Wang Y, Wu Q, Li WX, Gasciolli V, Vaucheret H, Ding SW (2011) The 21-nucleotide, but not 22-nucleotide, viral secondary small interfering RNAs direct potent antiviral defense by two cooperative argonautes in *Arabidopsis* thaliana. Plant Cell 23:1625–1638

Wani SH, Sanghera GS, Singh NB (2010) Biotechnology and plant disease control-role of RNA interference. Am J Plant Sci 1:55–68

Waterhouse PM, Fusaro AF (2006) Plant science. Viruses face a double defense by plant small RNAs. Science 313:54–55

Waterhouse PM, Graham MW, Wang MB (1998) Virus resistance and gene silencing in plants can be induced by simultaneous expression of sense and antisense RNA. Proc Natl Acad Sci 95:13959–13964

Wesley SV, Helliwell CA, Smith NA (2001) Construct design for efficient, effective and high throughput gene silencing in plants. Plant J 27:581–590

Whitehurst AW, Bodemann BO, Cardenas J, Ferguson D, Girard L, Peyton M, Minna JD, Michnoff C, Hao W, Roth MG, Xie XJ, White MA (2007) Synthetic lethal screen identification of chemosensitizer loci in cancer cells. Nature 446:815–819. https://doi.org/10.1038/nature05697

Xia H, Mao Q, Eliason S, Harper SQ, Martins IH, Orr HT, Paulson HL, Yang L, Kotin RM, Davidson BL (2004) RNAi suppresses polyglutamine-induced neurodegeneration in a model of spinocerebellar ataxia. Nat Med 10:816–820

Xu C, Gui Q, Chen W, Wu L, Sun W, Zhang N, Xu Q, Wang J, Fu X (2011) Small interference RNA targeting tissue factor inhibits human lung adenocarcinoma growth in vitro and in vivo. J Exp Clin Cancer Res 30:63

Yang Y, Sherwood TA, Patte CP, Hiebert E, Polston JE (2004) Use of tomato yellow leaf curl virus (TYLCV) Rep Gene Sequences to engineer TYLCV resistance in tomato. Phytopathology 94:490–496. https://doi.org/10.1094/PHYTO.2004.94.5.490

Yoo BC, Kragler F, Varkonyi-Gasic E, Haywood V, Archer Evans S, Lee YM, Lough TJ, Lucas WJ (2004) A systemic small RNA signaling system in plants. Plant Cell 16:1979–2000

Zamore PD, Tuschl T, Sharp PA, Bartel DP (2000) (2000) RNAi: double-stranded RNA directs the ATP-dependent cleavage of mRNA at 21 to 23 nucleotide intervals. Cell 101:25–33. https://doi.org/10.1016/S0092-8674(00)80620-0

Zhang X, Sato S, Ye X, Dorrance AE, Morris TJ, Clemente TE, Qu F (2011) Robust RNAi-based resistance to mixed infection of three viruses in soybean plants expressing separate short hairpins from a single transgene. Phytopathology 101:1264–1269

Zhang ZY, Wang YG, Shen XJ, Li L, Zhou SF, Li WC, Fu FL (2013) RNA interference-mediated resistance to maize dwarf mosaic virus. Plant Cell Tissue Organ Cult 113:571–578

Zhang L, Xiang Y, Chen S, Shi M, Jiang X, He Z, Gao S (2022) Mechanisms of MicroRNA biogenesis and stability control in plants. Front Plant Sci 13:844149. https://doi.org/10.3389/fpls.(2022)844149

Zhao Y, Yang X, Zhou G, Zhang T (2020) Engineering plant virus resistance: from RNA silencing to genome editing strategies. Plant Biotechnol J 18:328

Epigenomics for Nutrient Use Efficiency: Status and Future Applications for Improving Crop Nitrogen Response

9

Lekshmy Sathee, M. S. Nimmy, Archana Singh, Sweta Kumari, T. Vinutha, Kumar Durgesh, and Shailendra K. Jha

Abstract

An epigenome is a collective name for the biochemical alterations to nuclear DNA, changes in noncoding RNA biogenesis and histone protein modification. Without altering the underlying nucleotide sequence, these modifications frequently cause gene expression variations. The variations in chromatin structure brought on by the modifications could also alter how the genome functions or behaves. Epigenomic mechanisms regulate gene expression patterns and are influenced by various environmental factors, including nutrient availability. Nutrient use efficiency (NtUE) is the ability of an organism to acquire, assimilate, and utilize nutrients effectively, with nitrogen being one of the most essential components for the growth of plants. Efficient nitrogen uptake is crucial for optimizing crop yield, reducing fertilizer usage, and minimizing environmental impact caused by nitrogen depletion. Both endogenous and external stimuli can create epigenomic alterations and phenotypic plasticity in plants. Plant epigenomics has undergone a revolution owing to the advancement of techniques, such as next-generation sequencing methods. The initial studies concentrated on genes and DNA methylation at cell level. However, with the advent of technology, the focus is shifting towards mapping the entire epigenome of an organism. Apart from their functions in RNA degradation and translational repression,

L. Sathee (✉)
Division of Plant Physiology, ICAR-Indian Agricultural Research Institute, New Delhi, India

M. S. Nimmy
ICAR-National Institute of Plant Biotechnology, New Delhi, India

A. Singh · S. Kumari · T. Vinutha
Division of Biochemistry, ICAR-Indian Agriculture Research Institute, New Delhi, India

K. Durgesh · S. K. Jha
Division of Genetics, ICAR-Indian Agriculture Research Institute, New Delhi, India

K. Goswami et al. (eds.), *Non-coding RNAs for Crop Improvement*,
https://doi.org/10.1007/978-981-97-7213-1_9

small RNAs are potent in altering chromatin and target gene expression through RNA interference (RNAi). Some epigenetic modifications are heritable and can thus determine evolutionary adaptation for stress resilience. Understanding the responsible epigenetic machinery that maybe used to improve resource use efficiency to address nutritional/food security challenges is imperative.

Keywords

NUE · Nutrient · Nitrogen · Epigenome · RNAi · Small RNAs

9.1 Introduction

Understanding the epigenetic mechanisms underlying nutrient use efficiency, particularly the response to macronutrient nitrogen is mordanting developing strategies for crop improvement. The expression of nitrogen-responsive genes is modulated by DNA methylation, histone changes, and non-coding RNAs. The current status of epigenomic studies for nutrient use efficiency, specifically nitrogen response, involves both experimental and observational approaches. Experimental approaches include perturbation of the nutrient availability, particularly nitrogen, and investigating the resulting epigenetic alterations and their effect on gene expression and plant performance. Observational studies aim to identify correlations between NUE-related traits and epigenetic marks across the crop varieties and species. By manipulating nitrogen availability and studying the associated epigenomic patterns, one can gain insights towards the regulatory mechanisms involved in nitrogen response and identify potential epigenetic marks, landscapes, or targets for the crop improvement.

Similar to animal systems, epigenomics can be used as a source of novel variation for trait improvement in plants too (Gallusci et al. 2017). "Epigenetics" is a term that has been in prevalence since 1942, and refers to heritable yet reversible changes in gene expression. In the vernalization process, these epigenetic alterations can be passed down through mitosis. However, imprinting, which includes passing epigenetic variants straight through meiosis to the progeny, is also another possibility. According to Lee et al. (2019), epigenetic changes influence various plant developmental processes, including the ability to adapt to abiotic stressors. These variations also influence plant phenotypic plasticity.

CG/CHG (symmetric) or CHH (asymmetric) Cytosine methylation of genomic DNA is commonly seen in higher plants (Jin et al. 2011). Epigenetic changes also include histone modifications, RNAi by small RNAs (sRNAs), etc. (Fragou et al. 2011). The DNA methyltransferase 1 (MET1), catalyzes CG methylation while, chromo methylase 2 (CMT2), CMT3 and domains rearranged methyltransferase 2 (DRM2), catalyzes CG and CHH methylation (Fig. 9.1). Epialleles, which can control gene expression, are created due to the segregation of genomes with different methylation patterns and create stable hereditary epigenetic variations in genes (Gahlaut et al. 2020).

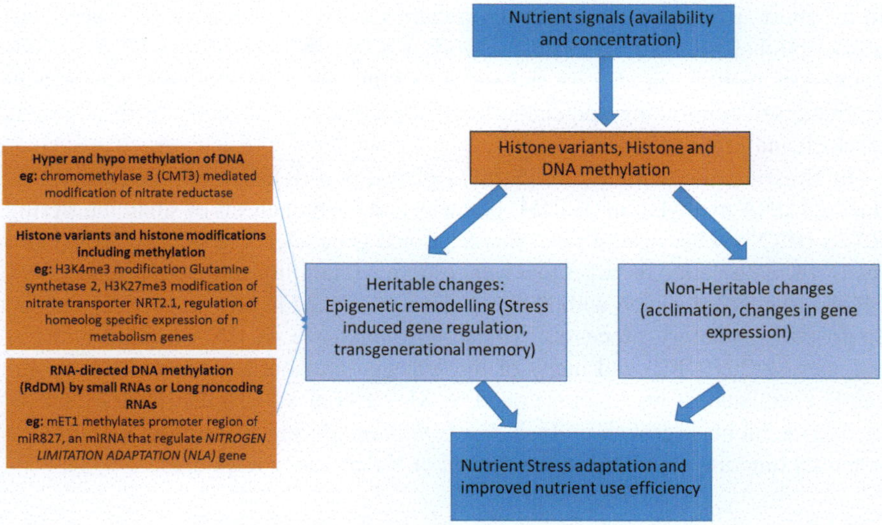

Fig. 9.1 Plant epigenetic changes triggered by nutrient signals

Different physiological and agronomic traits of crop plants are influenced by variations in methylation patterns (Wang et al. 2015). Plants subjected to stress can establish epigenetic memory, leading to epiallele formation (Bartels et al. 2018). The development of "epigenetic recombinant inbred lines (epiRILs) and epigenetic hybrids (epi hybrids)" uses epialleles, used for epi-breeding. Both chemical modifiers and epigenetic quantitative trait loci (epiQTLs) are used in epibreeding. Epigenetic alterations and epigenome editing can be used to develop climate-resilient plants, be applied (Singh et al. 2021). Epi-interventions can help produce crops ready for future challenges. These include epi-priming for retaining stress memory, epiQTL mapping for complex agronomic traits like yield, seed development, and plant growth, and epi-fingerprinting for assessing product quality. In order to change the epigenome specifically, epigenome editing uses cutting-edge techniques like clustered regularly interspaced short palindromic repeats (CRISPR)/CRISPR-associated protein 9 (Cas9) (Miglani 2017). The majority of such studies are focused on the model plant *Arabidopsis*, for change in development, or late flowering and for tolerance to drought (Gallego-Bartolomé et al. 2018; Papikian et al. 2019; Roca Paixão et al. 2019). Epialleles have an important role in determining plant development, shoot/root architecture, and nutrient uptake (Singh et al. 2021).

Different types of non-coding RNAs, that range in length from short to long, are known to regulate gene expression, defense against foreign genetic material and genome integrity. Small RNAs direct Argonaute-protein complex towards RNA scaffolds and thereby recruit DNA and histone methyltransferases, alter chromatin structure, and inhibit transcription (Holoch and Moazed 2015). Furthermore, new findings imply that chromatin-associated long non-coding RNA scaffolds, apart from short RNAs, also attract chromatin-modifying complexes. These co-transcriptional silencing mechanisms work together to create potent RNA surveillance systems that can identify and suppress improper transcription events as well as create self-reinforcing epigenetic loops that serve as a memory of these events (Holoch and Moazed 2015).

In Brassica and *Arabidopsis thaliana* plants, a correlation between the RNA-directed DNA methylation (RdDM) pathway and the biogenesis of small interfering RNAs (siRNAs) because of post-transcriptional gene silencing (PTGS) along with the DNA methylation of target loci were reported (Hamilton and Baulcombe 1999). Many proteins associated with N metabolism are linked to post-transcriptional and epigenetic regulatory processes. Transcript abundance of RNA-dependent RNA polymerase 2 (RDR2) and many of the nitrogen metabolism genes were significantly correlated. It was demonstrated by Meyer et al. (2019) that RDR2 contributed to the biomass growth in *Arabidopsis thaliana* during N shortage, signifying a potential function of RdDM in the control of N deficiency responses.

9.2 Modes of Epigenome Modifications

Both genetic and non-genetic factors can contribute to plant epigenetic changes (Hofmeister et al. 2017) (Fig. 9.1). Epigenetic diversity is created by the interaction of environment and epigenetic processes, the interaction of genotype and epigenetic processes and can be induced by chemicals. Chromatin methylation pattern alterations and siRNA generation are examples of genetic origins of epialleles that result in RdDM. The primary epigenetic mark for the epigenetic regulation of gene expression in plants and animals is DNA methylation. DNA methylation is present in plants at the position of cytosine in symmetric C.G. and C.H.G. bases as well as asymmetric CHH nucleotide sequences. Intriguingly, methylation patterns that differ significantly from those found in mammals are preferred in transposons and other repetitive DNA sequences in plants. Domains Rearranged Methyltransferase 2 (DRM2) catalyzes de novo DNA methylation in plants. However, three separate mechanisms work together to maintain DNA methylation. The conserved methyl transferase (MET1) performs symmetric methylation in collaboration with the VIM cofactor family (Johnson et al. 2014). However, chromo methylase 3 (CMT3), a DNA methyl transferase peculiar to plants, performs asymmetric CHH methylation with the aid of Kryptonite (KYP)-dependent histone-3 lysine-9 dimethylation (H3K9me2) (González et al. 2013). On the other hand, the RdDM machinery, which uses the siRNA-argonaute 4 (AGO4) complex, establishes C.H.G. methylation (Matzke et al. 2015). The ability of DNA methylation (hypo-and hypermethylation) to calm down the various environmental stresses is crucial for improving the capacity of plants for rapid acclimatization. For instance, the offspring of rice treated with azacytidine, a DNA methyltransferase inhibitor, exhibited acquired resistance to *Xanthomonas oryzae* (Akimoto et al. 2007).

Plants can use the RdDM pathway, frequently called chromatin modification, to modify DNA methylation under the control of siRNAs produced from dsRNA precursors (Zhang and Zhu 2011). Plant-specific Pol IV, Pol V, RNA-dependent RNA polymerase 2 (RDR2), and several auxiliary components like dicer-like 3 (DCL3) produce short dsRNAs. DCL3 cleaves these short dsRNAs into 24 nucleotide-long siRNAs. These siRNAs are then broken down and loaded into the AGO4 or AGO6 effector complex to carry out the chromatin modification through targeted methylation (Chellappan et al. 2010). For instance, Phytoene Desaturase (PDS) gene chromatin silencing in *Arabidopsis* led to abnormalities in flower development in the fca and fpa flowering time regulator mutants. The endogenous P.D.S. locus's asymmetric methylation occurred along with this repression (Deleris et al. 2016). Additionally, reports have demonstrated the role of miRNAs in DNA methylation. According to (Bao et al. 2019), miR165/166 regulates DNA methylation and controls PHB and PHV expression. (Chellappan et al. 2010) found that the miR2883 and miR2328 produced two sets of sRNA (21 and 27 nt), with the longer sRNAs controlling the DNA methylation of target genes. Additionally, it was shown that rice produces two sRNAs (21 and 24 nucleotide long miRNA), with the longer one suppressing its target gene by DNA methylation. This 24-nucleotide long miRNA played a direct role in the cytosine methylation of the loci from where it was derived. Interestingly, miRNAs can methylate their target genes to silence them (Wu et al. 2009). The discovery of miRNA-dependent DNA methylation in the extinct species *P. patens* (Khraiwesh et al. 2010) shows that this gene regulation is old and conserved. A chromatin modification complex called PRC2 interacts with two lncRNAs called COLDAIR and COOLAIR to repress the Flowering Locus C (FLC) by trimethylating histone H3 lysine (Kehr 2013) of chromatin at its locus in plants, which has been linked to the epigenetic regulation of vernalization (Heo and Sung 2011). Mangrove plants were denser and higher along rivers than they were in salt marsh areas. This different morphological variation was caused by the riverside mangrove's hypermethylation as opposed to the salt marsh mangrove (Lira-Medeiros et al. 2010).

Additionally, studies imply that DNA methylation through epigenetically controlled transgenerational inheritance significantly reduces the effects of the numerous environmental perturbations. For instance, even in unstressed settings, transgenerational inheritance of altered DNA methylation was observed in plants exposed to heavy metals (Cong et al. 2019). Additionally, locus-specific DNA methylation alternations were discovered in rice leaves subjected to N deficiency. Interestingly, 50% of the methylation pattern changes were recovered in the offspring, resulting in increased tolerance to N deprivation (Kou et al. 2011). Methylation of the H3 histone K4 and K36 sites often stimulates transcription, whereas methylation of the H3K9 and H3K27 residues typically represses transcription (Volpe et al. 2002). Histone lysine methyltransferases, or HKMTs, are responsible for methylating histones.

A similar histone alteration that has received attention and is connected to higher transcriptional activity is the acetylation of the H3 tail. Chromatin is decondensed as a result of HATs, or histone acetyltransferases, acetylating H2B and H3 to neutralize the positively charged lysine residues, allowing transcription factors to access DNA more readily (Eberharter and Becker 2002). The removal of acetyl groups from lysine residues by histone deacetylases (HDACs), performs the opposite of what HATs do by strengthening the interaction between lysine residues and phosphate group of the DNA backbone, rendering DNA inaccessibility for transcription (Fig. 9.1) (Wolffe 1996). In a mutant screen, a mutant of *Arabidopsis* appeared with enhanced expression of auxin-responsive genes even when auxin was absent. Surprisingly, in this mutant, none of the auxin-responsive genes had changed. A histone deacetylase named HDA6 was found to be altered, suggesting that HDA6 is involved in the gene silencing of auxin-responsive genes (Murfett et al. 2001).

Interestingly, HDA6 aids siRNA and improves DNA methylation while serving as an essential part of the RdDM mechanism. Thus, by altering the structure of chromatin and its interactions with transcriptional regulatory proteins, histone changes control general transcriptional processes. These covalent changes may follow a heritable epigenetic coding to affect gene expression during various abiotic or biotic stimuli. For instance, trimethylation of H3K4 and H3 acetylation has been linked to the up-regulation of stress-responsive respiratory genes in submerged rice seedlings, such as pyruvate decarboxylase 1 (PDC1) and alcohol dehydrogenase 1 (ADH1) (Tsuji et al. 2006). The *Arabidopsis* plant has also undergone similar changes that have boosted the expression of drought-response (Kim et al. 2008). The FLC (Flowering Locus) gene in *Arabidopsis* was also shown to be silenced by cold stress by changing the levels of H3 acetylation and H3K9/K27 methylation (Bastow et al. 2004). Similar to salt and cold shocks, H3 and H4 acetylation as well as other histone modifications, such as the phosphorylation of H3's S10 residue, were shown to be induced at the locus of stress-responsive genes (Sokol et al. 2007).

In addition to histone modification, nucleosome assembly protein 1 (NAP1) and chromatin assembly factor 1 (CAF1) control nucleosome maintenance and establishment as well as transcriptional regulation (Sudarsanam and Winston 2000). In *Arabidopsis*, FASCIATA1, FASCIATA2, and MS1 make up CAF1. Because the *fascia* mutants have faulty shoot and root apical meristem development, CAF1 is thought to be involved in plant cellular differentiation. Many chromatin remodelling enzyme complexes, including SWI2/SNF2, ISWI, and CHD have an ATPase domain. These complexes catalyse the chromatin remodelling activity while assisting transcriptional responses at the expense of ATP (Reina and Cavalieri 2020). For instance, a member of the CHD family, the PICKLE gene, encodes an ATP-dependent chromatin remodeler. Due to a reduction in meristematic activity, the *Arabidopsis* mutant (pkl) displayed a "pickle root" phenotype with opaque, thick, and, green primary roots with no or adventitious or lateral branches (Ogas et al. 1999; Aichinger et al. 2011). The pickle gene promotes trimethylation of H3 lysine 27, which represses several tissue-specific and development-related genes (Aichinger et al. 2011).

It is widely established that plants exposed to various conditions may have changes in their epigenomes (Gallusci et al. 2017). These epigenetic differences offer a theoretical underpinning for stress memory, which -enables plants to react effectively to reiterating stress events. Crop plants may benefit from using epigenetic modifications to maintain stress memory. sRNAs, RdDM components, histone variations, histone modification enzymes, and chromatin-remodelling factors are examples of epigenetic regulators affected by abiotic stress signals regarding their expression and function (Chinnusamy and Zhu 2009). While some of these alterations are temporary, others are heritable epigenetic modifications. The acclimatization response is mediated by transient chromatin changes. The generational and transgenerational stress memory that is produced by heritable epigenetic alterations is durable and can be used for crop breeding. In plants, transitory acclimation brought on by stress-induced signals, genes, and plant metabolites results in stress memory retention for a brief period against abiotic stress(es). The duration how long the short-term stress memory lasts depends on the half-life of these stress-induced biomolecules.

On the other hand, stable stress memory is produced by procedures including phenology and morphological reprogramming. Only a few natural epialleles are available in literature (Quadrana and Colot 2016; Song et al. 2019). Natural epialleles are bearable sources of epigenetic changes in plants. A reservoir of these variations linked to agronomic features and variations brought on by abiotic stress offers an intriguing source for climate-resilient epi-breeding. Targeted epigenetic changes call for techniques for editing the epigenome that are not well-developed in plant systems and require prior knowledge of the target sequences.

Additionally, it requires transformation protocols, which are constrained by the absence of induced modification inheritance (Kungulovski et al. 2015). EpiRIL production can result in non-targeted epigenetic variability. Using mutants with different epigenomes, two distinct *Arabidopsis* epiRILs were produced, laying the groundwork for non-targeted epigenetic diversity (Agarwal et al. 2020) (Singh et al. 2021). However, chemically induced epigenetic modifications (such as DNA methylase inhibitors like zebularine or azacytidine) can be created in the parental epigenome of crop plants where no such mutants exist (Séré and Martin 2020). According to (Hewezi 2020), the three primary sources of epigenetic diversity include (1) environmental variations brought on by the interaction of the environment and epigenetic processes, (2) genetic variations brought on by genotype interaction and epigenetic processes, and (3) chemical mutation-induced variations. In general, both genetic and non-genetic factors can contribute to epigenetic changes in plants. Maintaining methylation statuses, stress-induced chromatin modifications, or sRNA off-target effects can all be affected by non-genetic causes, such as developmental and environmental variables.

9.3 Epigenomics of Environmental Adaptation

Model plants have been used to identify and characterize several genes that regulate mineral homeostasis. According to recent studies, epigenetic factors also play a role in regulating the root system's flexibility during development and regulating cellular metabolism in response to nutrient deficiency. Consequently, it is crucial to investigate the functions of epigenetic regulation and traditional gene regulation to understand the regulatory network of stress-responsive genes in nutrient homoeostasis. In plants, sRNA-mediated epigenetic regulation in response to various abiotic stresses has been studied at the whole-genome level. The microRNA profiles of Brachypodium, rice, maize, wheat, and *Arabidopsis* (Fahlgren et al. 2007; Lu et al. 2008; Zhang et al. 2009; Zhai et al. 2013; Cadavid et al. 2020; Das et al. 2023) under various stresses are available. These are excellent resources for comprehending the regulation of stress signalling by sRNAs. Moreover, chromatin modifications are essential in conferring stress tolerance (Ganguly et al. 2017). Using chromatin immune precipitation (ChIP), histone modifications were detected in *Arabidopsis*, revealing the activation of genes coding for RD29A, RD29B, RD20, and AP2 (Kim et al. 2008). COLD-REGULATED genes such as COR15A and ATGOLS3 which codes for galactinol synthase, showed association with histone modifications in response to low temperature (Baker et al. 1994). Salt-tolerant and salt-stressed callus cells of *Medicago sativa* respond to ionic shifts in the nucleus by altering histone protein acetylation levels. This adaptive mechanism allows chromatin to function effectively under saline intranuclear conditions (Waterborg et al. 1989). Notably, the processes of histone acetylation/deacetylation and histone methylation/demethylation are believed to modulate hormonal signals governed by stress-responsive genes (Chinnusamy et al. 2008; Chinnusamy and Zhu 2009). For example, chromatin modification induction has been associated with the upregulation of stress-responsive genes in *Arabidopsis* and tobacco in response to ABA (Sokol et al. 2007).

In addition to essential minerals, soil also contains minerals that are toxic to plants, such as lead (Pb) and cadmium (Cd) Plants often develop detoxification mechanisms in response to this duress because of extensive gene expression regulation. An array of these modifications is credited to epigenetic controls. In maize roots subjected to Pb stress, genes associated with root development were differentially methylated (Ding et al. 2014). In rice plants exposed to salinity stress, DNA cytosine methylation levels of stress associated genes plays important role in stress adaptability (Karan et al. 2012). Furthermore, the rice and maize epigenomes offer an opportunity to examine the mechanism of epigenetic regulation in cereals (He et al. 2014). The development of high NtUE crops can be achieved by CRISPR-Cas technology, for example dCas9-SunTag system (Sathee et al. 2022) can be employed to manoeuvre DNA methylation and targeted gene activation in plants.

9.4 Epigenomics of Nutrient Stress Adaptation

Root architectural changes are very important for plants in adapting to different nutrient regimes. In agreement with this, root epidermal meristems are epigenetically reprogrammed to thrive in changing environments (González et al. 2013). The transcription factors associated with maintenance of root quiescent centre (QC) like scarecrow (SCR) are epigenetically regulated by histone acetylation (Jia et al. 2015). ASET Domain Group (SDG) family gene SDG2 is associated with trimethylation of histone lysine residues at the genome level and interacts with chromatin assembly factor (CAF1) to maintain root cell chromatin structure. SDG2 activity is essential for primary root stem cell niche (SCN) and for creating lateral root SCN (Sirohi et al. 2016). Many of the phosphate starvation responses (PSR) genes are induced by the nucleosome remodelling activity of SWI/SNF and histone acetylation by GENERAL CONTROL NON-REPRESSIBLE 5 (GCN5) (Hu et al. 2015). In *Arabidopsis*, actin-related protein ARP6-dependent H2AZ placement on chromatin modulates the transcription of a subset of PSR genes. *Arabidopsis arp6* mutant exhibits several PSR phenotypes, like primary root growth arrest, starch deposition, root hair proliferation, and high phosphatase activity (Deal et al. 2007). Several chromatin remodelling genes were upregulated in P-deficient wheat, including SWI3C, SWI3D, and GCN5-related N-acetyltransferase (Smith et al. 2010). In a recent study by Kumar et al. (2022) changes in cytosine DNA methylation patterns were observed in P-deficient rice seedlings undergoing a time-course experiment. In rice genotypes Pusa-44 and NIL-23, the role of Pup1 on DNA (de)methylation-mediated regulation of gene expression under P-starvation has been deciphered. A total of 13–15% of cytosines were found to be methylated, which were significantly more under stress (Kumar et al. 2022). Additionally, Malinas et al. (2022) provide a comprehensive review of nitrogen dynamics in wheat, focusing on key traits involved in nitrogen uptake, assimilation, distribution, and utilization efficiency. The review also explores the interactions among genetics (G), environment (E), and management (M). The study found an abundance of differentially methylated regions (DMRs) in transposable elements near significantly induced PSR genes. In addition, hypermethylation was observed in rice plants exposed to low P conditions for an extended period. It was hypothesized that the RDR6–RdDM pathway was responsible for these alterations, independent of the DCL3a–RdDM pathway. Widiez et al. (2011) demonstrated the function of posttranslational chromatin modification in plant nitrogen uptake and discovered high nitrogen insensitive 9 (HNI9) inhibits NRT2.1 expression in roots. HNI9 encodes the IWS1, which acts as a mediator of methylation and is a component of the RNA polymerase II complex. Under nitrogen conditions, H3K27me3 levels increase on NRT2.1 chromatin in an HNI9-dependent manner, particularly at the cis-elements in the promoter region of NRT2.1, leading to its repression. Research by Kou et al. (2011) demonstrated that through transgenerational inheritance of heavy cytosine methylation in rice, the offspring of nitrogen-deficient plants exhibit greater tolerance to nitrogen deficiency. Similarly, the offspring of *Plantago lanceolata* exposed to nutrient-poor conditions produce more leaves and spikes, have longer leaves, and exhibit greater photosynthetic capacity under non-stressed conditions. Nutrient-deprived *Polygonum persicaria* plants also produce progeny with altered nutritional stress responses. Recent studies indicate that epigenetic markers of methylation and demethylation due to iron deficiency are maintained even after barley seedlings recover from iron deficiency, suggesting these modifications can be inherited by future generations (Mlynárová et al. 2007). Therefore, transgenerational epigenetic inheritance plays a crucial role in modulating the nutritional responses of progeny, enhancing their ability to acclimate to environmental disturbances.

9.5 Epigenomics of Plant Nitrogen Response

Exposure to N deficiency can stimulate epigenetic alterations in DNA, which can be passed down through "epigenetic memory" to succeeding generations. Numerous studies have sought to comprehend the memory mechanism for stress events in various cultivated species. However, whether and how this memory benefits plants in low-N environments remains to be determined. Epigenetic regulation of N-response via miRNAs and chromatin remodelling is reported in response to N treatments (Nischal et al. 2012; Séré and Martin 2020; Li et al. 2021; Islam et al. 2022). Epigenetic regulation involving a non-canonical structure known as a G4 sequence or G-quadruplex is relatively unexplored. In G-quadruplexes, Hoogsteen hydrogen bonding occurs in DNA and RNA sequences containing four brief segments of guanine (Burge et al. 2006; Sengupta et al. 2021). These G4 sequences are dispersed throughout the genome but are often abundant in promoter regions, gene UTRs, and telomeres (Griffin and Bass 2018). They significantly contribute to chromatin remodeling, gene control, epigenetic regulation, and genomic instability (Agarwal et al. 2020). Despite their importance, the occurrence, and functions of G4 DNA and G4 RNA in plant species have not been thoroughly investigated, with few reports on their role in stress responses (Kopec and Karlowski 2019). Cagirici et al. (2021) suggest that genome-wide studies of G-quadruplexes could advance the understanding of their biological implications and practical applications in plants. Recently, Sharma et al. (2023) explored the potential of G4 sequences in rice nitrogen use efficiency (NUE). They identified N-responsive genes containing G-quadruplex sequences, which are predicted to be involved in the epigenetic regulation of NUE through G-quadruplexes. Histone modifications, including methylation, phosphorylation, acetylation, and ubiquitination, are covalent posttranslational alterations of histone proteins. According to several studies (Dong et al. 2008; Cazzonelli et al. 2009; Wang et al. 2014; Li et al. 2015;), the histone methyltransferase AtSDG8 plays pleiotropic roles in plant growth and development, including pigment synthesis, flowering, shoot branching, defence, and carbon and light responses. In order to maintain proper H3K36me3 levels at gene loci encoding nitrate or nitrite reductases, AtSDG8 played a key role along with a regulatory role regarding the N assimilation gene AtASN1 (Li et al. 2015). N deficit raised H3K36me3 levels in 143 genes in wild-type plants compared to N-sufficient conditions, while H3K36me3 levels declined in 53 genes, with the numbers decreasing to

3 and 23 in the sdg8–5 mutants. Additionally, in sdg8–5 mutants, N responses of increased lateral root development and chlorophyll content were decreased. These findings imply that different histone alterations regulate N responses (Wang et al. 2014).

Nitrate transporters carry out nitrate absorption and translocation in plants. Nitrate transporters can be divided into two main categories: NRT1 (low-affinity transport system) and NRT2 (high-affinity transport system) (Wang et al. 2020; Wang et al. 2021a, b). While AtNRT2.1 is a critical component of the high-affinity transport system for nitrate absorption in the root, AtNRT1.1/CHL1 is a dual-affinity nitrate transporter in *Arabidopsis* that also functions as a transceptor sensing nitrate availability (Guo et al. 2003; Liu and Tsay 2003; Wang et al. 2020). The optimum N availability and the total plant N demand impacted the transcript level of AtNRT2.1 (Forde 2002). AtNRT2.1 expression was significantly downregulated in conditions of high N supply (Zhuo et al. 1999). AtHNI9 encodes a part of the RNA polymerase II complex (IWS1), which was implicated in the regulation of transcription by recruiting chromatin-remodelling agents or positioning H3K36me3 on histone proteins. 2348 genes showed differential expression due to AtHNI9's loss of function. Of these, 686 gene overlap with 2253 N-regulated genes, indicating that AtHNI9 is crucial for controlling the N response. Intriguingly, it was discovered that AtNRT2.1 significantly accumulated H3K27me3, a transcription repression marker, under high N conditions compared to low N conditions (Widiez et al. 2011). Repression of AtNRT2.1 by high N was broken by the loss-of-function of AtHNI9, which reduced the amount of H3K27me3 (Widiez et al. 2011). However, a different study (Bellegarde et al. 2017) discovered that the H3K27me3 level of AtNRT2.1 was higher under low N conditions than in high N conditions. AtHNI9 is also associated with ROS homeostasis by controlling the accumulation of H3K4me3 on a number of ROS-detoxifying genes. It is possible that AtHNI9 indirectly downregulates AtNRT2.1 by lowering cellular ROS levels (Bellegarde et al. 2017, 2019; Jacquot et al. 2020). *ZmCHB101*, found in maize, encodes a key member of the *SWI/SNF*-type ATP-dependent chromatin remodelling complex and plays a role in leaf development, dehydration, and ABA response. Under low N circumstances, ZmCHB101-RNAi lines showed faster root development and more biomass than WT plants. The promoter of ZmNRT2.1/2.2 contains nitrate-responsive cis-elements to which ZmCHB101 binds, thus preventing ZmNLP3.1 from attaching and activating ZmNRT2.1/2.2. The binding of ZmCHB101 is significantly reduced in the presence of N, helping ZmNLP3.1 to bind with ZmNRT2.1/2.2 leading it to up-regulation (Meng et al. 2020).

The availability of N also determines yield. For instance, rice yield can primarily be affected by the number of tillers, the number of grains per panicle, and the weight of the grains. Of these, the number of tillers responds most noticeably to N and shows an incremental trend along with increased N fertilizer. OsD14 (Guo et al. 2013; Zhang et al. 2013; Chen et al. 2018). OsNGR5 and AOsSPL14 are branch-inhibitory genes that can be activated by N supply at the mRNA and protein levels. OsNGR5 then attracts PRC2 to add H3K27me3 on these genes, inhibiting their expression and encouraging tiller bud development. Additionally, under varied N fertilizer concentrations, over-expression of OsNGR5 considerably boosted rice yield and tiller number, demonstrating its considerable potential for breeding novel rice varieties with higher NUE (Wang et al. 2013). DNA methylation typically inhibits the transcription of genes. More than 95% of the differentially methylated regions (DMRs) between low nitrogen and control conditions were hypomethylated. Consequently, nitrogen deprivation reduced the overall methylated CG content in maize roots from 26.6% to 12.0% (Mager and Ludewig 2018). In contrast, the total methylation level in rice leaves was not significantly altered by nitrogen deprivation, with locus-specific hypomethylation and hypermethylation occurring at comparable rates. This disparity may result from differing DNA methylation reactions to N in other species or from distinct experimental approaches.

In the plastid, nitrite is transported and converted to ammonium by nitrite reductase. This ammonium is then incorporated into glutamate to produce glutamine via the glutamine synthetase/glutamine-2-oxoglutarate aminotransferase (GS/GOGAT) cycle (Wang et al. 2021a, b). *AtCMT3* encodes a methyltransferase that is essential for embryonic development and whose activity may be increased by AtSIZ1 through sumoylation (Kim et al. 2015). Nitrate reductase gene AtNIA2's promoter and gene sequence can be bound by AtCMT3, which methylates CHG sequences. It is interesting to note that ammonium treatment can totally stop this process, which increases the transcript level of AtNIA2 (Kim et al. 2015). The three TaGS2 homoeologs (TaGS2-A, TaGS2-B, and TaGS2-D) are primarily expressed in wheat leaves and to a lesser extent in roots, with TaGS2-B showing higher expression levels than TaGS2-A and TaGS2-D. Notably, the expression patterns of these homoeologs correlate with H3K4me3 deposition levels and DNA methylation levels. This suggests that DNA methylation and histone modifications may influence the tissue-or homoeolog-specific expression patterns of genes involved in the nitrogen response (Sabari et al. 2017).

To summarise, epigenetics research, particularly nitrogen response, is an emerging and rapidly evolving domain. By deciphering the epigenetic mechanisms that influence gene expression and plant adaptation to nitrogen availability, this field holds great potential for advancing sustainable agriculture and improving crop productivity in the face of changing environmental conditions.

9.6 Conclusions

Since not many epigenetic changes happen naturally, there are several ways to create them. Targeted epigenetic changes call for techniques for editing the epigenome that is still in its infancy in plant systems. Applications of epigenomics in the context of nutrient use efficiency and nitrogen response hold immense promise. Firstly, understanding the epigenetic regulation of nitrogen response can directly help in developing high NUE genotypes through targeted breeding or genetic engineering approaches. Secondly, epigenomic studies can provide valuable information for optimizing nitrogen fertilizer application strategies, ensuring efficient nutrient utilization while minimizing environmental impacts. Additionally, epigenomics can contribute to precision agriculture by enabling the development of epigenetic biomarkers or signatures that can predict NUE performance in crops. Such biomarkers could facilitate the selection of high-yielding and nitrogen-efficient cultivars, helping farmers to make informed decisions regarding fertilizer application and crop management.

References

Agarwal G, Kudapa H, Ramalingam A, Choudhary D, Sinha P, Garg V, Singh VK, Patil GB, Pandey MK, Nguyen HT, Guo B, Sunkar R, Niederhuth CE, Varshney RK (2020) Epigenetics and epigenomics: underlying mechanisms, relevance, and implications in crop improvement. Funct Integr Genomics 20:739–761

Aichinger E, Villar CBR, di Mambro R, Sabatini S, Köhler C (2011) The CHD3 chromatin remodeler PICKLE and polycomb group proteins antagonistically regulate meristem activity in the Arabidopsis root. Plant Cell 23(3):1047–1060

Akimoto K, Katakami H, Kim HJ, Ogawa E, Sano CM, Wada Y, Sano H (2007) Epigenetic inheritance in rice plants. Ann Bot 100(2):205–217

Baker SS, Wilhelm KS, Thomashow MF (1994) The 5′-region of *Arabidopsis thaliana* cor15a has cis-acting elements that confer cold-, drought-and ABA-regulated gene expression. Plant Mol Biol 24:701–713

Bao S, Owens RA, Sun Q, Song H, Liu Y, Eamens AL, Feng H, Tian H, Wang MB, Zhang R (2019) Silencing of transcription factor encoding gene StTCP23 by small RNA.S derived from the virulence modulating region of potato spindle tuber viroid is associated with symptom development in potato. PLoS Pathog 15(12):e1008110

Bartels A, Han Q, Nair P, Stacey L, Gaynier H, Mosley M, Huang QQ, Pearson JK, Hsieh TF, An YQC, Xiao W (2018) Dynamic DNA methylation in plant growth and development. Int J Mol Sci 19(7):2144

Bastow R, Mylne JS, Lister C, Lippman Z, Martienssen RA, Dean C (2004) Vernalization requires epigenetic silencing of FLC by histone methylation. Nature 427(6970):164–167

Bellegarde F, Gojon A, Martin A (2017) Signals and players in the transcriptional regulation of root responses by local and systemic N signalling in *Arabidopsis thaliana*. J Exp Bot 68(10):2553–2565

Bellegarde F, Maghiaoui A, Boucherez J, Krouk G, Lejay L, Bach L, Gojon A, Martin A (2019) The chromatin factor HNI9 and ELONGATED HYPOCOTYL5 maintain R.O.S. Homeostasis under high nitrogen provision. Plant Physiol 180(1):582–592

Burge S, Parkinson GN, Hazel P, Todd AK, Neidle S (2006) Quadruplex DNA: sequence, topology and structure. Nucleic Acids Res 34(19):5402–5415

Cadavid IC, Guzman F, de Oliveira-Busatto L, de Almeida RMC, Margis R (2020) Transcriptional analyses of two soybean cultivars under salt stress. Mol Biol Rep 47(4):2871–2888

Cagirici HB, Budak H, Sen TZ (2021) Genome-wide discovery of G-quadruplexes in barley. Sci Rep 11(1):7876

Cazzonelli CI, Cuttriss AJ, Cossetto SB, Pye W, Crisp P, Whelan J, Finnegan EJ, Turnbull C, Pogson BJ (2009) Regulation of carotenoid composition and shoot branching in Arabidopsis by a chromatin modifying histone methyltransferase, SDG8. Plant Cell 21(1):39–53

Chellappan P, Xia J, Zhou X, Gao S, Zhang X, Coutino G, Vazquez F, Zhang W, Jin H (2010) siRNAs from miRNA sites mediate DNA methylation of target genes. Nucleic Acids Res 38(20):6883–6894

Chen L, Zhao Y, Xu S, Zhang Z, Xu Y, Zhang J, Chong K (2018) OsMADS57, together with OsTB1, coordinates transcription of its target OsWRKY94 and D14 to switch its organogenesis to defence for cold adaptation in rice. New Phytol 218(1):219–231

Chinnusamy V, Zhu JK (2009) Epigenetic regulation of stress responses in plants. Curr Opin Plant Biol 12(2):133–139

Chinnusamy V, Gong Z, Zhu JK (2008) Abscisic acid-mediated epigenetic processes in plant development and stress responses. J Integr Plant Biol 50(10):1187–1195

Cong W, Miao Y, Xu L, Zhang Y, Yuan C, Wang J, Zhuang T, Lin X, Jiang L, Wang N, Ma J, Sanguinet KA, Liu B, Rustgi S, Ou X (2019) Transgenerational memory of gene expression changes induced by heavy metal stress in rice (*Oryza sativa* L.). BMC Plant Biol 19:1–14

Das S, Singh D, Meena HS, Jha SK, Kumari J, Chinnusamy V, Sathee L (2023) Long-term nitrogen deficiency alters the expression of miRNAs and alters nitrogen metabolism and root architecture in Indian dwarf wheat (*Triticum sphaerococcum* perc.) genotypes. Sci Rep 13(1):5002

Deal RB, Topp CN, McKinney EC, Meagher RB (2007) Repression of flowering in Arabidopsis requires activation of FLOWERING LOCUS C expression by the histone variant H2A.Z. Plant Cell 19(1):74–83

Deleris A, Halter T, Navarro L (2016) DNA methylation and demethylation in plant immunity. Annu Rev Phytopathol 54:579–603

Ding H, Gao J, Qin C, Ma H, Huang H, Song P, Luo X, Lin H, Shen Y, Pan G, Zhang Z (2014) The dynamics of DNA methylation in maize roots under Pb stress. Int J Mol Sci 15(12):23537–23554

Dong G, Ma DP, Li J (2008) The histone methyltransferase SDG8 regulates shoot branching in Arabidopsis. Biochem Biophys Res Commun 373(4):659–664

Eberharter A, Becker PB (2002) Histone acetylation: a switch between repressive and permissive chromatin. Second in review series on chromatin dynamics. EMBO Rep 3:224–229

Fahlgren N, Howell M, Kasschau K, Chapman E, Sullivan C (2007) High-throughput sequencing of Arabidopsis microRNAs: evidence for frequent birth and death of MIRNA genes. PLoS One 2(2):e219

Forde B (2002) Local and long-range signalling pathways regulating plant responses to nitrate. Annu Rev Plant Biol 53(1):203–224

Fragou D, Fragou A, Kouidou S, Njau S, Kovatsi L (2011) Epigenetic mechanisms in metal toxicity. Toxicol Mech Methods 21(4):343–352

Gahlaut V, Zinta G, Jaiswal V, Kumar S (2020) Quantitative epigenetics: a new avenue for crop improvement. Epigenomes 4(4):25

Gallego-Bartolomé J, Gardiner J, Liu W, Papikian A, Ghoshal B, Kuo HY, Zhao JMC, Segal DJ, Jacobsen SE (2018) Targeted DNA demethylation of the Arabidopsis genome using the human TET1 catalytic domain. Proc Natl Acad Sci USA 115(9):E2125–E2134

Gallusci P, Dai Z, Génard M, Gauffretau A, Leblanc-Fournier N, Richard-Molard C, Vile D, Brunel-Muguet S (2017) Epigenetics for plant improvement: current knowledge and modeling avenues. Trends Plant Sci 22(7):610–623

Ganguly DR, Crisp PA, Eichten SR, Pogson BJ (2017) The arabidopsis DNA methylome is stable under transgenerational drought stress. Plant Physiol 175(4):1893–1912

González RM, Ricardi MM, Iusem ND (2013) Epigenetic marks in an adaptive water stress-responsive gene in tomato roots under normal and drought conditions. Epigenetics 8:864–872

Griffin BD, Bass HW (2018) Plant G-quadruplex (G4) motifs in DNA and RNA.; abundant, intriguing sequences of unknown function. Plant Sci 269:143–147

Guo F-Q, Young J, Crawford NM (2003) The nitrate transporter AtNRT1. 1 (CHL1) functions in stomatal opening and contributes to drought susceptibility in Arabidopsis. Plant Cell 15(1):107–117

Guo S, Xu Y, Liu H, Mao Z, Zhang C, Ma Y, Zhang Q, Meng Z, Chong K (2013) The interaction between OsMADS57 and OsTB1 modulates rice tillering via DWARF14. Nat Commun 4(1):1566

Hamilton AJ, Baulcombe DC (1999) A species of small antisense RNA in posttranscriptional gene silencing in plants. Science 286(5441):950–952

He S, Yan S, Wang P, Zhu W, Wang X, Shen Y, Shao K, Xin H, Li S, Li J (2014) Comparative analysis of genome-wide chromosomal histone modification patterns in maize cultivars and their wild relatives. PLoS One 9:e97364

Heo JB, Sung S (2011) Encoding memory of winter by noncoding RNAs. Epigenetics 6(5):544–547

Hewezi T (2020) Epigenetic mechanisms in nematode-plant interactions. Annu Rev Phytopathol 58:119–138

Hofmeister BT, Lee K, Rohr NA, Hall DW, Schmitz RJ (2017) Stable inheritance of DNA methylation allows creation of epigenotype maps and the study of epiallele inheritance patterns in the absence of genetic variation. Genome Biol 18:1–16

Holoch D, Moazed D (2015) RNA-mediated epigenetic regulation of gene expression. Nat Rev Genet 16(2):71–84

Hu Z, Song N, Zheng M, Liu X, Liu Z, Xing J, Ma J, Guo W, Yao Y, Peng H, Xin M, Zhou DX, Ni Z, Sun Q (2015) Histone acetyltransferase GCN5 is essential for heat stress-responsive gene activation and thermotolerance in Arabidopsis. Plant J 84(6):1178–1191

Islam W, Tauqeer A, Waheed A, Zeng F (2022) MicroRNA mediated plant responses to nutrient stress. Int J Mol Sci 23(5):2562

Jacquot A, Chaput V, Mauries A, Li Z, Tillard P, Fizames C, Bonillo P, Bellegarde F, Laugier E, Santoni V, Hem S, Martin A, Gojon A, Schulze W, Lejay L (2020) NRT2.1 C-terminus phosphorylation prevents root high-affinity nitrate uptake activity in *Arabidopsis thaliana*. New Phytol 228(3):1038–1054

Jia Y, Tian H, Li H, Yu Q, Wang L, Friml J, Ding Z (2015) The *Arabidopsis thaliana* elongator complex subunit 2 epigenetically affects root development. J Exp Bot 66(15):4631–4642

Jin B, Li Y, Robertson KD (2011) DNA methylation: superior or subordinate in the epigenetic hierarchy? Genes Cancer 2(6):607–617

Johnson LM, Du J, Hale CJ, Bischof S, Feng S, Chodavarapu RK, Zhong X, Marson G, Pellegrini M, Segal DJ, Patel DJ, Jacobsen SE (2014) SRA-and SET-domain-containing proteins link RNA. Polymerase v occupancy to DNA methylation. Nature 507(7490):124–128

Karan R, DeLeon T, Biradar H, Subudhi PK (2012) Salt stress-induced variation in DNA methylation pattern and its influence on gene expression in contrasting rice genotypes. PLoS One 7:e40203. https://doi.org/10.1371/JOURNAL.PONE.0040203

Kehr J (2013) Systemic regulation of mineral homeostasis by micro RNAs. Front Plant Sci 4:145

Khraiwesh B, Arif MA, Seumel GI, Ossowski S, Weigel D, Reski R, Frank W (2010) Transcriptional control of gene expression by microRNAs. Cell 140(1):111–122

Kim JM, To, T.K, Ishida J, Morosawa T, Kawashima M, Matsui A, Toyoda T, Kimura H, Shinozaki K, Seki M (2008) Alterations of lysine modifications on the histone H3 N-tail under drought stress conditions in *Arabidopsis thaliana*. Plant Cell Physiol 49(10):1580–1588

Kim DY, Han YJ, Kim SI, Song JT, Seo HS (2015) Arabidopsis CMT3 activity is positively regulated by AtSIZ1-mediated sumoylation. Plant Sci 239:209–215

Kopec PM, Karlowski WM (2019) Sequence dynamics of pre-mRNA G-quadruplexes in plants front. Plant Sci 10:435550

Kou HP, Li Y, Song XX, Ou XF, Xing SC, Ma J, Von Wettstein D, Liu B (2011) Heritable alteration in DNA methylation induced by nitrogen-deficiency stress accompanies enhanced tolerance by progenies to the stress in rice (*Oryza sativa* L.). J Plant Physiol 168(14):1685–1693

Kumar S, Seem K, Kumar S, Vinod KK, Chinnusamy V, Mohapatra T (2022) *Pup1* QTL regulates gene expression through epigenetic modification of DNA under phosphate starvation stress in rice. Front Plant Sci 13:871890

Kungulovski G, Nunna S, Thomas M, Zanger UM, Reinhardt R, Jeltsch A (2015) Targeted epigenome editing of an endogenous locus with chromatin modifiers is not stably maintained. Epigenetics Chromatin 8:1–11

Lee JE, Neumann M, Duro DI, Schmid M (2019) CRISPR-based tools for targeted transcriptional and epigenetic regulation in plants. PLoS One 14(9):e0222778

Li Y, Mukherjee I, Thum KE, Tanurdzic M, Katari MS, Obertello M, Edwards MB, McCombie WR, Martienssen RA, Coruzzi GM (2015) The histone methyltransferase SDG8 mediates the epigenetic modification of light and carbon-responsive genes in plants. Genome Biol 16:1–15

Li A, Hu B, Chu C (2021) Epigenetic regulation of nitrogen and phosphorus responses in plants. J Plant Physiol 258:153363

Lira-Medeiros CF, Parisod C, Fernandes RA, Mata CS, Cardoso MA, Ferreira PCG (2010) Epigenetic variation in mangrove plants occurring in contrasting natural environments. PLoS One 5:e10326

Liu KH, Tsay YF (2003) Switching between the two action modes of the dual-affinity nitrate transporter CHL1 by phosphorylation. EMBO J 22:1005–1013

Lu C, Jeong DH, Kulkarni K, Pillay M, Nobuta K, German R, Thatcher SR, Maher C, Zhang L, Ware D, Liu B, Cao X, Meyers BC, Green PJ (2008) Genome-wide analysis for the discovery of rice microRNAs reveals natural antisense microRNAs (nat-miRNAs). Proc Natl Acad Sci USA 105(12):4951–4956

Mager S, Ludewig U (2018) Massive Loss of DNA Methylation in Nitrogen-, but Not in Phosphorus-Deficient Zea mays Roots Is Poorly Correlated With Gene Expression Differences. Front Plant Sci 9:497. https://doi.org/10.3389/fpls.2018.00497

Malinas A, Vidican R, Rotar I, Malinas C, Moldovan CM, Proorocu M (2022) Current status and future prospective for nitrogen use efficiency in wheat (*Triticum aestivum* L.). Plan Theory 11(2):217

Matzke MA, Kanno T, Matzke AJM (2015) RNA-directed DNA methylation: the evolution of a complex epigenetic pathway in flowering plants. Annu Rev Plant Biol 66:243–267

Meng X, Yu X, Wu Y, Kim DH, Nan N, Cong W, Wang S, Liu B, Xu ZY (2020) Chromatin remodelling protein ZmCHB101 regulates nitrate-responsive gene expression in maize. Front Plant Sci 11:52

Meyer RC, Gryczka C, Neitsch C, Müller M, Bräutigam A, Schlereth A, Schön H, Weigelt-Fischer K, Altmann T (2019) Genetic diversity for nitrogen use efficiency in *Arabidopsis thaliana*. Planta 250:41–57

Miglani GS (2017) Genome editing in crop improvement: present scenario and future prospects. J Crop Improv 31(4):453–559

Mlynárová L, Nap JP, Bisseling T (2007) The SWI/SNF chromatin-remodelling gene AtCHR12 mediates temporary growth arrest in *Arabidopsis thaliana* upon perceiving environmental stress. Plant J 51(5):874–885

Murfett J, Wang XJ, Hagen G, Guilfoyle TJ (2001) Identification of Arabidopsis histone deacetylase HDA6 mutants that affect transgene expression. Plant Cell 13(5):1047–1061

Nischal L, Mohsin M, Khan I, Kardam H, Wadhwa A, Abrol YP, Iqbal M, Ahmad A (2012) Identification and comparative analysis of MicroRNAs associated with low-N tolerance in Rice genotypes. PLoS One 7(12):e50261

Ogas J, Kaufmann S, Henderson J, Somerville C (1999) PICKLE is a CHD3 chromatin-remodelling factor that regulates the transition from embryonic to vegetative development in Arabidopsis. Proc Natl Acad Sci USA 96(24):13839–13844

Papikian A, Liu W, Gallego-Bartolomé J, Jacobsen SE (2019) Site-specific manipulation of Arabidopsis loci using CRISPR-Cas9 SunTag systems. Nat Commun 10(1):729

Quadrana L, Colot V (2016) Plant transgenerational epigenetics. Annu Rev Genet 50:467–491

Reina C, Cavalieri V (2020) Epigenetic modulation of chromatin states and gene expression by g-quadruplex structures. Int J Mol Sci 21(11):1–22

Roca Paixão JF, Gillet F-X, Ribeiro TP, Bournaud C, Lourenço-Tessutti IT, Noriega DD, de Melo BP, de Almeida-Engler J, Grossi-de-Sa MF (2019) Improved drought stress tolerance in Arabidopsis by CRISPR/dCas9 fusion with a histone AcetylTransferase. Sci Rep 9(1):8080

Sabari BR, Zhang D, Allis CD, Zhao Y (2017) Metabolic regulation of gene expression through histone acylations. Nat Rev Mol Cell Biol 18(2):90–101

Sathee L, Jagadhesan B, Pandesha PH, Barman D, Adavi BS, Nagar S, Krishna GK, Tripathi S, Jha SK, Chinnusamy V (2022) Genome editing targets for improving nutrient use efficiency and nutrient stress adaptation. Front Genet 13:900897

Sengupta A, Roy SS, Chowdhury S (2021) Non-duplex G-Quadruplex DNA structure: a developing story from predicted sequences to DNA structure-dependent epigenetics and beyond. Acc Chem Res 54(1):46–56

Séré D, Martin A (2020) Epigenetic regulation: another layer in plant nutrition. Plant Signal Behav 15(1):1–6

Sharma N, Madan B, Khan MS, Sandhu KS, Raghuram N (2023) Weighted gene co-expression network analysis of nitrogen (N)-responsive genes and the putative role of G-quadruplexes in N use efficiency (N.U.E.) in rice. Front. Plant Sci 14:1135675

Singh D, Chaudhary P, Taunk J, Kumar Singh C, Sharma S, Singh VJ, Singh D, Chinnusamy V, Yadav R, Pal M (2021) Plant epigenomics for extenuation of abiotic stresses: challenges and future perspectives. J Exp Bot 72(20):6836–6855

Sirohi G, Pandey BK, Deveshwar P, Giri J (2016) Emerging trends in epigenetic regulation of nutrient deficiency response in plants. Mol Biotechnol 58:159–171

Smith AP, Jain A, Deal RB, Nagarajan VK, Poling MD, Raghothama KG, Meagher RB (2010) Histone H2A.Z regulates the expression of several classes of phosphate starvation response genes but not as a transcriptional activator. Plant Physiol 152(1):217–225

Sokol A, Kwiatkowska A, Jerzmanowski A, Prymakowska-Bosak M (2007) Up-regulation of stress-inducible genes in tobacco and Arabidopsis cells in response to abiotic stresses and A.B.A. Treatment correlates with dynamic changes in histone H3 and H4 modifications. Planta 227:245–254

Song X, Li Y, Cao X, Qi Y (2019) MicroRNAs and Their Regulatory Roles in Plant–Environment Interactions. Annu Rev Plant Biol 70:489–525

Sudarsanam P, Winston F (2000) The SWI/SNF family nucleosome-remodeling complexes and transcriptional control. Trends Genet 16(8):345–351

Tsuji H, Saika H, Tsutsumi N, Hirai A, Nakazono M (2006) Dynamic and reversible changes in histone H3-Lys4 methylation and H3 acetylation occurring at submergence-inducible genes in rice. Plant Cell Physiol 47(7):995–1003

Volpe TA, Kidner C, Hall IM, Teng G, Grewal SIS, Martienssen RA (2002) Regulation of heterochromatic silencing and histone H3 lysine-9 methylation by RNAi. Science 297(5588):1833–1837

Wang AH, Zare H, Mousavi K, Wang C, Moravec CE, Sirotkin HI, Ge K, Gutierrez-Cruz G, Sartorelli V (2013) The histone chaperone Spt6 coordinates histone H3K27 demethylation and myogenesis. EMBO J 32(8):1075–1086

Wang X, Chen J, Xie Z, Liu S, Nolan T, Ye H, Zhang M, Guo H, Schnable PS, Li Z, Yin Y (2014) Histone lysine methyltransferase SDG8 is involved in brassinosteroid-regulated gene expression in *Arabidopsis thaliana*. Mol Plant 7(8):1303–1315

Wang F, Polydore S, Axtell MJ (2015) More than meets the eye? Factors that affect target selection by plant miRNAs and heterochromatic siRNAs. Curr Opin Plant Biol 27:118–124

Wang W, Hu B, Li A, Chu C (2020) NRT1.1s in plants: functions beyond nitrate transport. J Exp Bot 71(15):4373–4379

Wang X, Feng C, Tian LL, Hou C, Tian W, Hu B, Zhang Q, Ren Z, Niu Q, Song J, Kong D, Liu L, He Y, Ma L, Chu C, Luan S, Li L (2021a) A transceptor–complex channel couples nitrate sensing to calcium signalling in Arabidopsis. Mol Plant 14(5):774–786

Wang W, Li A, Zhang Z, Chu C (2021b) Posttranslational modifications: regulation of nitrogen utilization and signaling. Plant Cell Physiol 62(4):543–552

Waterborg JH, Harrington RE, Winicov I (1989) Differential histone acetylation in alfalfa (*Medicago sativa*) due to growth in NaCl: responses in salt-stressed and salt tolerant callus cultures. Plant Physiol 90(1):237–245

Widiez T, El Kafafi ES, Girin T, Berr A, Ruffel S, Krouk G, Vayssières A, Shen WH, Coruzzi GM, Gojon A, Lepetit M (2011) High nitrogen insensitive 9 (HNI9)-mediated systemic repression of root NO3-uptake is associated with changes in histone methylation. Proc Natl Acad Sci USA 108(32):13329–13334

Wolffe AP (1996) Histone deacetylase: a regulator of transcription. Science 272(5260):371–371

Wu L, Zhang Q, Zhou H, Ni F, Wu X, Qi Y (2009) Rice MicroRNA effector complexes and targets. Plant Cell 21(11):3421–3435

Zhai L, Liu Z, Zou X, Jiang Y, Qiu F, Zheng Y, Zhang Z (2013) Genome-wide identification and analysis of microRNA responding to long-term waterlogging in crown roots of maize seedlings. Physiol Plant 147(2):181–193

Zhang H, Zhu JK (2011) RNA-directed DNA methylation. Curr Opin Plant Biol 14(2):142–147

Zhang J, Xu Y, Huan Q, Chong K (2009) Deep sequencing of Brachypodium small RNA.S at the global genome level identifies microRNAs involved in cold stress response. BMC Genomics 10:1–16

Zhang YC, Yu Y, Wang CY, Li ZY, Liu Q, Xu J, Liao JY, Wang XJ, Qu LH, Chen F, Xin P, Yan C, Chu J, Li HQ, Chen YQ (2013) Overexpression of microRNA OsmiR397 improves rice yield by increasing grain size and promoting panicle branching. Nat Biotechnol 31(9):848–852

Zhuo D, Okamoto M, Vidmar JJ, Glass ADM (1999) Regulation of a putative high-affinity nitrate transporter (Nrt2;1At) in roots of *Arabidopsis thaliana*. Plant J 17:563–568

Role of miRNAs in Regulating Response to High Temperature Stress

10

Ambreen Khan and Neeti Sanan-Mishra

Abstract

Plants are significantly influenced by different environmental cues that negatively impact their growth and yields. This necessitates the improvement of their adaptation to stressful conditions to achieve better yields. Global climate change, particularly high temperatures resulting from greenhouse gas production, has lead to various biochemical, morphological, and physiological changes in plants. At the cellular level, the responses involve transmission of signals, osmolyte production, and activation of defense mechanisms, culminating in the expression of genes responsive to high temperature stress (HTS). The plant heat stress transcription factors (HSFs) and heat shock proteins (HSPs) perform a significant role in managing responses to HTS. Several studies in the recent years have demonstrated the regulation of HTS-responsive gene expression through microRNAs (miRs), a major class of small noncoding RNAs (sncRNAs). Several of these miRs regulate heat shock activators and repressors in plants. Such interplay between HSFs/HSPs and miRs forms a distinctive strategy to enhance crop performance under stress conditions. It has been observed that pre-exposing plants to shorter durations of high temperature, known as thermopriming, generally leads to improved adaptation of plants when subsequently exposed to HTS. A role for miRs in thermopriming response has also been implicated. This chapter introduces the mechanisms underlying HTS tolerance and extends insights into the role of miRNA in this process.

Keywords

microRNAs · High temperature stress · HSPs · HSFs · sncRNAs

A. Khan · N. Sanan-Mishra (✉)
Plant RNAi Biology Group, International Centre for Genetic Engineering and Biotechnology, New Delhi, India
e-mail: neeti@icgeb.res.in

K. Goswami et al. (eds.), *Non-coding RNAs for Crop Improvement*,
https://doi.org/10.1007/978-981-97-7213-1_10

10.1 Introduction

The development and growth of plants are significantly influenced by environmental cues, and when a plant is well adapted to its environment, it has the potential to achieve high yields. Temperature, light, nutrient availability, and water content play vital roles in directing plant development (Kooke and Keurentjes 2012). Seasonal changes bring about alterations in light and temperature, which are key environmental indicators of energy availability and the circadian clock. These factors play essential roles in the emergence, germination, maturation, and overall growth of plants (Lorenzo et al. 2016). The coordination of growth and development in plants, transitioning from the vegetative phase to the reproductive phase, relies on a well-connected signaling network involving various members. It is essential for this inter-connected network to converge appropriately to effectively regulate the molecular machinery (Heggie and Halliday 2005). Therefore, any deviations from the ideal growth conditions negatively affect the net rate of crop production.

Temperature is a significant environmental factor that exerts a substantial influence on the overall developmental process of plants. The escalating temperatures and the accompanying rise in greenhouse gases have become a global concern due to their diverse impact on plants (Hall 2000; Wahid et al. 2007; Janni et al. 2020). In natural settings, plants are subjected to numerous other simultaneous stress signals along with high temperatures. For example, high temperatures accelerate the rate of transpiration in plants (Jacob et al. 2017), which can result in water deficiency and subsequent soil salinity (Mathur et al. 2013). The elevated soil salinity, in turn, diminishes water absorption, leading to physiological drought in plants (Lisar et al. 2012). Drought conditions disrupt the balance between light capture and utilization, resulting in the inhibition of photosynthesis (Lisar et al. 2012). Furthermore, the combination of warmer and wetter conditions negatively impacts crop productivity by providing favorable conditions for the proliferation of fungi, weeds, and pests (Shekhar et al. 2017). Considering these factors, global climate changes bring about variations in the required ambient temperature conditions, as well as alterations in various biotic and abiotic factors, ultimately impeding plant yield (Pandey et al. 2017).

10.2 High Temperature Stress and Its Effects

Each plant has a preference for a particular temperature at which it shows optimum growth. Above this lies the range of threshold temperatures, which signifies a figure of daily mean temperature at which a discernible slowdown in growth occurs. Any upsurge in temperature over the threshold level for a period of time adequate enough to cause irrevocable harm to plant development and growth is generally explicated as HTS. Therefore, HTS is an elaborate function of the duration and rate of temperature severity (in degrees). The level of HTS tolerance is described by the plant's capacity to thrive and produce considerable yield. Excessive HTS accompanied by high light is the most limiting variable impacting plant development and eventually the crop output.

According to various global circulation models, the greenhouse gases such as chlorofluorocarbons, CO_2, methane, and nitrous oxides will steadily raise the worldwide average ambient temperature. According to the Intergovernmental Panel on Climate Change (IPCC), the global mean temperature would climb 0.3 °C every decade, reaching 1 °C and 3 °C over the current levels by 2025 and 2100, respectively; ultimately culminating in global warming (Jones et al. 1999). It is predicted that for every 1 °C increase in the world's mean temperature, the yields of wheat, rice, maize, and soybeans will decrease by 6.0%, 3.2%, 7.4%, and 3.1%, respectively (Zhao et al. 2017).

According to FAO data, the major cereal crops are seeing a decline in their relative rates of production. To maintain food security, crop yield must rise as the population grows, and it is predicted that an approximately 70% increase in agricultural production would be required to meet the needs and requirements of a population of 9 billion people in 2050 (Bita and Gerats 2013). This necessitates an in-depth understanding of the detrimental effects of HTS on plant growth and development.

The effects of HTS have been the subject of extensive studies over the past few decades (Haslbeck and Vierling 2015). It is acknowledged that different plant species have inculcated diverse strategies for adaptation, acclimation or avoidance to survive HTS. Some avoid harsh conditions by increasing transpiration; some escape by early anthesis while others tolerate high temperatures by means of genetic and metabolic mechanisms (Haslbeck and Vierling 2015).

10.2.1 Morphological Effects of HTS

Plant sensitivity to HTS differs depending on the species and the developmental stage. Although all plant tissues are sensitive to HTS at nearly all growth and developmental phases, the germination and reproductive stages are the most sensitive. Major effects of HTS have been observed during the germination stage in a variety of cultivated plant species. The effects include decreased germination rate, which causes diminished emergence and seedling establishment, reduced radicle and plumule growth, malformed seedlings or low seedling vigor (Piramila et al. 2012). In mature plants, HTS can lead to scorching and sunburn of aerial parts, causing blistering and drying of leaves, leaf senescence and defoliation, inhibition of root and shoot growth, fruit discoloration and damage (Kilasi et al. 2018; Vollenweider and Günthardt-Goerg 2005). An elevation in temperature during the flowering stage results in a significant reduction of floral buds and, in extreme cases, flower abortion, which may result in no flowers or sterile flowers. In cereal crops, such as rice and sorghum, a decreased number of spikes or florets and reduced seed-set have been reported (Fahad et al. 2017).

HTS also has a significant impact on reproductive processes in plants, causing impairment in fertilization and post-fertilization processes, and hence lower crop yield. This is mainly due to impaired cell division in both male and female organ systems leading to altered anther shape, decreased anther dehiscence, poor pollen viability and number, decreased pollen retention and germination on the stigma, decreased pollen swelling, impeded elongation of pollen tubes, abnormal stigma and style positions, and ovule viability (Fu et al. 2016; Shi et al. 2018). In rice plants, HTS affects both mega-sporogenesis and micro-sporogenesis resulting in greater spikelet sterility and reduced rice seed-set (Qu et al. 2013).

10.2.2 Physiological Effects of HTS

HTS has a significant impact on a number of physiological parameters in plants like intercellular CO_2 concentration, leaf stomatal conductance, and leaf water status (Greer and Weedon 2012). The effect on water potential and photosynthetic rate is detailed below.

Water Potential: The water status of plants is the most essential variable under changing temperatures. In general, HTS increases transpiration throughout the day resulting in water deficit in plants by reducing the water potential and causing several physiological processes to be disrupted (Tsukaguchi et al. 2003). When moisture is plentiful, plants prefer to maintain consistent tissue water status regardless of temperature; but when water is scarce, HTS significantly impedes this propensity. HTS is also typically linked to a reduction in water availability in the field (Simões-Araújo et al. 2003). HTS appears to promote higher water loss in plants during the day than at night. In response to HTS, plants change their metabolism in a variety of ways to regulate cellular homeostasis. This includes synthesizing and accumulating suitable molecules that can maintain proteins and cellular structures by guiding their assembly, maintaining cell turgor pressure by osmotic adjustments, and restoring redox balance by modifying the antioxidant system (Munns and Tester 2008). Tolerance to HTS is closely connected with the plant's ability to maintain leaf gas exchange and CO_2 assimilation rates (Yang et al. 2006).

Photosynthetic rate: Photosynthesis exhibits a direct relationship with growth, so changes in photosynthetic characteristics under HTS affect the plant's thermotolerance ability. HTS limits the rate of photosynthesis and thus impedes plant development (Wang et al. 2009). When exposed to HTS, there is both a reduction in chlorophyll biosynthesis and an increase in its degradation (Fahad et al. 2017). Rice, like other C3 plants, is more sensitive to HTS compared to C4 plants in terms of photosynthetic capability (Yang et al. 2006). Carbon metabolism in the stroma and photochemical reactions in thylakoids are the major susceptible sites. High temperatures disrupt the electron transport and inactivate the oxygen-evolving enzymes of PSII, thus modifying the energy distribution and the activities of carbon metabolism enzymes, most notably Rubisco (Salvucci and Crafts-Brandner 2004). In the dark, HTS decreases the number of photosynthetic pigments, rubisco binding proteins (RBP), soluble proteins, and large-(LS) and small-subunits (SS) of rubisco (Demirevska-Kepova et al. 2005). Compared to wild-type (WT) rice plants, transgenic rice plants with greater Rubisco levels thrived more successfully in HTS and had faster photosynthetic rates (Wang et al. 2010). In wheat, cotton, and barley different components of PSII were damaged when exposed to high temperatures. Furthermore, starch or sucrose synthesis is strongly impacted by HTS, as evidenced by lower activities of sucrose phosphate synthase (Chaitanya et al. 2001), ADP glucose pyrophosphorylase, and invertase. The lower production and translocation of photoassimilates further impact the seed or grain.

10.2.3 Biochemical Effects of HTS

Immediate effects of high temperature include increased fluidity of the cell membrane, peroxidation of lipids, and denaturation of protein, thereby decreasing the thermostability of cells (Narayanan et al. 2016). Indirect effects include increased production and accumulation of carbohydrates, changes in the metabolism and phytohormone levels, along with synthesis of proteins involved in defense, like the heat shock transcription factors (HSTFs) and heat shock proteins (HSPs). Different amino acids and their derivatives, acyclic and cyclic polyols, sugars, fructans, ectoine, quaternary amino, and sulfonium compounds act as compatible solutes under stress conditions. These processes are mainly initiated to maintain proper cellular homeostasis and protect plants under HTS.

A major sign of cellular damage brought on by HTS is the production of activated oxygen species (AOS)/reactive oxygen species (ROS), such as superoxide radical (O^{2-}), singlet oxygen ($^1O_2$), hydroxyl radical (OH^-), and hydrogen peroxide (H_2O_2) (Liu and Huang 2000). The prevailing consensus is that low ROS rates are essential secondary messengers for cellular metabolism. The mitochondria and chloroplast routinely generate superoxide radicals, with reasonable amounts also synthesized in microbodies. The HTS-induced altered membrane fluidity results in the influx of Ca^{+2}, thus activating respiratory burst oxidase homolog (RBOH). By activating the RBOHs and controlling ABA signals, they also contribute significantly to cross-tolerance to biotic and abiotic stress (Xia et al. 2014). For instance, H_2O_2 has a variety of functions including that of a second messenger in cross-tolerance to various abiotic stressors (Mittler and Blumwald 2015). However, stress-induced high levels of ROS can damage DNA, protein, lipids, and chlorophyll, along with other vital macromolecules. They can lead to alteration in the functionalities of membranes by causing the autocatalytic peroxidation of membrane lipids, along with the pigments, enzyme inhibition, and carbohydrate oxidation, which ultimately results in programmed cell death (Suzuki et al. 2012). Accumulation of ROS may influence plant responses mediated by phytohormones as they presumably serve a variety of functions in phytohormone production, transport, and signaling (Xia et al. 2015). This has a devastating impact on plants and restricts their growth and productivity. Therefore, the accumulation of ROS during HTS itself acts as a signal for the activation of stress-responsive pathways in plants (Baxter et al. 2014).

Protection against oxidative stress is a key factor in determining whether the plant will survive under HTS. To combat ROS and safeguard cells from oxidative damage, plants have evolved a variety of enzymatic and non-enzymatic detoxifying processes (Sairam and Tyagi 2004). The major ROS scavenging enzymes include Superoxide dismutase (SOD), catalase (CAT), ascorbate peroxidase (APX), glutathione reductase (GR), glutathione peroxidase (GPX), dehydroascorbate reductase

(DHAR), monodehydroascorbate reductase (MDHAR), and peroxiredoxin (PrxR). The main non-enzymatic antioxidants are reduced glutathione (GSH), ascorbic acid (ASA), carotenoids, flavonoids, tocopherol, and phenolic compounds (Mittler and Blumwald 2015). Overexpression of SOD in plants results in oxidation of toxic reductants, removal of H_2O_2, defensive reactions to injury, biosynthesis, and degradation of lignin in cell walls, defense against pathogen or insect attack (Scandalios 1993). APX expression and activation more precisely, are connected to the onset of physiological damage brought on by HTS in plants.

10.2.4 Molecular Responses Governing HTS Tolerance

The ability of plants to quickly transmit the detected signals for commencing the proper biochemical, physiological and molecular adaptations determines their response to HTS (Hasanuzzaman et al. 2013). The stress signals are enhanced by several second messengers such as calcium ions (Ca^{+2}) and ROS. The cascade culminates in changes in gene expression, which are aided by a network of transcription factors (TFs) and miRs (Kinoshita and Seki 2014). Given the growing number of regulators engaged in gene networks, it is intriguing to comprehend how miRs, TFs, HSPs and mRNAs interact dynamically (Fig. 10.1).

There have been numerous studies providing insights into the temperature-sensing process, but the authentic temperature receptors have not been identified as yet (Miura and Furumoto 2013). This may be so, as the thermodynamic effects profoundly influence the molecular and supramolecular structures. There are several suggestions on the potential temperature sensors, including calcium ion channels and the thermodynamic variation in fluidity of the plasma membrane.

The Ca^{+2} ions are thought to be an essential component of temperature perception and signaling in plants, in addition to acting as second messengers (Steinhorst and Kudla 2014). The cell wall is the first line of defense against the HTS response

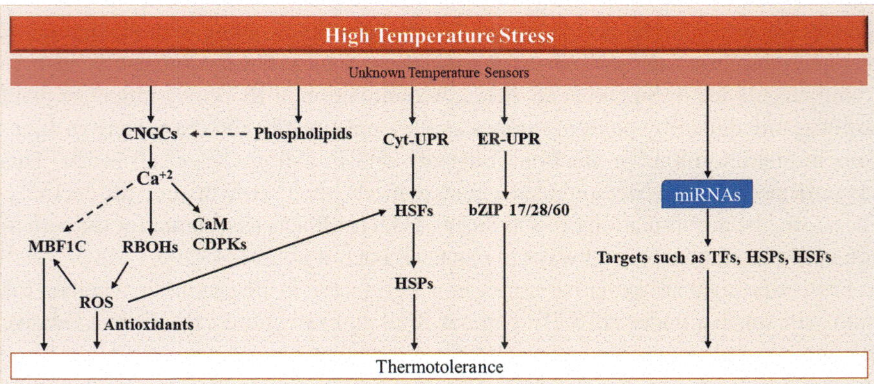

Fig. 10.1 Schematic representation showing the molecular networks during high temperature stress tolerance in plants.

(HSR) in plants; affecting the release of apoplastic Ca^{+2} and raising the concentration of free cytosolic Ca^{+2}. HTS-induced changes in membrane structures and properties, such as fluidity, permeability, and thickness, can affect the functions of membrane proteins and initiate intracellular HTS-responsive signaling cascades (Niu and Xiang 2018). The plasma membrane can directly connect HTS signals to intracellular signaling molecules. Calcium-conducting CNGC (cyclic nucleotide-gated channel) found in plasma membranes are well known for their roles in HTS sensing (Ohama et al. 2017). An increase in temperatures above the threshold limit leads to changes in membrane fluidity through opening of these Ca^{+2} ion channels, thereby promoting activation of HTS response. Extracellular Ca^{+2} can enter the cytoplasm by transiently activated Ca^{+2} channels that are triggered by increased membrane fluidity (Ohama et al. 2017). There are currently 16 fully characterized CNGC genes in rice, 13 of which are projected to be located in the plasma membrane, and the majority of them are temperature-responsive. OsCNGC16 and OsCNGC14 are positive regulators of rice's cytosolic Ca^{+2} increase caused by HT (Cui et al. 2020). Annexin 1 (ANN1), a membrane-associated Ca^{+2} channel, is crucial for thermotolerance and also contributes to the rise in cytosolic Ca^{+2} caused by HTS (Wang et al. 2015). CNGC2 has been shown to have a thermo-sensory role in the induction of *HSP* expression in response to high temperatures. The Ca^{+2} dependent protein kinases (CDPKs) are also associated with stress resistance upon exposure to HTS (Hake and Romeis 2019).

Likewise, lipid signaling is activated by HTS-induced changes in membrane fluidity. HTS causes changes in the tertiary as well as quaternary structures of membrane proteins, membrane stability and integrity. The motion of molecules within membranes and hence their kinetic energy is increased under HTS leading to the disruption of their bonds. These result in either an increase in unsaturated fatty acids or denaturation of proteins; thereby making the lipid bilayer of biological membranes highly fluid (Savchenko et al. 2002). HTS-induced changes in membrane fluidity can cause certain plasma membrane-located phosphoinositide-specific phospholipase C (PLCs) to become active (Niu and Xiang 2018). The other components of lipid signaling, namely Phospholipase D (PLD) and phosphatidylinositol-4-phosphate 5-kinase (PIP5K), have been shown to be stimulated by HTS (Yao et al. 2011).

Under HTS the cell's ability to maintain protein homeostasis is disrupted as protein synthesis, folding, sub-cellular localization and quality control are affected. The unfolded or improperly folded proteins are harmful to plant cells and their build-up in cells also acts as a signal to activate the HSTFs and HSP (Fragkostefanakis et al. 2016).

High temperatures trigger changes in the membrane structure and transmit the stress signals through the Ca^{+2} and lipid-dependent pathways to regulate HTS-responsive gene expression. The Ca^{+2} ions can be conducted through the CDPK or CaM pathways to activate the ROS and HSFs. The HSFs bind to their respective cis-elements to promote the transcription of HSPs and ROS activates the antioxidant enzymes. HSPs inhibit the thermal denaturation of cellular proteins. miRs act as negative regulators of their target genes, some of which include HSPs, HSFs, and TFs, to trigger positive or negative feedback loops for retention of thermotolerance memory.

1. Heat shock proteins: An increase in temperatures above the threshold value causes a signature rise in the expression of HSPs. This activates a chaperone cascade which aids in averting the denaturation, misfolding and anomalous aggregation of proteins (Dafny-Yelin et al. 2008). According to their molecular weight, HSPs are divided into six classes, including HSP100, HSP90, HSP70, HSP60, HSP40, and small HSPs (sHSPs) having low molecular weights.

   The expression levels of five HSPs: HSP16.9A, HSP17.4, HSP26.7, HSP23.2, and HSP17.9A are upregulated in response to HTS. In drought- or thermo-primed plants, HSPs and HOP (Hsp70-Hsp90 organizing protein) proteins are strongly expressed, which help in increasing grain filling during HTS in wheat (Zhang et al. 2016) and rice (Kushawaha et al. 2021).The two largest families of HSPs, HSP90, and HSP70, are controlled by the stress-induced HOP (Carrigan et al. 2006). Under HTS there is an increase in the abundance of HSP90, HSP101, and HSP70 proteins in rice, and these proteins probably enhance thermotolerance. Their levels are higher in heat-tolerant rice cultivars than in heat-susceptible rice cultivars. HSP90 enhances the transcription factors such as Auxin Response Factors (ARFs) and suppresses the build-up of transcriptional repressors like Auxin/indole-3-acetic acid (Aux/IAA) concentration, and thus participates in hormonal cross-talk to regulate plant growth under HTS (Watanabe et al. 2016).

   The ability of plants to tolerate HTS is also mediated by an interaction between the Arabidopsis Hikeshi-Like Protein1 (HLP1) and HSP70 (Koizumi et al. 2014). The retention of HTS memory and thermotolerance depends on the induction of the *Heat Stress Associated 32* (*HSA32*) gene by HSP101 (Wu et al. 2013). The HSA32 interacts with Brushy1/Tonsoku/Mgoun3 (BRU1/TSK/MGO3) to cause changes in chromatin structure and enable the prolonged activation of HTS memory genes (Brzezinka et al. 2019).
2. Heat stress transcription factors are a conserved family of transcription factors that interact directly with heat shock elements (HSEs) to regulate the expression of genes relevant to HTS response (Ohama et al. 2017). The HSEs are characterized by the pentameric consensus sequence nGAAn (nTTCnnGAAnnTTCn) and AGGGG motifs, which comprise tandem inverted repeats (von Koskull-Döring et al. 2007). HSFs trigger a transcriptional cascade by activating other HTS-induced transcription factors, metabolic enzymes, ROS-scavenging enzymes, and HSPs.

   HSFs are classified into three classes: HSF-A, HSF-B, and HSF-C. Among the Class-A HSF transcriptional network, HSFA1 is regarded as a chief regulator, working as an early response gene for activating other HTS response factors, like HSFA2 and DREB2A (Guo and Wang 2011; Ohama et al. 2017). Another direct target of HSFA1 is the multiprotein bridging factor 1c (MBF1c), which encodes a conserved transcription co-activator. HSFA1b directly controls its expression by binding to the HSE in the MBF1c promoter, resulting in accumulation of MBF1c under HTS. MBF1c then interacts with the CTAGA motif in the promoters of its target genes DREB2A, HSFB2a, and HSFB2b (Bechtold et al. 2013). When TaMBF1c was overexpressed in rice, the plants exhibited an increase in thermotolerance. Under normal temperatures, HSP70 and HSP90 interact with HSFA1 to impair its transactivation activity and nuclear localization (Ohama et al. 2017). However, HTS activates HsfA1 by allowing it to dissociate from HSP70 and HSP90 (Hahn et al. 2011). The rice HSF-binding proteins OsHSBP1 and OsHSBP1 interact with activated HSFs and function as negative regulators in HSR; overexpressing these proteins leads to lower plant survival rates following HTS (Rana et al. 2012). Rice HSR and thermotolerance are also reliant on a number of HSFA1-independent transcription factors, such as OsMYB55, OsbZIP46, SNAC3, and OsWRKY11 (El-Kereamy et al. 2012; Fang et al. 2015).

HSFA2 is known to play a crucial role during HTS response in plants (Ohama et al. 2017). When rice plants are exposed to HTS, alternative splicing induces the transcriptionally active form of OsHSFA2d, enhancing the expression of genes such as HSP17.7, HSP18.2, HSP21, HSP83.1, and HSP101. HSFA2 performs chromatin modification to regulate long-term gene expression; in addition to activating HTS-regulated genes. For instance, HSFA2 is necessary to maintain high levels of two particular histone modifications (H3K4me2 and H3K4me3) in HSP22, HSP18.2 and APX2, among other proteins. These changes are essential for plants to sustain HTS memory; a process also known as prolonged acquired thermotolerance (Lämke et al. 2016). Since HSFA2 is not a DNA methyltransferase, it is unclear how DNA methyltransferases are recruited to certain loci in order to maintain the expression of all these HTS memory-related genes.

The expression of HSFA3, which is controlled by DREB2a and DREB2c, is necessary for thermotolerance (Sato et al. 2014). Other HSFs, such as HSFA4a and HSFA8, are also involved in detecting ROS and activating the plant's antioxidant defense mechanism. Only in the latter phases of seed development is HSFA9 expressed (Kotak et al. 2007). It is known that the promoters of various HSFs, including HSFA1b, HSFA6b, HSFA7a, and HSFC1, have NAC binding sites (Guan et al. 2014).

The primary function of the Class B HSFs, including HSFB1 and HSFB2b, is to suppress the transcription of genes that are activated by HTS. In conjunction with class-A HSFs, they function as co-activators. Class C HSFs' function in HTS is unclear. Early HTS treatment was found to negatively regulate TF-HSFC1a in rice (Fragkostefanakis et al. 2016), although overexpressing HSFC2a-B enhanced tolerance to HTS in wheat by increasing the expression of HTS-related genes.

10.3 Thermopriming

A plant can be primed by sub-lethal or moderately high temperatures to increase its tolerance to HTS that might otherwise be fatal to an unadapted plant. Thermopriming has gained significant research attention over the past 10 years since primed plants demonstrate superior adaptability and stress tolerance. The thermotolerance of primed plant is maintained for a considerable period after returning to non-stress temperatures and is known as HTS memory. The HTS memory is genetically distinct from thermopriming (Stief et al. 2014). Priming can have either a short-term or long-term effect, and it can be accomplished by exposing plants to HTS at three different developmental stages:

1. Priming of seeds enables them to avoid a variety of challenges during germination. Seed priming promotes cross-stress resistance to other abiotic stress and in general hastened the process of emergence and germination (Banerjee and Roychoudhury 2020).
2. Stress priming in the post-embryonic stage increases the probability of the mature plant surviving the HTS. It is sometimes referred to as an intergenerational, somatic or mitotic stress memory since its effects are mostly seen in the current generation. Intergenerational thermopriming causes changes in small RNA expression for modulation of their target transcripts, thus aiding in adaptation to HTS (Kushawaha et al. 2021).
3. Trans-generational stress priming is the process of influencing the parental population in a way that its effects are passed down to offspring. It is often referred to as meiotic stress memory since its effects have an impact on subsequent generations (Munné-Bosch and Alegre 2013). The memory for responding to environmental pressures is likely bundled into seeds together with other resources, and these resources supply vital information to the new seedling during germination and the beginning of plant growth (Herman and Sultan 2011). Studies on *Brassica rapa* revealed that HTS-induced alterations in non-coding RNA profiles are passed along to the next generation (Bilichak et al. 2015). Similar studies have been performed on wheat to see how HTS affects seed vigor and germination in the offspring of thermo-primed parents (Liu et al. 2020; Zhou et al. 2020).

In addition, priming may be divided into *cis*- and *trans*-priming depending on the kind of stress signals. "Cis-priming" refers to the situation where the stimulus that elicits the response and the stress that triggers it are of the same kind. Pre-exposure to sub-lethal temperatures up to 38 °C boost thermotolerance of rice plants both at the vegetative stage and anthesis stage due to upregulation of HSPs and various other HTS-related pathway proteins (Jagadish et al. 2010; Kushawaha et al. 2021). In "trans-priming" the stimulus eliciting the reaction is distinct from the stress that triggers the response (Hilker et al. 2016). Various reports suggest that thermopriming of plants improves their adaptability to biotic (*Xanthomonas oryzae*) and heavy metal (cadmium) stressors due to suppression of abscisic acid responsive genes and accumulation of higher antioxidant levels, respectively (Hsu and Kao 2007; Chou et al. 2012; Cohen et al. 2017).

The priming and memory of HTS may be advantageous for plant fitness and survival in natural settings. In order to explore the molecular underpinnings and evolution of priming and memory in response to abiotic stress, thermopriming and memory have been investigated using Arabidopsis and rice as model systems. It is hypothesized that HTS priming entails activating HSFs that trigger the production of HSPs, which in turn help maintain protein homeostasis through their function as chaperones (Finka et al. 2012; Haslbeck and Vierling 2015). In plants, mammals, and fungi, this HTS response is retained over a longer time period to confer thermotolerance (Richter et al. 2010). It is unclear why HTS memory offers a competitive advantage over de-novo acclimation. Even while remarkable progress has been achieved recently, a mechanistic explanation is still a long way off. The new understanding is that thermomemory is controlled at a variety of levels, including protein stability, miRNA-controlled mRNA stability, and transcriptional memory.

10.4 MicroRNAs and Their Role in Heat Stress Tolerance

The miRs constitute an endogenously coded sub-class of 20–24 nt noncoding RNAs. They negatively regulate gene expression at the post-transcriptional level (Bartel 2004; Thomson and Dinger 2016) by complementarily binding to the transcripts and targeting them for cleavage or inhibition of protein translation (Sanei and Chen 2015). Mature miRs are derived from precursor stem-loop forms (pre-miRNAs), which are processed from primary miR transcripts (pri-miRNAs). The pri-miRNAs are transcribed from their respective *MIR* genes via the RNA polymerase II-mediated transcription process (Yu et al. 2019). The miR targets include transcripts of a variety of TFs (Rhoades et al. 2002; Mitsuda and Ohme-Takagi 2009) and other proteins like enzymes, receptors, transporters and so on. The TFs can stimulate or inhibit the transcription of numerous genes, which can transcribe pri-miRNA and transcripts coding for proteins (Istrail et al. 2007; Chow et al. 2016;). The miR-TF circuitry indicates that both regulators have a significant influence on each other's expression variability to orchestrate a multitude of biological processes. Therefore, genetic control of biological processes operates through complex regulatory loops or circuits (Krol et al. 2010; Nazarov et al. 2013).

A number of miRs are deregulated when plants are subjected to HTS and/or thermopriming, suggesting that they play crucial roles in fine-tuning thermotolerance and response to HTS. The HTS-induced deregulations vary in plant species (Table 10.1).

Table 10.1 The HTS induced deregulations vary in plant species.

miR family	Target	Rice	Arabidopsis	Wheat	Barley	Maize	Tomato
miR156	SPL	↓	↑	↑		↓	
miR159	MYB	↓		↑		↓	
miR160	ARF	↓	↑	↑	↑		↓
miR164	NAC	↓	↑			↓	
miR166	HD-ZIPIII, HSP	↓	↑	↑	↑	↓	↓
miR167	ARF, UDPG	↓		↑	↑		↑
miR168	AGO1	↓		↑			↓
miR169	NF-YA	↑	↑	↑		↓	
miR171	SCL		↑				
miR172	AP2		↓	↓		↑	
miR319	TCP, MYB			↓			↓
miR390	RLK, TAS3	↓					
miR393	TIR, AFB		↓	↑			↓
miR394	F-box						
miR395	APS, SULTR			↑			
miR396	GRF, bHLH, WRKY	↓				↓	↑
miR397	Laccase	↑	↓				↓
miR398	CSD	↓	↑	↑			↓
miR399	PHO2			↑			
miR408	Plastocyanin						
miR528	DCL	↓					
miR529	SBP-box						
miR827	SPX-MFS protein			↑			
References		Jeong et al. (2011), Sailaja et al. (2014), Liu and Liu (2017), Mangrauthia et al. (2017)	Guan et al. (2013), Lu et al. (2013), May et al. (2013), Mahale et al. (2014), Stief et al. (2014)	Xin et al. (2010), Wang et al. (2012); Kumar et al. (2015)	Kruszka et al. (2014)	Zhang et al. (2019)	Pan et al. (2017), Zhou et al. (2020)

10.5 Conclusion

HTS leads to a range of detrimental effects, such as increased accumulation of ROS, diminished germination, premature leaf senescence, inhibited root growth, decreased photosynthesis rate, reduced vigor, lower seed-set, protein denaturation, spikelet sterility, anther indehiscence, and heightened lipid peroxidation rates. In order to attain sustainable agriculture, it is imperative to augment plant responses and enhance their ability to withstand these stresses. The concept of thermopriming by moderately high temperatures holds great potential to enhance the long-term sustainability of agriculture and boost crop yields in the current context of global warming. The concept can be employed to enable them to withstand extreme temperatures. Our current understanding of the effects of priming and induced stress memory in plants remains uncertain. There is a need to investigate the maintenance of memory and its transmission. Understanding the duration and regulation of the memory phase is crucial.

Advances in molecular biology have made substantial progress in the identification of genomic regions and underlying mechanisms that influence transcriptional and post-transcriptional regulation of abiotic stress pathways. The miRs have emerged as fine tuners of response to HTS and mitigation of the detrimental effects related to HTS. They act by regulating the abundance of transcription factors and proteins associated with the metabolism of ROS, as well as in several mitogen-activated protein kinases signaling pathway. Having a thorough understanding of the functions of the small, noncoding RNA molecules holds the potential to advance our knowledge and aid in the development of plant tolerance toward HTS. It is likely that there are still numerous undiscovered miRs and their associated target genes with functions in modulating plant response to HTS. It is thus crucial to explore and delve deeper into the functioning of miR-mediated regulatory networks.

10.6 Future Perspectives

Agricultural research will continue to focus on the function of miRs in plant growth and stress responses in the upcoming years. The identification of novel miRs and their target genes will be facilitated by developments in next-generation sequencing, molecular biology, and bioinformatics, providing precise tools for agricultural improvement. The engineering of stress-tolerant crops will be made possible by biotechnological methods including targeted degradation and miR editing. By incorporating such strategies into sustainable agricultural practices, food security in the face of climate change issues will eventually be improved by increased crop productivity and resilience. All things considered, there is a lot of hope for creating next-generation crops that can resist a variety of stressors while maintaining sustainable farming practices if we can fully comprehend and utilize miRs.

References

Banerjee A, Roychoudhury A (2020) Seed priming as a method to generate heat-stress tolerance in plants: a minireview. In: Heat stress tolerance in plants: physiological, molecular and genetic perspectives. Wiley, Hoboken, pp 23–32

Bartel DP (2004) MicroRNAs: genomics, biogenesis, mechanism, and function. Cell 116(2):281–297

Baxter A, Mittler R, Suzuki N (2014) ROS as key players in plant stress signalling. J Exp Bot 65(5):1229–1240

Bechtold U, Albihlal WS, Lawson T, Fryer MJ, Sparrow PA, Richard F, Persad R, Bowden L, Hickman R, Martin C (2013) Arabidopsis heat shock transcription FACTORA1b overexpression enhances water productivity, resistance to drought, and infection. J Exp Bot 64(11):3467–3481

Bilichak A, Ilnytskyy Y, Wóycicki R, Kepeshchuk N, Fogen D, Kovalchuk I (2015) The elucidation of stress memory inheritance in Brassica rapa plants. Front Plant Sci 6:5

Bita CE, Gerats T (2013) Plant tolerance to high temperature in a changing environment: scientific fundamentals and production of heat stress-tolerant crops. Front Plant Sci 4:273

Brzezinka K, Altmann S, Bäurle I (2019) BRUSHY1/TONSOKU/MGOUN3 is required for heat stress memory. Plant Cell Environ 42(3):771–781

Carrigan PE, Sikkink LA, Smith DF, Ramirez-Alvarado M (2006) Domain: domain interactions within Hop, the Hsp70/Hsp90 organizing protein, are required for protein stability and structure. Protein Sci 15(3):522–532

Chaitanya K, Sundar D, Reddy AR (2001) Mulberry leaf metabolism under high temperature stress. Biol Plant 44:379–384

Chou T-S, Chao Y-Y, Kao CH (2012) Involvement of hydrogen peroxide in heat shock-and cadmium-induced expression of ascorbate peroxidase and glutathione reductase in leaves of rice seedlings. J Plant Physiol 169(5):478–486

Chow C-N, Zheng H-Q, Wu N-Y, Chien C-H, Huang H-D, Lee T-Y, Chiang-Hsieh Y-F, Hou P-F, Yang T-Y, Chang W-C (2016) PlantPAN 2.0: an update of plant promoter analysis navigator for reconstructing transcriptional regulatory networks in plants. Nucleic Acids Res 44(D1):D1154–D1160

Cohen SP, Liu H, Argueso CT, Pereira AV, Cruz C, Verdier V, Leach JE (2017) RNA-Seq analysis reveals insight into enhanced rice Xa7-mediated bacterial blight resistance at high temperature. PLoS One 12(11):e0187625

Cui J, Pottosin I, Lamade E, Tcherkez G (2020) What is the role of putrescine accumulated under potassium deficiency? Plant Cell Environ 43(6):1331–1347

Dafny-Yelin M, Tzfira T, Vainstein A, Adam Z (2008) Non-redundant functions of sHSP-CIs in acquired thermotolerance and their role in early seed development in Arabidopsis. Plant Mol Biol 67:363–373

Demirevska-Kepova K, Holzer R, Simova-Stoilova L, Feller U (2005) Heat stress effects on ribulose-1,5-bisphosphate carboxylase/oxygenase, rubisco binding protein and rubisco activase in wheat leaves. Biol Plant 49(4):521–525

El-Kereamy A, Bi Y-M, Ranathunge K, Beatty PH, Good AG, Rothstein SJ (2012) The rice R2R3-MYB transcription factor OsMYB55 is involved in the tolerance to high temperature and modulates amino acid metabolism. PLoS One 7(12):e52030

Fahad S, Bajwa AA, Nazir U, Anjum SA, Sadia S, Nasim W, Adkins S, Saud S, Ihsan MZ (2017) Crop production under drought and heat stress: plant responses and management options. Front Plant Sci 8:265598

Fang Y, Liao K, Du H, Xu Y, Song H, Li X, Xiong L (2015) A stress-responsive NAC transcription factor SNAC3 confers heat and drought tolerance through modulation of reactive oxygen species in rice. J Exp Bot 66(21):6803–6817

Finka A, Cuendet AFH, Maathuis FJ, Saidi Y, Goloubinoff P (2012) Plasma membrane cyclic nucleotide gated calcium channels control land plant thermal sensing and acquired thermotolerance. Plant Cell 24(8):3333–3348

Fragkostefanakis S, Mesihovic A, Simm S, Paupière MJH, Paul Y, Mishra P, Tschiersch SK, Theres B, K. & Bovy, A. (2016) HsfA2 controls the activity of developmentally and stress-regulated heat stress protection mechanisms in tomato male reproductive tissues. Plant Physiol 170(4):2461–2477

Fu G, Yang Y, Jin Q, Tao L (2016) Heat stress is more damaging to superior spikelets than inferiors of rice (Oryza sativa L.) due to their different organ temperatures. Front Plant Sci 7:217932

Greer DH, Weedon MM (2012) Modelling photosynthetic responses to temperature of grapevine (Vitis vinifera cv. Semillon) leaves on vines grown in a hot climate. Plant Cell Environ 35(6):1050–1064

Guan Q, Lu X, Zeng H, Zhang Y, Zhu J (2013) Heat stress induction of mi R 398 triggers a regulatory loop that is critical for thermotolerance in Arabidopsis. Plant J 74(5):840–851

Guan Q, Yue X, Zeng H, Zhu J (2014) The protein phosphatase RCF2 and its interacting partner NAC019 are critical for heat stress–responsive gene regulation and thermotolerance in Arabidopsis. Plant Cell 26(1):438–453

Guo J, Wang M-H (2011) Expression profiling of the DREB2 type gene from tomato (Solanum lycopersicum L.) under various abiotic stresses. Hortic Environ Biotechnol 52:105–111

Hahn A, Bublak D, Schleiff E, Scharf K-D (2011) Crosstalk between Hsp90 and Hsp70 chaperones and heat stress transcription factors in tomato. Plant Cell 23(2):741–755

Hake K, Romeis T (2019) Protein kinase-mediated signalling in priming: immune signal initiation, propagation, and establishment of long-term pathogen resistance in plants. Plant Cell Environ 42(3):904–917

Hall AE (2000) Crop responses to environment. CRC Press, Boca Raton

Hasanuzzaman M, Nahar K, Alam MM, Roychowdhury R, Fujita M (2013) Physiological, biochemical, and molecular mechanisms of heat stress tolerance in plants. Int J Mol Sci 14(5):9643–9684

Haslbeck M, Vierling E (2015) A first line of stress defense: small heat shock proteins and their function in protein homeostasis. J Mol Biol 427(7):1537–1548

Heggie L, Halliday KJ (2005) The highs and lows of plant life: temperature and light interactions in development. Int J Dev Biol 49(5/6):675

Herman JJ, Sultan SE (2011) Adaptive transgenerational plasticity in plants: case studies, mechanisms, and implications for natural populations. Front Plant Sci 2:18242

Hilker M, Schwachtje J, Baier M, Balazadeh S, Bäurle I, Geiselhardt S, Hincha DK, Kunze R, Mueller-Roeber B, Rillig MC (2016) Priming and memory of stress responses in organisms lacking a nervous system. Biol Rev 91(4):1118–1133

Hsu YT, Kao CH (2007) Toxicity in leaves of rice exposed to cadmium is due to hydrogen peroxide accumulation. Plant Soil 298:231–241

Istrail S, De-Leon SB-T, Davidson EH (2007) The regulatory genome and the computer. Dev Biol 310(2):187–195

Jacob P, Hirt H, Bendahmane A (2017) The heat-shock protein/chaperone network and multiple stress resistance. Plant Biotechnol J 15(4):405–414

Jagadish S, Muthurajan R, Oane R, Wheeler TR, Heuer S, Bennett J, Craufurd P (2010) Physiological and proteomic approaches to address heat tolerance during anthesis in rice (Oryza sativa L.). J Exp Bot 61(1):143–156

Janni M, Gullì M, Maestri E, Marmiroli M, Valliyodan B, Nguyen HT, Marmiroli N (2020) Molecular and genetic bases of heat stress responses in crop plants and breeding for increased resilience and productivity. J Exp Bot 71(13):3780–3802

Jeong D-H, Park S, Zhai J, Gurazada SGR, Paoli E, Meyers BC, Green PJ (2011) Massive analysis of rice small RNAs: mechanistic implications of regulated microRNAs and variants for differential target RNA cleavage. Plant Cell 23(12):4185–4207

Jones PD, New M, Parker DE, Martin S, Rigor IG (1999) Surface air temperature and its changes over the past 150 years. Rev Geophys 37(2):173–199

Kilasi NL, Singh J, Vallejos CE, Ye C, Jagadish S, Kusolwa P, Rathinasabapathi B (2018) Heat stress tolerance in rice (Oryza sativa L.): identification of quantitative trait loci and candidate genes for seedling growth under heat stress. Front Plant Sci 9:373853

Kinoshita T, Seki M (2014) Epigenetic memory for stress response and adaptation in plants. Plant Cell Physiol 55(11):1859–1863

Koizumi S, Ohama N, Mizoi J, Shinozaki K, Yamaguchi-Shinozaki K (2014) Functional analysis of the Hikeshi-like protein and its interaction with HSP70 in Arabidopsis. Biochem Biophys Res Commun 450(1):396–400

Kooke R, Keurentjes J (2012) Multi-dimensional regulation of metabolic networks shaping plant development and performance. J Exp Bot 63(9):3353–3365

Kotak S, Larkindale J, von Lee U, Koskull-Döring P, Vierling E, Scharf K-D (2007) Complexity of the heat stress response in plants. Curr Opin Plant Biol 10(3):310–316

Krol J, Loedige I, Filipowicz W (2010) The widespread regulation of microRNA biogenesis, function and decay. Nat Rev Genet 11(9):597–610

Kruszka K, Pacak A, Swida-Barteczka A, Nuc P, Alaba S, Wroblewska Z, Karlowski W, Jarmolowski A, Szweykowska-Kulinska Z (2014) Transcriptionally and post-transcriptionally regulated microRNAs in heat stress response in barley. J Exp Bot 65(20):6123–6135

Kumar RR, Pathak H, Sharma SK, Kala YK, Nirjal MK, Singh GP, Goswami S, Rai RD (2015) Novel and conserved heat-responsive microRNAs in wheat (Triticum aestivum L.). Funct Integr Genomics 15:323–348

Kushawaha AK, Khan A, Sopory SK, Sanan-Mishra N (2021) Priming by high temperature stress induces microRNA regulated heat shock modules indicating their involvement in thermopriming response in rice. Life 11(4):291

Lämke J, Brzezinka K, Altmann S, Bäurle I (2016) A hit-and-run heat shock factor governs sustained histone methylation and transcriptional stress memory. EMBO J 35(2):162–175

Lisar S, Motafakkerazad R, Hossain MM, Rahman I (2012) Causes, effects and responses. Water Stress 25(1):33

Liu X, Huang B (2000) Heat stress injury in relation to membrane lipid peroxidation in creeping bentgrass. Crop Sci 40(2):503–510

Liu Q, Liu B (2017) Integrating small RNA sequencing with QTL mapping for identification of miRNAs and their target genes associated with heat tolerance at the flowering stage in rice. Front Plant Sci 8:243607

Liu H, Able AJ, Able JA (2020) Transgenerational effects of water-deficit and heat stress on germination and seedling vigour—new insights from durum wheat microRNAs. Plants 9(2):189

Lorenzo CD, Sanchez-Lamas M, Antonietti MS, Cerdán PD (2016) Emerging hubs in plant light and temperature signaling. Photochem Photobiol 92(1):3–13

Lu X, Guan Q, Zhu J (2013) Downregulation of CSD2 by a heat-inducible miR398 is required for thermotolerance in Arabidopsis. Plant Signal Behav 8(8):e24952

Mahale BM, Fakrudin B, Ghosh S, Krishnaraj P (2014) LNA mediated in situ hybridization of miR171 and miR397a in leaf and ambient root tissues revealed expressional homogeneity in response to shoot heat shock in Arabidopsis thaliana. J Plant Biotechnol 23:93–103

Mangrauthia SK, Bhogireddy S, Agarwal S, Prasanth VV, Voleti S, Neelamraju S, Subrahmanyam D (2017) Genome-wide changes in microRNA expression during short and prolonged heat stress and recovery in contrasting rice cultivars. J Exp Bot 68(9):2399–2412

Mathur S, Mehta P, Jajoo A (2013) Effects of dual stress (high salt and high temperature) on the photochemical efficiency of wheat leaves (Triticum aestivum). Physiol Mol Biol 19:179–188

May P, Liao W, Wu Y, Shuai BR, McCombie W, Zhang MQ, Liu QA (2013) The effects of carbon dioxide and temperature on microRNA expression in Arabidopsis development. Nat Commun 4(1):2145

Mitsuda N, Ohme-Takagi M (2009) Functional analysis of transcription factors in Arabidopsis. Plant Cell Physiol 50(7):1232–1248

Mittler R, Blumwald E (2015) The roles of ROS and ABA in systemic acquired acclimation. Plant Cell 27(1):64–70

Miura K, Furumoto T (2013) Cold signaling and cold response in plants. Int J Mol Sci 14(3):5312–5337

Munné-Bosch S, Alegre L (2013) Cross-stress tolerance and stress "memory" in plants. Environ Exp Bot 94:1–88

Munns R, Tester M (2008) Mechanisms of salinity tolerance. Annu Rev Plant Biol 59:651–681

Narayanan S, Tamura PJ, Roth MR, Prasad PV, Welti R (2016) Wheat leaf lipids during heat stress: I. High day and night temperatures result in major lipid alterations. Plant Cell Environ 39(4):787–803

Nazarov PV, Reinsbach SE, Muller A, Nicot N, Philippidou D, Vallar L, Kreis S (2013) Interplay of microRNAs, transcription factors and target genes: linking dynamic expression changes to function. Nucleic Acids Res 41(5):2817–2831

Niu Y, Xiang Y (2018) An overview of biomembrane functions in plant responses to high-temperature stress. Front Plant Sci 9:366067

Ohama N, Sato H, Shinozaki K, Yamaguchi-Shinozaki K (2017) Transcriptional regulatory network of plant heat stress response. Trends Plant Sci 22(1):53–65

Pan C, Ye L, Zheng Y, Wang Y, Yang D, Liu X, Chen L, Zhang Y, Fei Z, Lu G (2017) Identification and expression profiling of microRNAs involved in the stigma exsertion under high-temperature stress in tomato. BMC Genomics 18:1–16

Pandey P, Irulappan V, Bagavathiannan MV, Senthil-Kumar M (2017) Impact of combined abiotic and biotic stresses on plant growth and avenues for crop improvement by exploiting physio-morphological traits. Front Plant Sci 8:237767

Piramila B, Prabha A, Nandagopalan V, Stanley A (2012) Effect of heat treatment on germination, seedling growth and some biochemical parameters of dry seeds of black gram. Int J Pharm Phytopharmacol Res 1:194–202

Qu A-L, Ding Y-F, Jiang Q, Zhu C (2013) Molecular mechanisms of the plant heat stress response. Biochem Biophys Res Commun 432(2):203–207

Rana RM, Dong S, Tang H, Ahmad F, Zhang H (2012) Functional analysis of OsHSBP1 and OsHSBP2 revealed their involvement in the heat shock response in rice (Oryza sativa L.). J Exp Bot 63(16):6003–6016

Rhoades MW, Reinhart BJ, Lim LP, Burge CB, Bartel B, Bartel DP (2002) Prediction of plant microRNA targets. Cell 110(4):513–520

Richter K, Haslbeck M, Buchner J (2010) The heat shock response: life on the verge of death. Mol Cell 40(2):253–266

Sailaja B, Voleti S, Subrahmanyam D, Sarla N, Prasanth VV, Bhadana V, Mangrauthia S (2014) Prediction and expression analysis of miRNAs associated with heat stress in Oryza sativa. Rice Sci 21(1):3–12

Sairam RK, Tyagi A (2004) Physiology and molecular biology of salinity stress tolerance in plants. Curr Sci 10:407–421

Salvucci ME, Crafts-Brandner SJ (2004) Relationship between the heat tolerance of photosynthesis and the thermal stability of Rubisco activase in plants from contrasting thermal environments. Plant Physiol 134(4):1460–1470

Sanei M, Chen X (2015) Mechanisms of microRNA turnover. Curr Opin Plant Biol 27:199–206

Sato H, Mizoi J, Tanaka H, Maruyama K, Qin F, Osakabe Y, Morimoto K, Ohori T, Kusakabe K, Nagata M (2014) Arabidopsis DPB3-1, a DREB2A interactor, specifically enhances heat stress-induced gene expression by forming a heat stress-specific transcriptional complex with NF-Y subunits. Plant Cell 26(12):4954–4973

Savchenko G, Klyuchareva E, Abramchik L, Serdyuchenko E (2002) Effect of periodic heat shock on the inner membrane system of etioplasts. Russ J Plant Physiol 49:349–359

Scandalios JG (1993) Oxygen stress and superoxide dismutases. Plant Physiol 101(1):7

Shekhar S, Gambhir G, Padaria JC (2017) Bioprospecting of genes from microbes for stress management in agricultural crops. In: Agro-environmental sustainability: volume 1: managing crop health. Springer, Cham, pp 127–147

Shi W, Li X, Schmidt RC, Struik PC, Yin X, Jagadish SK (2018) Pollen germination and in vivo fertilization in response to high-temperature during flowering in hybrid and inbred rice. Plant Cell Environ 41(6):1287–1297

Simões-Araújo JL, Rumjanek NG, Margis-Pinheiro M (2003) Small heat shock proteins genes are differentially expressed in distinct varieties of common bean. Braz J Plant Physiol 15:33–41

Steinhorst L, Kudla J (2014) Signaling in cells and organisms—calcium holds the line. Curr Opin Plant Biol 22:14–21

Stief A, Altmann S, Hoffmann K, Pant BD, Scheible W-R, Bäurle I (2014) Arabidopsis miR156 regulates tolerance to recurring environmental stress through SPL transcription factors. Plant Cell 26(4):1792–1807

Suzuki N, Koussevitzky S, Mittler R, Miller G (2012) ROS and redox signalling in the response of plants to abiotic stress. Plant Cell Environ 35(2):259–270

Thomson DW, Dinger ME (2016) Endogenous microRNA sponges: evidence and controversy. Nat Rev Genet 17(5):272–283

Tsukaguchi T, Kawamitsu Y, Takeda H, Suzuki K, Egawa Y (2003) Water status of flower buds and leaves as affected by high temperature in heat-tolerant and heat-sensitive cultivars of snap bean (Phaseolus vulgaris L.). Plant Prod Sci 6(1):24–27

Vollenweider P, Günthardt-Goerg MS (2005) Diagnosis of abiotic and biotic stress factors using the visible symptoms in foliage. Environ Pollut 137(3):455–465

von Koskull-Döring P, Scharf K-D, Nover L (2007) The diversity of plant heat stress transcription factors. Trends Plant Sci 12(10):452–457

Wahid A, Gelani S, Ashraf M, Foolad MR (2007) Heat tolerance in plants: an overview. Environ Exp Bot 61(3):199–223

Wang J, Cui L, Wang Y, Li J (2009) Growth, lipid peroxidation and photosynthesis in two tall fescue cultivars differing in heat tolerance. Biol Plant 53:237–242

Wang D, Li XF, Zhou ZJ, Feng XP, Yang WJ, Jiang DA (2010) Two Rubisco activase isoforms may play different roles in photosynthetic heat acclimation in the rice plant. Physiol Plant 139(1):55–67

Wang Y, Sun F, Cao H, Peng H, Ni Z, Sun Q, Yao Y (2012) TamiR159 directed wheat TaGAMYB cleavage and its involvement in anther development and heat response. PLoS One 7(11):e48445

Wang X, Ma X, Wang H, Li B, Clark G, Guo Y, Roux S, Sun D, Tang W (2015) Proteomic study of microsomal proteins reveals a key role for Arabidopsis annexin 1 in mediating heat stress-induced increase in intracellular calcium levels. Mol Cell Prot 14(3):686–694

Watanabe CK, Yamori W, Takahashi S, Terashima I, Noguchi K (2016) Mitochondrial alternative pathway-associated photoprotection of photosystem II is related to the photorespiratory pathway. Plant Cell Physiol 57(7):1426–1431

Wu T-Y, Juan Y-T, Hsu Y-H, Wu S-H, Liao H-T, Fung RW, Charng Y-Y (2013) Interplay between heat shock proteins HSP101 and HSA32 prolongs heat acclimation memory posttranscriptionally in Arabidopsis. Plant Physiol 161(4):2075–2084

Xia XJ, Gao CJ, Song LX, Zhou YH, Shi K, Yu JQ (2014) Role of H_2O_2 dynamics in brassinosteroid-induced stomatal closure and opening in Solanum lycopersicum. Plant Cell Environ 37(9):2036–2050

Xia X-J, Zhou Y-H, Shi K, Zhou J, Foyer CH, Yu J-Q (2015) Interplay between reactive oxygen species and hormones in the control of plant development and stress tolerance. J Exp Bot 66(10):2839–2856

Xin M, Wang Y, Yao Y, Xie C, Peng H, Ni Z, Sun Q (2010) Diverse set of microRNAs are responsive to powdery mildew infection and heat stress in wheat (Triticum aestivum L.). BMC Plant Biol 10:1–11

Yang X, Chen X, Ge Q, Li B, Tong Y, Zhang A, Li Z, Kuang T, Lu C (2006) Tolerance of photosynthesis to photoinhibition, high temperature and drought stress in flag leaves of wheat: a comparison between a hybridization line and its parents grown under field conditions. Plant Sci 171(3):389–397

Yao J, Liu B, Qin F (2011) Modular thermal sensors in temperature-gated transient receptor potential (TRP) channels. Proc Natl Acad Sci 108(27):11109–11114

Yu Y, Zhang Y, Chen X, Chen Y (2019) Plant noncoding RNAs: hidden players in development and stress responses. Annu Rev Cell Dev Biol 35:407–431

Zhang X, Zhou Q, Wang X, Cai J, Dai T, Cao W, Jiang D (2016) Physiological and transcriptional analyses of induced post-anthesis thermo-tolerance by heat-shock pretreatment on germinating seeds of winter wheat. Environ Exp Bot 131:181–189

Zhang M, An P, Li H, Wang X, Zhou J, Dong P, Zhao Y, Wang Q, Li C (2019) The miRNA-mediated post-transcriptional regulation of maize in response to high temperature. Int J Mol Sci 20(7):1754

Zhao C, Liu B, Piao S, Wang X, Lobell DB, Huang Y, Huang M, Yao Y, Bassu S, Ciais P (2017) Temperature increase reduces global yields of major crops in four independent estimates. Proc Natl Acad Sci 114(35):9326–9331

Zhou R, Yu X, Li XD, Santos TM, Rosenqvist E, Ottosen C-O (2020) Combined high light and heat stress induced complex response in tomato with better leaf cooling after heat priming. Plant Physiol Biochem 151:1–9

Exploring RNAi's Prospective Role in Curtailing Virus Transmission in Crop Plants

11

Abreham Chebte Alemu

Abstract

RNA interference (RNAi) is an innate antiviral plant immunity that inhibits viral genes and reduces viral infection. To activate this systemic protection, double-stranded RNA (dsRNA) derived from the viral genome and artificially synthesized can be used. Proteins such as Dicer-like ribonucleases (DCLs), Argonaute (AGO) proteins, dsRNA-binding proteins (DRBPs), and RNA-dependent RNA polymerase (RDR) are at the heart of RNAi-based antiviral plant system. Noncoding RNAs (ncRNAs) protect plants from virus-induced stress by either directly targeting the virus's genetic material or indirectly modulating plant defense pathways. Small noncoding RNAs (sncRNAs) are the most varied category, having a length of 20–31 nucleotides and including small interfering RNAs (siRNAs), microRNAs (miRNAs), and trans-acting siRNA (tas-iRNA). By capitalizing on this system, experts have been able to develop transgenic crops that can resist multiple virus and viroid types, such as rice stripe virus (RSV), tobacco mosaic virus (TMV), and citrus exocortis viroid (CEVd). While dsRNAs have been used successfully to generate resistance, there are still barriers to the application, such as dsRNA delivery reliability and off-target effects. This chapter covers the potential use of RNAi to inhibit viral infection in crops, including mechanism and virus-resistant crops produced by RNAi approach.

A. C. Alemu (✉)
Department of Biotechnology, College of Agriculture and Natural Resource Sciences, Debre Birhan University, Debre Birhan, Ethiopia

Festetics Doctoral School, Hungarian University of Agriculture and Life Sciences, Georgikon Campus, Keszthely, Hungary
e-mail: AbrehamChebte@dbu.edu.et

K. Goswami et al. (eds.), *Non-coding RNAs for Crop Improvement*,
https://doi.org/10.1007/978-981-97-7213-1_11

Keywords

Plant innate immunity · RNAi · Noncoding RNAs · Virus-resistances

11.1 Introduction

Viruses are a main problem in agriculture worldwide in the twenty-first century, causing global economic losses of approximately US$30 billion annually (Nicaise 2014). The problem is further exacerbated by the issue of global climate change adding significance to epidemics and prevention efforts of plant viruses (Trebicki and Finlay 2019; Trebicki 2020). The most common solution for the management of plant-virus diseases lies in the application of chemicals that have grave consequences on human health and environment (Damalas and Eleftherohorinos 2011). Various traditional methods, on the other hand, have been used to control plant viral diseases. These contain natural predators, reflective mulch and UV-absorbing sheets, early planting, weed suppression, crop-free periods, virus-free propagation material as well as discarding of affected plants. Nevertheless, all those approaches come with their costs and sometimes with ineffectiveness (Legg et al. 2014; Zaidi et al. 2016).

Combating viruses through the improvement of virus-resistant varieties appears to be the most economically viable and effective solution (Duan et al. 2012; Zhao et al. 2020). Recent discoveries in sequence information are expected to aid in generating novel, safe, selective, low-cost, and quickly deployable means of controlling plant viruses (Mahas and Mahfouz 2018; Yin and Qiu 2019). One such powerful tool that is coming up to help induce resistance to viruses is RNA interference (RNAi) (Simon-Mateo and Garcia 2011; Duan et al. 2012). This phenomenon was first investigated in plants as early as 1990 (Napoli et al. 1990), where it has been widely studied in fungi, animals, and plants (Fire et al. 1998; Baulcombe 2005). In 2006 Andrew Z. Fire and Craig C. Mello were granted a joint Nobel Prize in Physiology or Medicine for their recognition of double-stranded RNA silencing (Fire et al. 1998; Mello and Conte Jr. 2004).

The RNAi phenomena in plants is initiated by the existence of double strands of RNA (dsRNA) (Duan et al. 2012). Antiviral RNA silencing involves four core components; Dicer-like ribonucleases (DCLs), dsRNA-binding proteins (DRBs), Argonaute (AGO) proteins, and RNA-dependent RNA polymerase (RDR) proteins (Liu et al. 2009; Ding 2010; Carbonell and Carrington 2015). It takes places in four main steps: (1) transcription and cut of dsRNA into miRNA or siRNA, reaching from 21 to 24 nucleotides (nt) in length (Liu et al. 2009; Garcia-Ruiz et al. 2010); (2) loading of miRNA/siRNA into an RNA induced silencing complex (RISC) (Carbonell and Carrington 2015); (3) complementary bases pairing between a target gene with miRNA/siRNA (Ding and Voinnet 2007; Liu et al. 2009; Ding 2010); and, finally, (4) disruption or inhibition of the target gene via AGO protein(s) with cleavage activity (Borges and Martienssen 2015; Zvereva and Pooggin 2012).

ncRNAs have a pivotal role in antiviral defense mechanisms that plants employ against viral infections (Hou et al. 2019). There are three major classes of ncRNAs based on length, such as small (<31 nt), medium (32–200 nt), long (>200 nt)

(Wang et al. 2021). sncRNAs are the most diverse category, 20-30 nucleotides in length, and include miRNAs, Piwi-interacting RNAs (piRNAs), and siRNAs (Borges and Martienssen 2015; Budak and Akpinar 2015). piRNA is the most abundant type of sncRNA with a size typically between 26–31 nucleotides, and it can be derived from single-stranded precursor molecules (Lu et al. 2005). miRNAs are small, endogenously produced RNAs made of 20–24 nt (Borges and Martienssen 2015). They are derived from precursors of imperfectly paired "hairpin" RNA molecules and processed through Dicer-like proteins (Voinnet 2009; Axtell 2013). siRNAs are also composed of similar-sized nucleotides and require Dicer-like proteins for their production. However, siRNAs are derived from perfectly matched double-stranded trigger RNA molecules that can be endogenous or derived from introduced RNA, such as transgenes or viruses, and affect multiple sites in the target RNA molecule (Bartel 2004). Additionally, plant antiviral defense systems comprise other types of ncRNAs; trans-acting small interfering RNAs (ta-siRNAs), small nuclear RNAs (snRNAs), and small nucleolar RNAs (snoRNAs) (Qu 2010; Song et al. 2021).

Numerous studies show the effectiveness of RNAi against a variety of economically important plant viruses. These include those affecting citrus trees (Fagoaga et al. 2006), Papaya (Saxena et al. 2011, 2013), Wheat (Fahim et al. 2012; Kis et al. 2016), Soybean (Kim et al. 2013), Potato (Ntui et al. 2013), Tobacco (Hu et al. 2011), Barley (Wang et al. 2000), Rice (Shimizu et al. 2012), Tomato (Bucher et al. 2006; Lin et al. 2011; Saxena et al. 2011), Maize (Gan et al. 2010; Zhang et al. 2011), and cotton plants (Hashmi et al. 2011; Ali et al. 2013). Efficiency in target gene selection, reliable delivery of dsRNA, and off-target effects are the major challenges for RNAi-mediated virus control (Das and Sherif 2020). This chapter outlines the potential of RNAi to inhibit viral infection in crops focusing on mechanisms of antiviral plant RNAi, sncRNAs types and their mechanisms of action, virus-resistant crops produced by this approach, and limiting barriers.

11.2 RNAi as Innate and Adaptive Antiviral Immunity in Plants

RNAi is a well-maintained molecular mechanism that controls gene expression and functions as an antiviral agent upon virus invasion of various eukaryotes (Ratcliff et al. 1997; Kasschau and Carrington 1998; Li et al. 2013; Maillard et al. 2013). Evidence for the effect of regulated RNA activity on virus defeat was first discovered when transgenic expression of viral sequences protected plants from homologous viruses (Lindbo and Dougherty 1992; Dougherty et al. 1994). In 1990, the finding of RNA silencing in plants was made when a petunia was introduced with a chalcone synthase gene, causing the inhibition of its homolog (Napoli et al. 1990; van der Krol et al. 1990). In 1998, Fire and Mello found that dsRNA from *Caenorhabditis elegans* could result in sequence-specific inhibition of protein expression. This led to their coining of the term "RNA interference" (Fire et al. 1998). As a result of this ground-breaking discovery, Fire and Mello were granted

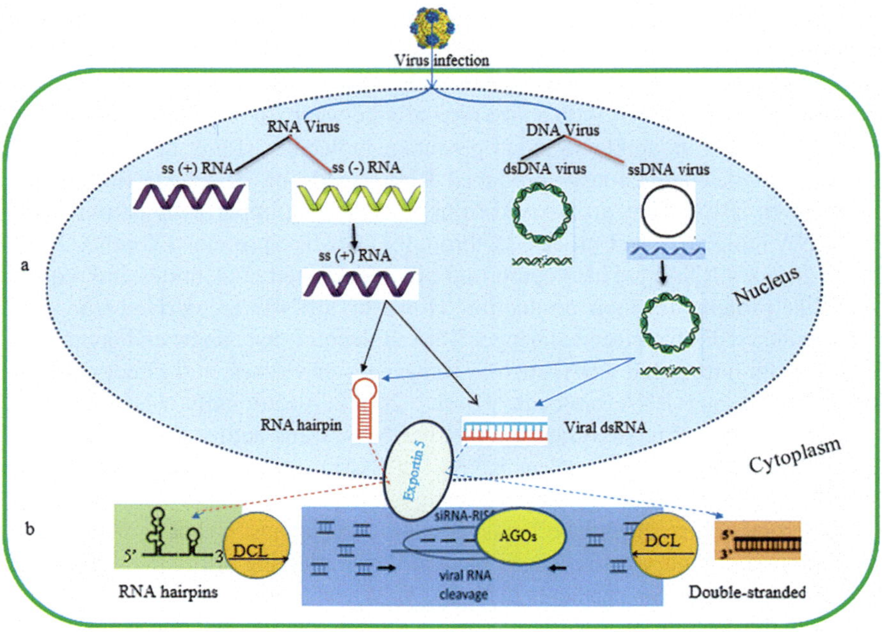

Fig. 11.1 (**a**) Schematic representation of basic viral double-stranded RNA substrates processing for Dicer (Pantaleo 2011), (**b**) a diagram illustrating how viral RNAs are processed and how gene silencing takes place in plant antiviral RNAi system (Taliansky et al. 2021).

the Nobel Prize in Physiology or Medicine in 2006 (Mello and Conte Jr. 2004), which had a seismic impact on biology.

Research has shown that RNAi-inducing molecules are short double-stranded RNAs (dsRNAs) or a hairpin structure that are processed by the Dicer enzyme (Ding 2010; Guo et al. 2019). When a virus infects a cell, it produces dsRNAs and also hairpins (Fig. 11.1a). These dsRNAs are then identified and cut by the host protein Dicer in to siRNAs, which are about 21–24 nucleotides long (Fig. 11.1b) (Liu et al. 2009; Garcia-Ruiz et al. 2010). The vsiRNA then enters an effector complex known as RISC, where it can bind to complementary nucleotides of their target gene via base pairing (Ding and Voinnet 2007; Liu et al. 2009; Ding 2010). The AGO proteins in this complex then proceed to cut out the target gene from the mRNA (Zvereva and Pooggin 2012; Borges and Martienssen 2015).

11.3 Plants Recruit Key Proteins Involved in RNAi-Antiviral Defense Pathway

Among the key enzymes and components of the RNAi-based plant response to viruses are DCL, AGO, DRB, and RNA-RDR, some of these proteins initiate the process and others act as effectors, enhancers, and transporters (Table 11.1).

Table 11.1 Key proteins involved in plant antiviral defense pathway.

Protein	Size (kDa)	Sub-class	Functional domain	Major function	References
Dicer	~220	DCL1 DCL2 DCL3 DCL4	N-terminal RNA helicase C/ ATPase	Unwinding of dsRNA	Cenik et al. (2011), Bologna and Voinnet (2014)
			PAZ	Recognizes the end of dsRNA	Zhang et al. (2004)
			Two tandem RNase III (a, b)	Cuts one of the strands of dsRNA	Zhang et al. (2004), Carmell and Hannon (2004)
			DUF283	Promote RNA–RNA base pairing and RNA-binding	Qin et al. (2010), Vergani-Junior et al. (2021)
			dsRBDs	Provides dsRNA binding and protein distribution in the nucleocytoplasmic	Banerjee and Barraud (2014)
AGO	~100	AGO 1 AGO 2 AGO 3 AGO 4 AGO 5 AGO 6 AGO 7 AGO 8 AGO 9 AGO 10	N-terminal (N)	Blocks guide-target duplexoutside position 16	Wang et al. (2000, 2009), Mallory and Vaucheret (2010)
			PAZ	Recognizes the 3′ end of sRNA	Wang et al. (2000), Mallory and Vaucheret (2010), Simon-Mateo and Garcia (2011)
			MID (middle)	Anchors the 5′ phosphate of sRNA	Mallory and Vaucheret (2010)
			PIWI	Cleavage of targeted mRNA	
			DUF1785	Destroys perfectly matched siRNA and miRNA duplexes	Derrien et al. (2018)
RDR	–	RDR1 RDR2 RDR3 RDR4 RDR5 RDR6	RDR1, RDR2 and RDR6 share the C-terminal canonical DLDGD catalytic motif of eukaryotic RDRs and orthologs in many plant species, whereas RDR3, RDR4 and RDR5 share an atypical DFDGD amino acid motif in their catalytic domains	RDR1, RDR2, and RDR6 are essential proteins for vsiRNA re-amplification, RdDM, and epigenetic silencing in plants. It also spreads antiviral defense signals to other cells. However, the ability to repeat RDR3, RDR4, and RDR5 consecutively has not been precisely established	Borges and Martienssen (2015)

(continued)

Table 11.1 (continued)

Protein	Size (kDa)	Sub-class	Functional domain	Major function	References
DRB	35–40	DRB1/HYL1 DRB2 DRB3 DRB4 DRB5	Contain 1 to 3 dsRBM of approximately 70 amino acids, which forms an α-β-β-β-α fold and two α-helices for interaction with dsRNA	Interact with DCLs for the effective production of siRNA, miRNA, nat-siRNA, hcsiRNAs, ta-siRNAs	Eamens et al. (2012), Fukudome and Fukuhara (2017), Incarbone et al. (2021)
HSP	~69	HSP70-90	–	Role in RISC formation and siRNA loading	Iki et al. (2010)
Other factors	–	AVI/ALA2 AVI2 ALA1	ALA1/ALA2 contain the typical P4-type ATPase structure	Cooperates with RDR1 and RDR6 to stimulate vsiRNA secondary biogenesis	Guo et al. (2017, 2019)
		CAMTA3	–	It activates BN2 to stabilize AGO1/2 and DCL1 and triggers RDR6 for vsiRNA amplification	Wang et al. (2021)

PAZ Piwi-Argonaute-Zwille, *283* domain of unknown function, *dsRBDs* dsRNA-binding domains, *RdDM* RNA-directed DNA methylation, *nat-siRNAs* natural antisense transcript-derived siRNAs, *HSP* heat shock protein, *AVI1* antiviral RNAi-defective 1; *ALA2* aminophospholipid transporting ATPase 2; *CAMTA3* calmodulin-binding transcription activator-3; *BN2* bifunctional nuclease-2

11.4 Noncoding RNAs in Antiviral Plant RNAi System

A summary of the types, mechanisms and biogenesis of key ncRNA molecules involved in plant antiviral RNAi systems is presented as follows.

11.4.1 miRNAs

miRNAs are part of several biological processes, such as development and stress responses (Borges and Martienssen 2015). They are small, noncoding RNAs produced by DCL1 from single-stranded RNA precursors transcribed by RNA polymerase II (Pol II) (Zhou and Tzanetakis 2019). Apart from their well-known role in siRNAs, increasing evidence shows miRNAs's significant role in plant antiviral defense (Du et al. 2011). For example, the miRNA expression level is altered when viruses infect plants. In the case of rice streak virus infection, the abundance of certain guide strands increased without any change to the corresponding guide strands (Du et al. 2011). Studies have also revealed a role for certain miRNAs in plant antiviral immunity. miR472 was negatively regulated by viral infections and targeted transcripts from resistance genes essential for immunity. Additionally, some other miRNAs have been documented to be involved with this process: miR16848, miR52847, miR31964, and miR44452 (Guo et al. 2015; Lian et al. 2016).

The posttranscriptional roles played by miRNAs have been characterized in Arabidopsis AGO1. AGO1 mRNA is regulated by miR168 at post-transcription levels before progressing to its translation activities which AGO1 itself explores while under virus attack; its expression is then boosted accompanied by up-regulation of miR168 compromising its defense activity against the virus (Várallyay et al. 2010). But this happens differently in rice where AGO18 is recruited competently via miR168 blocking off the formation of AGO1-miR548 RISC thereby releasing the AGO1 so it can function as an effector controlling viral silencing (Du et al. 2011; Wu et al. 2015). Meanwhile differently affecting diverse hosts where monocots possess exclusively miR528. Regulation of AO gene enhancing reactive oxygen species ROS leading to defense response upon certain virus infection hence increasing understanding on possible functions exerted by endogenous miRNA pathway, especially regarding regulation of candidate genes involved with silencing pathway (Wu et al. 2017).

11.4.2 siRNAs

siRNAs have gained much attention as they play a vital role in the natural antiviral defense known as RNAi, targeting viral RNAs for degradation (Baulcombe 2005). The production of siRNAs in plants involves a DCL that cuts virus-specific dsRNAs generated during virus replication into fragments of 21–25 nucleotides (Morozov et al. 2019). These fragments produce one strand, known as the guide strand, which is incorporated into a multi-protein complex composed of AGO, an endonuclease. This complex helps match the guide strand with complementary sequences within viral RNAs and then cleave them (Yang and Li 2018). Once the viral RNA cleaved, these aberrant products are amplified by endogenous RdRP leading to systemic siRNA spread throughout the plant; essential for efficient antiviral protection. To improve the siRNA's stability and lifespan methylation can be done through HEN1 (Guo et al. 2017).

For DNA plant viruses, such as geminiviruses, Pol II mediates bidirectional transcription in both sense and antisense orientations giving rise to perfect dsRNAs which are further cut by DCL into siRNAs (Aregger et al. 2012) (Fig. 11.1a). Viruses often evolve suppressors to overcome this defense system and infect plants far more easily; they do so by targeting different points in the siRNA pathway such as inhibiting production, sequestering siRNAs, or preventing their spread (Csorba et al. 2015). Others use structures that enlist incompatibility between host machinery and the virus itself (Li and Ding 2006). siRNA triggers powerful defenses against plant viruses that have been proven in many species and crops (Gaffar et al. 2019), enabling its utilization for the creation of innovative technologies aiming to control plant viruses.

11.4.3 ta-siRNA

ta-siRNAs play a vital role during plant-virus interactions. ta-siRNA, belonging to the family of phasiRNAs, is generated through mRNA cleavage in the trans (Fei et al. 2013). This process begins with miRNA-mediated clipping of TAS transcripts, followed by copying of resultant RNAs into dsRNAs by RDR6, and lastly processing of dsRNAs through type III ribonuclease to create several ta-siRNA strands in a phased manner (Axtell et al. 2006). These ta-siRNAs are present in the Arabidopsis genome in four families, *TAS1, TAS2, TAS3*, and *TAS4*, with eight loci altogether (Chen 2009). When viruses such as ToLCDNV infect plants, ta-siRNAs are created and transported to induce systemic acquired resistance (SAR) (Sahu et al. 2010). However, some viruses have developed strategies to suppress this ta-siRNAs generation through small peptides or proteins that interact with certain proteins associated with ta-siRNA biogenesis (Singh et al. 2015). Viruses also use different approaches to suppress ta-siRNAs generation, for example, the small peptide VISP1 has been reported to impair antiviral immunity by degrading SGS3 and hence inhibiting both SGS3/RDR6 dependent siRNA amplification as well as endogenous ta-siRNAs biogenesis in CMV infection (Tong et al. 2021).

11.4.4 snRNA

snRNAs are a common and highly abundant class of ncRNAs, involved in the post-transcriptional modification of pre-mRNA molecules to generate mature mRNAs (Valadkhan et al. 2009; Hudzik et al. 2020). snRNAs have been proposed to be important players within antiviral RNAi defense when assisting in the recruitment of siRNAs to specific viral targets. Multiple transcription factors initiate the production of snRNAs which undergo several steps: capping, splicing, 3′ end cleavage, processing, and transportation from the nucleus to cytoplasm, before maturation into functional snRNP complexes (Price et al. 1998; Will et al. 2001; Valadkhan et al. 2009). Alongside their role in splicing pre-mRNA, snRNAs can act as support molecules for metabolic processes such as antiviral defense (Singh and Cooper 2012). This is believed to work by allowing access to target viruses or infected cells due to a higher binding affinity than siRNAs and by recruiting AGO proteins at target sites. Evidence from recent research suggests that proteolysis complexes associated with snRNP may have regulatory effects on antiviral immunity reactions, introducing new interplays between plants and invading organisms that need further investigation (Ye 2013).

11.4.5 snoRNA

snoRNAs play an important role in eukaryote posttranscriptional modifications of other RNAs (Maxwell and Fournier 1995; Huang et al. 2022). In plants, they have several of functions, including controlling gene expression, RNA processing, and epigenetic control (Liang et al. 2002; Brown et al. 2003; Kruszka et al. 2003). Moreover, the biogenesis and maturation of double-stranded viral RNA which is essential for trans-generational antiviral defense are guided by snoRNAs (Lange and Gerbi 2000; Kiss 2002). To start an antiviral response against viral infection, these snoRNAs get transcribed from their genomic sites by RNA polymerase II to become pri-snoRNAs, which possess a 2′-O methyltransferase activity or endoribonuclease cleavage sites that allow them to target and modify particular RNAs with signaling recognition elements (Rakitina et al. 2011; Loza-Muller et al. 2015). Additionally, through targeting mRNA or miRNA levels in connection with ribosomal protein genes or other core cellular pathways, like splicing and elongation, snoRNAs can modulate antiviral responses such as CBF/DREB mediated defense signaling performed.

11.5 Viral Pathogens Affecting Crops

Plant diseases are mostly caused by viruses, which are estimated to have an annual economic impact of about $30 billion (Nicaise 2014; Muniyappa 2016). Nearly 50% of the pathogens that cause emerging plant diseases globally are viruses, which affect both cultivated and wild plants (Cooper and Jones 2006; Jones 2009). Viruses can infect a wide range of crop species, causing yield losses, reduced quality, and even complete crop failure (Reddy and Thirumala-Devi 2003; Anderson et al. 2004). Table 11.2 provides a comprehensive overview of major viral pathogens affecting major crop species, including their genome, symptom, and impact on crop yield and quality.

Table 11.2 Major viruses influencing important crop species.

Crop	Virus	Genome	Symptom and impact	Transmission mode	References
Wheat	*Wheat streak mosaic virus* (WSMV)	ssRNA+	Mosaic symptoms on leaves and stunted growth	Wheat curl mite	Singh et al. (2018)
	Wheat dwarf virus (WDV)	Circular ssDNA	Dwarfing, leaf chlorosis, reduced spike number, and significant yield loss	Leafhopper (*Psammotettixalienus*)	Nygren et al. (2015)
Rice	*Rice dwarf virus* (RSV)	dsRNAs	Severe stunting and shorter crown roots	Green rice leafhopper (*Nephotettixcincticeps*)	Pu et al. (2011), Jin et al. (2016)
	Rice tungro spherical virus (RTBV)	Circular dsDNA	Severe stunting, yellowing of the leaves, and reduced tillering		Azzam and Chancellor (2002), Budot et al. (2014)
Maize	*Maize dwarf mosaic virus* (MDMV)	ssRNA+	Mosaic symptom and 70% yield loss	Aphid species	Kannan et al. (2018)
	Maize streak virus (MSV)	Circular ssDNA	Chlorotic streaks on leaves, severe stunting, and failure to produce complete cobs or seed	Leafhopper species	Harrison et al. (1977), Shepherd et al. (2010)
Soybeans	*Soybean mosaic virus* (SMV)	ssRNA+	Severe leaf and seed mottling, distortion of leaves, necrosis, stunting, and death of infected plants	Aphid species	Hajimorad et al. (2018)
	Soybean vein necrosis virus (SVNV)	ssRNA+	Necrotic, stunting and reduced growth	Soybean thrips (*Neohydatothripsvariabilis*)	Zhou and Tzanetakis (2019), Suarez et al. (2021)
Potatoes	*Potato virus Y* (PVY)	ssRNA	Leaf mosaic, mottle and crinkling, vein necrosis, leaf drop and stunting of the plants	Aphids	Quenouille et al. (2013)
	Potato leaf roll virus (PLRV)	ssRNA+	Smaller tubers with internal net necrosis and 33–50% yield loss		Jones et al. (1982), Petrov et al. (2023)
Sugarcane	*Sugarcane mosaic virus*(SCMV)	ssRNA	Chlorotic streaks on the leaves, reduced growth, and stunted stems	Aphid (*Rhopalosiphummaidis Fitch*)	Lu et al. (2021)
	Sugarcane yellow leaf virus (SCYLV)	ssRNA+	Yellowing of the leaves, stunting, uneven growth, and reduced sugar content in the stalks	Aphid (*Melanaphissacchari*)	Holkar et al. (2020)

Barley	*Barley yellow dwarf viruses* (BYDVs)	ssRNA+	Stunting, reduced tillering, suppressed heading, sterility, and failure to fill the kernels	Aphids	Shen et al. (2020)
	Barley stripe mosaic virus (BSMV)	ssRNA+	Yellow streaks or spots, mosaic leaves, and stunted growth	Mechanical	Jackson et al. (2009), Zhang et al. (2017)
Tomatoes	*Tomato yellow leaf curl virus*(TYLCV)	Circular ssDNA	Curling, yellowing, and downward rolling of leaves, and stunted growth	Whiteflies (*Bemisiatabaci*)	El-Sappah et al. (2022), Yang et al. (2019)
	Tomato mosaic virus (ToMV)	ssRNA	Mosaic patterns, leaf distortion, wilting, and necrotic spots on fruit	Mechanical	Kumar et al. (2011), Lyu et al. (2023)
Bananas	*Banana bunchy top virus*(BBTV)	Circular ssDNA	Stunted growth, yellowing of leaves, and deformed or small fruit	Banana aphids (*Pentalonianigronervosa*)	Thomas et al. (2021), Guyot et al. (2022)
	Banana streak virus(BSV)	dsDNA	Streaking or chlorotic patterns on leaves	Mealybugs	Gayral et al. (2008), Côte et al. (2010)
Apples	*Apple mosaic virus*(ApMV)	ssRNA+	Yellow discolorations, line or oak-leaf patterns, and vein banding of leaves, andup to 25% yield reduction	Propagative material and grafting	Sutic et al. (1999), Kinoti et al. (2018)
	Apple stem pitting virus(ASPV)	ssRNA+	Quince fruit deformation and pear vein yellows		Jelkmann (1994), Mathioudakis et al. (2021)

11.6 RNAi Techniques for Crop Protection, Application and Limiting Barriers

The main applications of RNAi-based techniques have been in the generation of transgenic plants and Host-Induced Gene Silencing (HIGS) (Waterhouse et al. 1998; Qi et al. 2019). However, due to concerns about the effects on the environment and human health, acceptance of RNAi-based transgenic plants has always been difficult (Puyam and Kaur 2020). siRNAs, dsRNAs, and hpRNAs have all recently gained attention as potential exogenous uses of RNAs that may be more acceptable in terms of social, environmental, and economic factors compared to transgenic (Dubrovina and Kiselev 2019; Das and Sherif 2020). It is reported that various RNA interference pathways are used to artificially silence viral genes. These consist of Spray-Induced Gene Silencing (SIGS), Virus-Induced Gene Silencing (VIGS), and Host-Induced Gene Silencing (HIGS) combinations (Qi et al. 2019, 2023; Vetukuri et al. 2021).

According to Roth et al. 2004, the majority of viruses that infect plants have single-stranded RNA and reproduce via dsRNA intermediates. Plants can be genetically modified to express and carry new genes, giving them resistance to these viruses. RNAi-based antivirus resistance has shown efficacy in treating plant-virus infections in agricultural systems like papaya ring spot virus-resistant papaya, tobacco mosaic virus, cucumber mosaic virus, and cassava black streak disease (Gonsalves 2006; Hu et al. 2011; Patil et al. 2015). Temporary HIGS induced by viruses have been shown to target numerous eukaryotic plant diseases and their insect vectors (Nowara et al. 2010; Tinoco et al. 2010; Jahan et al. 2015). *Tobacco rattle virus* (TRV) and *barley stripe mosaic virus* (BSMV) are two plant viruses that can be altered to express desired sequences in plants (Yin et al. 2011; Chen et al. 2016; Song and Thomma 2018). RNAi methods have shown promise for targeting insects, particularly those that harbor plant diseases (Rosa et al. 2018).

Exogenous application of dsRNA has primarily used spray or mechanical inoculation to apply RNAi molecules, with potential for other delivery mechanisms (Gan et al. 2010; Konakalla et al. 2016; Kaldis et al. 2018). Various genes are suitable targets, with dsRNA, hpRNA, siRNA, and ssRNA species being most effective (Tenllado and Díaz-Ruíz 2001). Protection typically lasts 2–7 days but can be extended up to 70 days with protective agents (Mitter et al. 2017a). This has been effective against plant viruses, viroids, fungal pathogens, and insect vectors of plant pathogens (Robinson et al. 2014; Mitter et al. 2017b; Worrall et al. 2019; Dalakouras et al. 2020; Das and Sherif 2020; Voloudakis et al. 2022). The study demonstrates that loading RNAi-inducing dsRNA into BioClay nanoparticles and spraying it on plant surfaces provides sustained viral protection for up to 20 days, a significant improvement over previous methods (Mitter et al. 2017b).

Bacteria have been engineered to induce RNAi effects in their hosts, with *Escherichia coli* being used to express and deliver dsRNAs in 2001 (Timmons et al. 2001). This led to a Nobel Prize for discovering the mechanism of RNAi. Bacteria can secrete noncoding RNAs, or extracellular RNAs, by producing outer membrane vesicles (Baldrich et al. 2019). Many bacteria are endosymbionts in plants and insect vectors of plant viruses, suggesting they could be used to express RNAi inducers to help control plant diseases (Whitten et al. 2016). One recent example is using bacterial endosymbiont-mediated RNAi targeting the tubulin gene transcripts of western flower thrips, an insect vector of the tomato spotted wilt virus (Whitten et al. 2016).

There are various obstacles to the successful implementation of RNAi in controlling plant viruses (Whitehead et al. 2009; Bennett et al. 2020). To be successful, efficient delivery of a suitable dsRNA molecule is required but this is hard to achieve due to physical barriers such as cell walls and endomembrane systems obstructing its diffusion (Fenner et al. 2006). Additionally, there may be off-target effects if the pre-designed siRNAs do not align with their target sequence accurately, leading to other non-target genes being impacted instead of just those related to silencing viral RNA replication (Christiaens et al. 2020; Papadopoulou et al. 2020). On top of this, extracellular degradation of the dsRNA and host immune suppression further creates prominent challenges.

The high variability among different viral isolates makes it complex to create efficient siRNAs that can cover existing virus variants in a species or genotype (Christiaens et al. 2020). Therefore, specific siRNAs or broader class-based approaches may be used, which could result in an increased cost and a faster design. Plant viruses are mostly spread from host to host, or horizontally through insect vectors, which can make them systemic. Treatments are difficult if not impossible when selective pressures act on an individual host and have limited results for a long distance (Escriu et al. 2000, 2003). Additionally, some plant viruses suppress their hosts' capability of recognizing foreign material and weaken the efficacy of transgenic constructs applied with the RNAi technology needed to control diseases in grain and crops (Díaz-Pendón and Ding 2008; Zhang et al. 2019; Pumplin and Voinnet 2013).

11.7 Viral Tolerant Crops Developed by RNAi

The method involves introducing specific pieces of dsRNA into the crop genome which targets and silences the viral genetic material infects plant. This silencing disrupts the replication process so that the viruses can no longer reproduce and are reduced. Currently, this technique is being used to develop virus-resistant crops for numerous commercial applications including potatoes, maize, cotton, rice, and tomatoes (Table 11.3) (Zhang et al. 2011; Shimizu et al. 2012; Ntui et al. 2013; Sohrab et al. 2016; Rego-Machado et al. 2020).

Table 11.3 List of agriculturally important transgenic crops produced by RNAi systems against viruses and viroids.

Crop	Virus/viroid	Target gene/domain	RNA production and delivery method	Size of dsRNA (bp)	ncRNAs	Gene silencing level (%)	References
Rice	RGDV	Pns9	Agrobacterium-based transformation	500	siRNA	95	Shimizu et al. (2012)
	RTBV	Cp1-3		300	siRNA	Infection decreased by 100-fold	Sharma et al. (2018)
	RTSV	Cp1-3		300	siRNA	Infection decreased by 500-fold	
Maize	SCMV	CP1, CP2	Bacterially expressed dsRNA/hpRNAs (crude extracts) spraying	147, 140	siRNA	80, 70	Gan et al. (2010)
	MDMV	CP	Agrobacterium-based transformation	404	–	10.1–25.0	Zhang et al. (2011)
Wheat	WSMV	HC-pro	Microparticle Bombardment	700	miR395	Resistance breaking down over time	Fahim et al. (2012)

Barely	WDV	–	Agrobacterium-based transformation	413	miRNA	–	Ntui et al. (2013)
	BYDV-PAV	CP	BYDV-PAV hpRNA construct	626	–	– (yes)	Wang et al. (2000)
Cowpea	CMV, PMMoV	RP	Direct application of in vitro synthesized or bacterially expressed dsRNA (crude extract) or loading into LDH	330, 977	siRNA		Gupta et al. (2021)
	GBNV	NSs	Foliar application of bacterially expressed dsRNA	1320	–	12.5–20	
		N, NSs	Agrobacterium-based transformation	3546, 410	–	Infection decreased by 900 folds	Prasad Babu et al. (2019)
	MYMIV	AC2, AC4		186, 197	siRNA	~100	Kumar et al. (2011)
	BCMV	NIB	Direct application of in vitro synthesized or bacterially expressed dsRNA	480	–	53	Worrall et al. (2019)
		CP		461	–	87	
Soybean	SMV	SMV-CP	Agrobacterium-based transformation	527	siRNA	– (yes)	Kim et al. (2013)
	MYMIV	AC2		160	shRNA	Infection decreased by 80-fold	Ramesh et al. (2019)
Pea	PSbMV	CP (PSB117CZ)	In vitro synthesized dsRNAsspraying	500	–	55	Šafářová et al. (2014)

(continued)

Table 11.3 (continued)

Crop	Virus/viroid	Target gene/ domain	RNA production and delivery method	Size of dsRNA (bp)	ncRNAs	Gene silencing level (%)	References
Watermelon	ZYMV	HC-Pro	Mechanical inoculation of in vitro synthesized dsRNAs	588	miRNA	50	Kaldis et al. (2018)
		CP		498	miRNA	43	
Potato	CMV	RNA2	Agrobacterium-based transformation	1128	siRNA	100	Ntui et al. (2013)
Squash	ZYMV	HC-Pro		588	miRNA	18	Kaldis et al. (2018)
		CP		498	miRNA	16	
Brassica	TuMV	TIR-NBS-LRR		377	MIR1885	Yes	He et al. (2008)
		BraCP24		320	MIR1885/ tasiRNA	–	Cui et al. (2020)
Tomato	CEVd	–		365	siRNA	50	Carbonell and Carrington (2015)
	PSTVd	–		353	siRNA	Delay in symptom	
	ToLCV	AC1/AC4, AV1/AV2		449, 432	siRNA	45, 60	Namgial et al. (2019)
	TMV	CP		456	siRNA	43–63	Rego-Machado et al. (2020)
	PVY	CP		481	siRNA	57	
	TYLCV	V2, C4		351, 294	siRNA	28, 64	Melita et al. (2021)
	TLCV	AV1, AC1	Agrobacterium-mediated transformation of chemically synthesized siRNA duplexes	421	siRNA	75	Saxena et al. (2011)
	TSWV	N-gene		375	siRNA	65	Lin et al. (2011)
	ToLCTWV	C2		110	siRNA	80	
	TSWV, GRSV, TCSV and WSMoV	N-gene	Agrobacterium-based transformation	150	siRNA	82	Bucher et al. (2006)

Cucumber	ZYMV	HC-Pro	In vitro synthesized dsRNAs mechanical inoculation	588	miRNA	82	Kaldis et al. (2018)
		CP		498	miRNA	70	
Citrus	CTV	p23	–	630	siRNA	– (yes)	Fagoaga et al. (2006)
Papaya	PRSV	HC-Pro, CP	Mechanical inoculation of in vitro synthesized dsRNAs	594, 714	siRNA	81, 94	Vadlamudi et al. (2020)
		CP	Bacterially expressed hpRNAs (crude extracts) mechanical inoculation	279	siRNA	65	Shen et al. (2020)
	PLCV	AC1, AV1	Agrobacterium-mediated transformation of chemically synthesized siRNA duplexes	421	siRNA	75	Saxena et al. (2011)
	Geminiviruses	AV2, AC2 and AC4	–	–	siRNA	–	Saxena et al. (2013)
Pepper	PMMoV, TEV, AMV		Mechanical inoculation of in vitro synthesized dsRNAs	–			Tenllado and Díaz-Ruíz (2001)
	CMV	2b		336		40	Borah et al. (2018)

(continued)

Table 11.3 (continued)

Crop	Virus/viroid	Target gene/ domain	RNA production and delivery method	Size of dsRNA (bp)	ncRNAs	Gene silencing level (%)	References
Tobacco	PMMoV, TEV, AMV		Mechanical inoculation of in vitro synthesized dsRNAs	977, 1483, 1493		–	Tenllado and Díaz-Ruíz (2001)
	TMV	CP	Mechanical inoculation of *E. coli* expressed dsRNA (crude extracts)	480	siRNA	50	Yin et al. (2011)
		CP, p126		480, 666	siRNA	50, 65	Konakalla et al. (2016)
		CP, MP, RP		480	siRNA	48, 66, 34	Yin et al. (2011), Sun et al. (2010)
	TEV	HC-Pro		1483	–	100	Tenllado and Díaz-Ruíz (2001)
	PVY	HC-Pro, NIb, CP		600	siRNA	46, 72, 56	Sun et al. (2010)
	CMV	CP, 2b	Mechanical inoculation of in vitro synthesized dsRNAs	657, 336	siRNA	85, 62	Holeva et al. (2021)
		2b, AV1/AV2	Mechanical inoculation of *E. coli* expressed dsRNA (crude extracts)	336	siRNA	33	Namgial et al. (2019)
		2b		330	siRNA	76	Mitter et al. (2017a, b)
		CMV-2b	Agrobacterium-based transformation	335	siRNA	40	Namgial et al. (2019)
	TSWV	NSm, N	Mechanical inoculation of in vitro synthesized dsRNAs	646, 717	siRNA	54, 84	Konakalla et al. (2016)
	TMV	MP	Agrobacterium-based transformation	–	siRNA	–	Hu et al. (2011)
	CMV	RP		–	siRNA	–	
Cotton	Monopartite-Begomoviruses	V2		21	miRNA-169a	–	Ali et al. (2013), Melita et al. (2021)
	CLCuV	5′AC1, 3′AC1		669, 783	–	72, 81	Hashmi et al. (2011)
		βC1		–	–	60–70	Sohrab et al. (2016)

RGDV rice gall dwarf virus, *RTBV* rice tungro bacilliform virus, *RTSV* rice tungro spherical virus, *SCMV* sugarcane mosaic virus, *MDMV* maize dwarf mosaic virus, *WSMV* wheat streak mosaic virus, *WDV* wheat dwarf virus, *BYDV-PAV* barley yellow dwarf virus–PAV; *CMV* cucumber mosaic virus, *PMMoV* pepper mild mottle virus, *GBNV* groundnut bud necrosis virus, *BCMV* bean common mosaic virus, *SMV* soybean mosaic virus, *MYMIV* mungbean yellow mosaic India virus, *PSbMV* pea seed-borne mosaic virus, *ZYMV* zucchini yellow mosaic virus, *TuMV* turnip mosaic virus, *CEVd* citrus exocortis viroid, *PSTVd* potato spindle tuber viroid, *ToLCV* tomato leaf curl virus, *ToMV* tomato mosaic virus, *PVY* tobacco mosaic virus potato virus Y, *ToLCTWV* tomato leaf curl Taiwan virus, *TSWV* tomato spotted wilt virus, *TLCV* tomato leaf curl virus, *GRSV* groundnut ring spot virus, *TCSV* tomato chlorotic spot virus, *WSMoV* watermelon silver mottle virus, *PRSV* papaya ringspot virus, *PLCV* papaya leaf curl virus, *TEV* tobacco etch virus, *AMV* alfalfa mosaic virus, *CTV* citrus tristeza virus, *CP* coat protein, *AO* L-Ascorbate oxidase, *LDH* layered double hydroxide, *2b* clay nano-sheets, *2b* suppressor of silencing; *AC1* open reading frame 1 from complementary (C) strand of DNA A, *AC2* open reading frame 2 from complementary (C) strand of DNA A, *AC4* open reading frame 4 from complementary (C) strand of DNA A, *AV1* open reading frame 1 from viral (V) strand of DNA A, *AV2* open reading frame 2 from viral (V) strand of DNA A, *C4* open reading frame 2 from complementary (C) strand, *HC-Pro* heleper component proteinase, *MV* movement protein, *N* nucleocapsid, *NIb* nuclear inclusion protein b, *RP* replicase protein, *V2* open reading frame 2 from viral (V) strand, *p23* 23 KDa protein

11.8 Current Research and Innovations

RNAi technology has been used in crop protection to enhance resistance against insects, nematodes, and viruses. Currently, research focuses on optimizing RNA delivery methods, identifying suitable target genes, understanding off-target effects, and addressing regulatory concerns (Bartoszewski and Sikorski 2019; Bharathi et al. 2023). Other cutting-edge technologies include CRISPR-Cas9 (Soyars et al. 2018; Li et al. 2023), Antisense Oligonucleotides (ASOs) (Hunter et al. 2021; Sandoval-Mojica et al. 2021; Tarn et al. 2021), zinc finger nucleases (ZFNs), and transcription activator-like effector nucleases (TALENs) (Bhuyan et al. 2023; Daniel et al. 2023). These technologies have the potential to correct disease-causing mutations in plants (Daniel et al. 2023). Epigenetics, which refers to heritable changes in gene expression, plays a critical role in regulating gene expression patterns. By understanding and modifying these marks, novel approaches to the management of plant diseases may be developed (Kumari et al. 2022).

11.9 Conclusions and Perspectives

The RNAi pathway has long been known to possess innate and adaptive defense mechanisms against viruses in plants. Recent studies have proved that various ncRNAs are critical components of this pathway and appear to show a major role in antiviral response in plants. Virus-resistant crops generated by RNAi-based approaches show potential for practical applications; however, such approaches often affect from limitations related to the delivery efficiency and off-target effects of dsRNA into the cell, providing major challenges in terms of timely and efficient regulation of virus replication. Despite these obstacles, an increased awareness of the role and complexity of the several layers involved in RNAi antiviral immunity opens up prospects for its effective exploitation in combating plant viruses and ultimately contributes significantly to food security worldwide. Research focuses on optimizing RNA delivery methods, identifying target genes, understanding off-target effects, and addressing regulatory concerns. Other cutting-edge technologies in plant disease management include CRISPR-Cas9, ASOs, small compounds, ZFNs, and TALENs.

References

Ali I, Amin I, Briddon RW, Mansoor S (2013) Artificial microRNA-mediated resistance against the monopartite begomovirus Cotton leaf curl Burewala virus. Virol J 10:231

Anderson PK, Cunningham AA, Patel NG, Morales FJ, Epstein PR, Daszak P (2004) Emerging infectious diseases of plants: pathogen pollution, climate change and agrotechnology drivers. Trends Ecol Evol 19(10):535–544

Aregger M, Borah BK, Seguin J, Rajeswaran R, Gubaeva EG, Zvereva AS, Windels D, Vazquez F, Blevins T, Farinelli L, Pooggin MM (2012) Primary and secondary siRNAs in geminivirus-induced gene silencing. PLoS Pathog 8:e1002941

Axtell MJ (2013) Classification and comparison of small RNAs from plants. Annu Rev Plant Biol 64:137–159

Axtell MJ, Jan C, Rajagopalan R, Bartel DP (2006) A two-hit trigger for siRNA biogenesis in plants. Cell 127:565–577

Azzam O, Chancellor TCB (2002) The biology, epidemiology, and management of rice tungro disease in Asia. Plant Dis 86(2):88–100

Baldrich P, Rutter BD, Karimi HZ, Podicheti R, Meyers BC, Innes RW (2019) Plant extracellular vesicles contain diverse small RNA species and are enriched in 10- to 17-nucleotide "Tiny" RNAs. Plant Cell 31(2):315–324

Banerjee S, Barraud P (2014) Functions of double-stranded RNA-binding domains in nucleo-cytoplasmic transport. RNA Biol 11(10):1226–1232. https://doi.org/10.4161/15476286.2014.972856

Bartel DP (2004) MicroRNAs: genomics, biogenesis, mechanism, and function. Cell 116(2):281–297

Bartoszewski R, Sikorski AF (2019) Editorial focus: understanding off-target effects as the key to successful RNAi therapy. Cell Mol Biol Lett 24:1–23

Baulcombe D (2005) RNA silencing. Trends Biochem Sci 30:290–293

Bennett M, Deikman J, Hendrix B, Iandolino A (2020) Barriers to Efficient Foliar Uptake of dsRNA and Molecular Barriers to dsRNA Activity in Plant Cells. Front Plant Sci 11. https://doi.org/10.3389/fpls.2020.00816

Bharathi JK, Anandan R, Benjamin LK, Muneer S, Prakash MAS (2023) Recent trends and advances of RNA interference (RNAi) to improve agricultural crops and enhance their resilience to biotic and abiotic stresses. Plant Physiol Biochem 194:600–618

Bhuyan SJ, Kumar M, Ramrao Devde P, Rai AC, Mishra AK, Singh PK, Siddique KHM (2023) Progress in gene editing tools, implications and success in plants: a review. Front Genome 5:1272678

Bologna NG, Voinnet O (2014) The diversity, biogenesis, and activities of endogenous silencing small RNAs in Arabidopsis. Annu Rev Plant Biol 65:473–503. https://doi.org/10.1146/annurev-arplant-050213-035728

Borah M, Berbati M, Reppa C, Holeva M, Nath P, Voloudakis A (2018) RNA-based vaccination of Bhut Jolokia pepper (Capsicum chinense Jacq.) against cucumber mosaic virus. VirusDisease 29. https://doi.org/10.1007/s13337-018-0452-6

Borges F, Martienssen RA (2015) The expanding world of small RNAs in plants. Nat Rev Mol Cell Biol 16(12):727–741

Brown JW, Echeverria M, Qu L-H (2003) Plant snoRNAs: functional evolution and new modes of gene expression. Trends Plant Sci 8(1):42–49

Bucher E, Lohuis D, Van Poppel P, Geerts-Dimitriadou C, Goldbach R, Prins M (2006) Multiple virus resistance at a high frequency using a single transgene construct. J Gen Virol 87(12):3697–3701

Budak H, Akpinar BA (2015) Plant miRNAs: biogenesis, organization and origins. Funct Integr Genomics 15:523–531

Budot BO, Encabo JR, Ambita ID, Atienza-Grande GA, Satoh K, Kondoh H, Ulat VJ, Mauleon R, Kikuchi S, Choi IR (2014) Suppression of cell wall-related genes associated with stunting of Oryza glaberrima infected with Rice tungro spherical virus. Front Microbiol 5:26

Carbonell A, Carrington JC (2015) Antiviral roles of plant argonautes. Curr Opin Plant Biol 27:111–117

Carmell MA, Hannon GJ (2004) RNase III enzymes and the initiation of gene silencing. Nat Struct Mol Biol 11(3):214–218. https://doi.org/10.1038/nsmb729

Cenik ES, Fukunaga R, Lu G, Dutcher R, Wang Y, Tanaka Hall TM, Zamore PD (2011) Phosphate and R2D2 restrict the substrate specificity of Dicer-2, an ATP-driven ribonuclease. Mol Cell 42(2):172–184. https://doi.org/10.1016/j.molcel.2011.03.002

Chen X (2009) Small RNAs and their roles in plant development. Annu Rev Cell Dev Biol 25:21–44

Chen W, Kastner C, Nowara D, Oliveira-Garcia E, Rutten T, Zhao Y, Deising HB, Kumlehn J, Schweizer P (2016) Host-induced silencing of Fusarium culmorum genes protects wheat from infection. J Exp Bot 67(17):4979–4991

Christiaens O, Whyard S, Velez AM, Smagghe G (2020) Double-stranded RNA technology to control insect pests: current status and challenges. Front Plant Sci 11:505962

Cooper I, Jones RA (2006) Wild plants and viruses: under-investigated ecosystems. Adv Virus Res 67:1–47

Côte FX, Galzi S, Folliot M, Lamagnère Y, Teycheney PY, Iskra-Caruana ML (2010) Micropropagation by tissue culture triggers differential expression of infectious endogenous Banana streak virus sequences (eBSV) present in the B genome of natural and synthetic interspecific banana plantains. Mol Plant Pathol 11(1):137–144

Csorba T, Kontra L, Burgyan J (2015) viral silencing suppressors: tools forged to fine-tune host-pathogen coexistence. Virology 479:85–103

Cui C, Wang JJ, Zhao JH, Fang YY, He XF, Guo HS, Duan CG (2020) A Brassica miRNA Regulates Plant Growth and Immunity through Distinct Modes of Action. Mol Plant 13(2):231–245. https://doi.org/10.1016/j.molp.2019.11.010

Dalakouras A, Wassenegger M, Dadami E, Ganopoulos I, Pappas ML, Papadopoulou K (2020) Genetically modified organism-free RNA interference: exogenous application of RNA molecules in plants. Plant Physiol 182(1):38–50

Damalas CA, Eleftherohorinos IG (2011) Pesticide exposure, safety issues, and risk assessment indicators. Int J Environ Res Public Health 8(5):1402–1419

Daniel MA, Sebastin R, Yu J-K, Soosaimanickam MP, Chung JW (2023) Enhancing horticultural crops through genome editing: applications, benefits, and considerations. Horticulture 9(8):884

Das PR, Sherif SM (2020) Application of exogenous dsRNAs-induced RNAi in agriculture: challenges and triumphs. Front Plant Sci 11:542801

Díaz-Pendón JA, Ding SW (2008) Direct and indirect roles of viral suppressors of RNA silencing in pathogenesis. Annu Rev Phytopathol 46:303–326

Derrien B, Clavel M, Baumberger N, Iki T, Sarazin A, Hacquard T, Genschik P (2018) A Suppressor Screen for AGO1 Degradation by the Viral F-Box P0 Protein Uncovers a Role for AGO DUF1785 in sRNA Duplex Unwinding. Plant Cell 30(6):1353–1374. https://doi.org/10.1105/tpc.18.00111

Ding SW (2010) RNA-based antiviral immunity. Nat Rev Immunol 10(9):632–644

Ding SW, Voinnet O (2007) Antiviral immunity directed by small RNAs. Cell 130(3):413–426

Dougherty WG, Lindbo JA, Smith HA, Parks TD, Swaney S, Proebsting WM (1994) RNA-mediated virus resistance in transgenic plants: exploitation of a cellular pathway possibly involved in RNA degradation. Mol Plant-Microbe Interact 7(5):544–552

Du P, Wu J, Zhang J, Zhao S, Zheng H, Gao G, Wei L, Li Y (2011) Viral infection induces expression of novel phased microRNAs from conserved cellular microRNA precursors. PLoS Pathog 7(8):e1002176

Duan CG, Wang CH, Guo HS (2012) Application of RNA silencing to plant disease resistance. Silence 3:1–8

Dubrovina AS, Kiselev KV (2019) Exogenous RNAs for gene regulation and plant resistance. Int J Mol Sci 20:2282

Eamens AL, Kim KW, Curtin SJ, Waterhouse PM (2012) DRB2 Is Required for MicroRNA Biogenesis in Arabidopsis thaliana. PLoS One 7(4):e35933. https://doi.org/10.1371/journal.pone.0035933

El-Sappah AH, Qi S, Huang Q, Wan L, Cheng GT, Liu J, Ihtisham M, Noor Z, Rouf Mir R, Zhao X, Yan K, Abbas M, Li J (2022) Natural resistance of tomato plants to Tomato yellow leaf curl virus. Front Plant Sci 13:1081549

Escriu F, Fraile A, García-Arenal F (2000) Evolution of virulence in natural populations of the satellite RNA of cucumber mosaic virus. Phytopathology 90(5):480–485

Escriu F, Fraile A, García-Arenal F (2003) The evolution of virulence in a plant virus. Evolution 57(4):755–765

Fagoaga C, Lopez C, De Mendoza AH, Moreno P, Navarro L, Flores R, Pena L (2006) Post-transcriptional gene silencing of the p23 silencing suppressor of Citrus tristeza virus confers resistance to the virus in transgenic Mexican lime. Plant Mol Biol 60:153–165

Fahim M, Millar AA, Wood CC, Larkin PJ (2012) Resistance to wheat streak mosaic virus generated by expression of an artificial polycistronic microRNA in wheat. Plant Biotechnol J 10(2):150–163

Fei Q, Xia R, Meyers BC (2013) Phased, secondary, small interfering RNAs in posttranscriptional regulatory networks. Plant Cell 25(7):2400–2415

Fenner BJ, Goh W, Kwang J (2006) Sequestration and protection of double-stranded RNA by the betanodavirus b2 protein. J Virol 80(14):6822–6833

Fire A, Xu S, Montgomery MK, Kostas SA, Driver SE, Mello CC (1998) Potent and specific genetic interference by double-stranded RNA in Caenorhabditis elegans. Nature 391(6669):806–811

Fukudome A, Fukuhara T (2017) Plant dicer-like proteins: double-stranded RNA-cleaving enzymes for small RNA biogenesis. J Plant Res 130(1):33–44. https://doi.org/10.1007/s10265-016-0877-1

Gaffar FY, Imani J, Karlovsky P, Koch A, Kogel KH (2019) Different components of the RNA interference machinery are required for conidiation, ascosporogenesis, virulence, deoxynivalenol production, and fungal inhibition by exogenous double-stranded RNA in the head blight pathogen Fusarium graminearum. Front Microbiol 10:1662

Gan D, Zhang J, Jiang H, Jiang T, Zhu S, Cheng B (2010) Bacterially expressed dsRNA protects maize against SCMV infection. Plant Cell Rep 29:1261–1268

Garcia-Ruiz H, Takeda A, Chapman EJ, Sullivan CM, Fahlgren N, Brempelis KJ, Carrington JC (2010) Arabidopsis RNA-dependent RNA polymerases and dicer-like proteins in antiviral defense and small interfering RNA biogenesis during turnip mosaic virus infection. Plant Cell 22(2):481–496

Gayral P, Noa-Carrazana JC, Lescot M, Lheureux F, Lockhart BE, Matsumoto T, Piffanelli P, Iskra-Caruana ML (2008) A single banana streak virus integration event in the banana genome as the origin of infectious endogenous pararetrovirus. J Virol 82(13):6697–6710

Gonsalves D (2006) Transgenic papaya: development, release, impact and challenges. Adv Virus Res 67:317–354

Guo C, Li L, Wang X, Liang C (2015) Alterations in siRNA and miRNA expression profiles detected by deep sequencing of transgenic rice with siRNA-mediated viral resistance. PLoS One 10(1):e0116175

Guo Z, Lu J, Wang X, Zhan B, Li W, Ding SW (2017) Lipid flippases promote antiviral silencing and the biogenesis of viral and host siRNAs in Arabidopsis. Proc Natl Acad Sci USA 114(6):1377–1382

Guo Z, Li Y, Ding SW (2019) Small RNA-based antimicrobial immunity. Nat Rev Immunol 19:31–44

Gupta D, Singh OW, YB B, Roy A, Mukherjee S, Mandal B (2021) Direct Foliar Application of dsRNA Derived From the Full-Length Gene of NSs of Groundnut Bud Necrosis Virus Limits Virus Accumulation and Symptom Expression. Front Plant Sci 12. https://doi.org/10.3389/fpls.2021.734618

Guyot V, Rajeswaran R, Chu HC, Karthikeyan C, Laboureau N, Galzi S, Mukwa LFT, Krupovic M, Kumar PL, Iskra-Caruana ML, Pooggin MM (2022) A newly emerging alphasatellite affects banana bunchy top virus replication, transcription, siRNA production and transmission by aphids. PLoS Pathog 18(4):e1010448

Hajimorad MR, Domier LL, Tolin SA, Whitham SA, Saghai Maroof MA (2018) Soybean mosaic virus: a successful potyvirus with a wide distribution but restricted natural host range. Mol Plant Pathol 19(7):1563–1579

Harrison BD, Barker H, Bock KR, Guthrie EJ, Meredith G, Atkinson M (1977) Plant viruses with circular single-stranded DNA. Nature 270:760–762

Hashmi JA, Zafar Y, Arshad M, Mansoor S, Asad S (2011) Engineering cotton (Gossypium hirsutum L.) for resistance to cotton leaf curl disease using viral truncated AC1 DNA sequences. Virus Genes 42:286–296

He X-F, Fang Y-Y, Feng L, Guo H-S (2008) Characterization of conserved and novel microRNAs and their targets, including a TuMV-induced TIR–NBS–LRR class R gene-derived novel miRNA in Brassica. FEBS Lett 582(16):2445–2452. https://doi.org/10.1016/j.febslet.2008.06.011

Holeva MC, Sklavounos A, Rajeswaran R, Pooggin MM, Voloudakis AE (2021) Topical application of double-stranded RNA targeting 2b and CP genes of Cucumber mosaic virus protects plants against local and systemic viral infection. Plants 10(5):963

Holkar SK, Balasubramaniam P, Kumar A, Kadirvel N, Shingote PR, Chhabra ML, Kumar S, Kumar P, Viswanathan R, Jain RK, Pathak AD (2020) Present status and future management strategies for sugarcane yellow leaf virus: a major constraint to the global sugarcane production. Plant Pathol J 36(6):536–557

Hou J, Lu D, Mason AS, Li B, Xiao M, An S, Fu D (2019) Non-coding RNAs and transposable elements in plant genomes: emergence, regulatory mechanisms and roles in plant development and stress responses. Planta 250:23–40

Hu Q, Niu Y, Zhang K, Liu Y, Zhou X (2011) Virus-derived transgenes expressing hairpin RNA give immunity to Tobacco mosaic virus and Cucumber mosaic virus. Virol J 8:41

Huang Z-H, Du Y-P, Wen J-T, Lu B-F, Zhao Y (2022) snoRNAs: functions and mechanisms in biological processes, and roles in tumor pathophysiology. Cell Death Dis 8(1):259

Hudzik C, Hou Y, Ma W, Axtell MJ (2020) Exchange of small regulatory RNAs between plants and their pests. Plant Physiol 182(1):51–62

Hunter WB, Cooper WR, Sandoval-Mojica AF, Mccollum G, Aishwarya V, Pelz-Stelinski KS (2021) Improving suppression of hemipteran vectors and bacterial pathogens of citrus and solanaceous plants: advances in antisense oligonucleotides (FANA). Front Agron 3:675247

Iki T, Yoshikawa M, Nishikiori M, Jaudal MC, Matsumoto-Yokoyama E, Mitsuhara I, Ishikawa M (2010) In vitro assembly of plant RNA-induced silencing complexes facilitated by molecular chaperone HSP90. Mol Cell 39(2):282–291. https://doi.org/10.1016/j.molcel.2010.05.014

Incarbone M, Clavel M, Monsion B, Kuhn L, Scheer H, Vantard É, Ritzenthaler C (2021) Immunocapture of dsRNA-bound proteins provides insight into Tobacco rattle virus replication complexes and reveals Arabidopsis DRB2 to be a wide-spectrum antiviral effector. Plant Cell 33(11):3402–3420. https://doi.org/10.1093/plcell/koab214

Jackson AO, Lim HS, Bragg J, Ganesan U, Lee MY (2009) Hordeivirus replication, movement, and pathogenesis. Annu Rev Phytopathol 47:385–422

Jahan SN, Åsman AK, Corcoran P, Fogelqvist J, Vetukuri RR, Dixelius C (2015) Plant-mediated gene silencing restricts growth of the potato late blight pathogen Phytophthora infestans. J Exp Bot 66(9):2785–2794

Jelkmann W (1994) Nucleotide sequences of apple stem pitting virus and of the coat protein gene of a similar virus from pear associated with vein yellows disease and their relationship with potex- and carlaviruses. J Gen Virol 75(7):1535–1542

Jin L, Qin Q, Wang Y, Pu Y, Liu L, Wen X, Ji S, Wu J, Wei C, Ding B, Li Y (2016) Rice dwarf virus P2 protein hijacks auxin signaling by directly targeting the rice OsIAA10 protein, enhancing viral infection and disease development. PLoS Pathog 12(9):e1005847

Jones RA (2009) Plant virus emergence and evolution: origins, new encounter scenarios, factors driving emergence, effects of changing world conditions, and prospects for control. Virus Res 141(12):113–130

Jones R, Fribourg C, Slack SA (1982) Potato virus and virus-like diseases. potato virus and virus-like diseases. Clemson University, Clemson

Kaldis A, Berbati M, Melita O, Reppa C, Holeva M, Otten P, Voloudakis A (2018) Exogenously applied dsRNA molecules deriving from the Zucchini yellow mosaic virus (ZYMV) genome move systemically and protect cucurbits against ZYMV. Mol Plant Pathol 19(4):883–895

Kannan M, Ismail I, Bunawan H (2018) Maize dwarf mosaic virus: from genome to disease management. Viruses 10(9):492

Kasschau KD, Carrington JC (1998) A counterdefensive strategy of plant viruses: suppression of posttranscriptional gene silencing. Cell 95(4):461–470

Kim HJ, Kim M-J, Pak JH, Jung HW, Choi HK, Lee Y-H, Baek I-Y, Ko J-M, Jeong S-C, Pack IS, Ryu KH, Chung Y-S (2013) Characterization of SMV resistance of soybean produced by genetic transformation of SMV-CP gene in RNAi. Plant Biotechnol Rep 7:425–433

Kinoti WM, Constable FE, Nancarrow N, Plummer KM, Rodoni B (2018) The incidence and genetic diversity of apple mosaic virus (ApMV) and prune dwarf virus (PDV) in prunus species in Australia. Viruses 10(3):136

Kis A, Tholt G, Ivanics M, Varallyay E, Jenes B, Havelda Z (2016) Polycistronic artificial miRNA-mediated resistance to wheat dwarf virus in barley is highly efficient at low temperature. Mol Plant Pathol 17(3):427–437

Kiss T (2002) Small nucleolar RNAs: an abundant group of noncoding RNAs with diverse cellular functions. Cell 109(2):145–148

Konakalla NC, Kaldis A, Berbati M, Masarapu H, Voloudakis AE (2016) Exogenous application of double-stranded RNA molecules from TMV p126 and CP genes confers resistance against TMV in tobacco. Planta 244:961–969

Kruszka K, Barneche F, Guyot R, Ailhas J, Meneau I, Schiffer S, Marchfelder A, Echeverría M (2003) Plant dicistronic tRNA-snoRNA genes: a new mode of expression of the small nucleolar RNAs processed by RNase Z. EMBO J 22:621–632

Kumar S, Udaya Shankar AC, Nayaka SC, Lund OS, Prakash HS (2011) Detection of tobacco mosaic virus and tomato mosaic virus in pepper and tomato by multiplex RT-PCR. Lett Appl Microbiol 53(3):359–363

Kumari P, Khan S, Wani IA, Gupta R, Verma S, Alam P, Alaklabi A (2022) Unravelling the role of epigenetic modifications in development and reproduction of angiosperms: a critical appraisal. Front Genet 13:819941

Lange TS, Gerbi SA (2000) Transient nucleolar localization of U6 small nuclear RNA InXenopus laevis oocytes. Mol Biol Cell 11(7):2419–2428

Legg JP, Shirima R, Tajebe LS, Guastella D, Boniface S, Jeremiah S, Nsami E, Chikoti P, Rapisarda C (2014) Biology and management of bemisia whitefly vectors of cassava virus pandemics in Africa. Pest Manag Sci 70(10):1446–1453

Li F, Ding SW (2006) Virus counterdefense: diverse strategies for evading the RNA-silencing immunity. Ann Rev Microbiol 60:503–531

Li Y, Lu J, Han Y, Fan X, Ding SW (2013) RNA interference functions as an antiviral immunity mechanism in mammals. Science 342(6155):231–234

Li T, Yang Y, Qi H, Cui W, Zhang L, Fu X, He X, Liu M, Li P-F, Yu T (2023) CRISPR/Cas9 therapeutics: progress and prospects. Signal Transduct Target Ther 8(1):36

Lian J, Guo J, Huang X, Yang XI, Huang G, Mao H, Sun HH, Ba Y, Zhou J (2016) miRNAs regulate hERG. J Cardiovasc Electrophysiol 27(12):1472–1482

Liang D, Zhou H, Zhang P, Chen YQ, Chen X, Chen CL, Qu LH (2002) A novel gene organization: intronic snoRNA gene clusters from Oryza sativa. Nucleic Acids Res 30(14):3262–3272

Lin CY, Ku HM, Tsai WS, Green SK, Jan FJ (2011) Resistance to a DNA and a RNA virus in transgenic plants by using a single chimeric transgene construct. Transgenic Res 20:261–270

Lindbo JA, Dougherty WG (1992) Pathogen-derived resistance to a potyvirus: immune and resistant phenotypes in transgenic tobacco expressing altered forms of a potyvirus coat protein nucleotide sequence. Mol Plant-Microbe Interact 5(2):144–153

Liu Q, Feng Y, Zhu Z (2009) Dicer-like (DCL) proteins in plants. Funct Integr Genomics 9:277–286

Loza-Muller L, Rodríguez-Corona U, Sobol M, Rodríguez-Zapata LC, Hozak P, Castano E (2015) Fibrillarin methylates H2A in RNA polymerase I trans-active promoters in Brassica oleracea. Front Plant Sci 6:976

Lu J, Getz G, Miska EA, Alvarez-Saavedra E, Lamb J, Peck D, Sweet-Cordero A, Ebert BL, Mak RH, Ferrando AA, Downing JR, Jacks T, Horvitz HR, Golub TR (2005) MicroRNA expression profiles classify human cancers. Nature 435(7043):834–838

Lu G, Wang Z, Xu F, Pan YB, Grisham MP, Xu L (2021) Sugarcane mosaic disease: characteristics, identification and control. Microorganisms 9(9):1984

Lyu J, Yang Y, Sun X, Jiang S, Hong H, Zhu X, Liu Y (2023) Genetic variability and molecular evolution of tomato mosaic virus populations in three Northern China Provinces. Viruses 15:1617

Mahas A, Mahfouz M (2018) Engineering virus resistance via CRISPR-Cas systems. Curr Opin Virol 32:1–8

Maillard PV, Ciaudo C, Marchais A, Li Y, Jay F, Ding SW, Voinnet O (2013) Antiviral RNA interference in mammalian cells. Science 342(6155):235–238

Mallory A, Vaucheret H (2010) Form, Function, and Regulation of ARGONAUTE Proteins. Plant Cell 22(12):3879–3889. https://doi.org/10.1105/tpc.110.080671

Mathioudakis MM, Maliogka VI, Candresse T, Nickel O, Fajardo TVM, Budzyńska D, Hasiów-Jaroszewska B, Katis NI (2021) Molecular characterization of the coat protein gene of greek apple stem pitting virus isolates: evolution through deletions, insertions, and recombination events. Plants 10(5):91

Maxwell ES, Fournier MJ (1995) The small nucleolar RNAs. Annu Rev Biochem 64(1):897–934

Melita O, Kaldis A, Berbati M, Reppa C, Holeva M, Lapidot M, Voloudakis A (2021) Topical application of double-stranded RNA molecules deriving from Tomato yellow leaf curl virus reduces cognate virus infection in tomato. Biol Plant 65:100–110. https://doi.org/10.32615/bp.2020.172

Mello CC, Conte D Jr (2004) Revealing the world of RNA interference. Nature 431(7006):338–342

Mitter N, Worrall EA, Robinson KE, Li P, Jain RG, Taochy C, Fletcher SJ, Carroll BJ, Lu GQ, Xu ZP (2017a) Clay nanosheets for topical delivery of RNAi for sustained protection against plant viruses. Nat Plants 3(2):16207

Mitter N, Worrall EA, Robinson KE, Xu ZP, Carroll BJ (2017b) Induction of virus resistance by exogenous application of double-stranded RNA. Curr Opin Virol 26:49–55

Morozov SY, Solovyev AG, Kalinina NO, Taliansky ME (2019) Double-stranded RNAs in plant protection against pathogenic organisms and viruses in agriculture. Acta Nat 11(43):13–21

Muniyappa V (2016) K. Subramanya Sastry and Thomas A. Zitter: plant virus and viroid diseases in the tropics, volume 2: epidemiology and management. Springer, Cham, p 489

Namgial T, Kaldis A, Chakraborty S, Voloudakis A (2019) Topical application of double-stranded RNA molecules containing sequences of Tomato leaf curl virus and Cucumber mosaic virus confers protection against the cognate viruses. Physiol Mol Plant Pathol 108:101432. https://doi.org/10.1016/j.pmpp.2019.101432

Napoli C, Lemieux C, Jorgensen R (1990) Introduction of a chimeric chalcone synthase gene into petunia results in reversible co-suppression of homologous genes in trans. Plant Cell 2(4):279–289

Nicaise V (2014) Crop immunity against viruses: outcomes and future challenges. Front Plant Sci 5:118474

Nowara D, Gay A, Lacomme C, Shaw J, Ridout C, Douchkov D, Hensel G, Kumlehn J, Schweizer P (2010) HIGS: host-induced gene silencing in the obligate biotrophic fungal pathogen Blumeria graminis. Plant Cell 22(9):3130–3141

Ntui VO, Kynet K, Azadi P, Khan RS, Chin DP, Nakamura I, Mii M (2013) Transgenic accumulation of a defective cucumber mosaic virus (CMV) replicase derived double stranded RNA modulates plant defence against CMV strains O and Y in potato. Transgenic Res 22:1191–1205

Nygren J, Shad N, Kvarnheden A, Westerbergh A (2015) Variation in susceptibility to Wheat dwarf virus among wild and domesticated wheat. PLoS One 10:e0121580

Pantaleo V (2011) Plant RNA silencing in viral defence. Adv Exp Med Biol 722:39–58

Papadopoulou N, Devos Y, Alvarez-Alfageme F, Lanzoni A, Waigmann E (2020) Risk assessment considerations for genetically modified RNAi plants: EFSA's activities and perspective. Front Plant Sci 11:522254

Patil BL, Legg JP, Kanju E, Fauquet CM (2015) Cassava brown streak disease: a threat to food security in Africa. J Gen Virol 96(5):956–968

Petrov NM, Stoyanova MI, Gaur RK (2023) Chapter 12: biodiversity and characterization of economically important viruses on potato cultivars. In: Plant RNA viruses. Academic, New York, pp 245–270

Prasad Babu K, Maligeppagol M, Asokan R, Krishna Reddy M (2019) Screening of a multi-virus resistant RNAi construct in cowpea through transient vacuum infiltration method. VirusDisease 30(2):269–278. https://doi.org/10.1007/s13337-018-00509-y

Price JL, Blau J, Rothenfluh A, Abodeely M, Kloss B, Young MW (1998) Double-time is a novel Drosophila clock gene that regulates PERIOD protein accumulation. Cell 94(1):83–95

Pu Y, Kikuchi A, Moriyasu Y, Tomaru M, Jin Y, Suga H, Hagiwara K, Akita F, Shimizu T, Netsu O, Suzuki N, Uehara-Ichiki T, Sasaya T, Wei T, Li Y, Omura T (2011) Rice dwarf viruses with dysfunctional genomes generated in plants are filtered out in vector insects: implications for the origin of the virus. J Virol 85(6):2975–2979

Pumplin N, Voinnet O (2013) RNA silencing suppression by plant pathogens: defence, counter-defence and counter-counter-defence. Nat Rev Microbiol 11(11):745–760

Puyam A, Kaur K (2020) Exploiting RNA interference mechanism in plants for disease resistance. In: Plant disease management strategies for sustainable agriculture through traditional and modern approaches. Springer, Cham, pp 217–236

Qi T, Guo J, Peng H, Liu P, Kang Z, Guo J (2019) Host-induced gene silencing: a powerful strategy to control diseases of wheat and barley. Int J Mol Sci 20:206

Qi X, Mo Q, Li J, Zi Z, Xu M, Yue S, Zhao H, Zhu H, Wang G (2023) Establishment of virus-induced gene silencing (VIGS) system in Luffa acutangula using Phytoene desaturase (PDS) and tendril synthesis related gene (TEN). Plant Methods 19(1):94

Qin H, Chen F, Huan X, Machida S, Song J, Yuan YA (2010) Structure of the Arabidopsis thaliana DCL4 DUF283 domain reveals a noncanonical double-stranded RNA-binding fold for protein–protein interaction. RNA 16(3):474–481

Qu F (2010) Antiviral role of plant-encoded RNA-dependent RNA polymerases revisited with deep sequencing of small interfering RNAs of virus origin. Mol Plant-Microbe Interact 23(10):1248–1252

Quenouille J, Vassilakos N, Moury B (2013) Potato virus Y: a major crop pathogen that has provided major insights into the evolution of viral pathogenicity. Mol Plant Pathol 14(5):439–452

Rakitina DV, Taliansky M, Brown JW, Kalinina NO (2011) Two RNA-binding sites in plant fibrillarin provide interactions with various RNA substrates. Nucleic Acids Res 39(20):8869–8880

Ramesh SV, Shivakumar M, Ramteke R, Bhatia VS, Chouhan BS, Goyal S, Chand S (2019) Quantification of a legume begomovirus to evaluate soybean genotypes for resistance to yellow mosaic disease. J Virol Methods 268:24–31. https://doi.org/10.1016/j.jviromet.2019.03.002

Ratcliff F, Harrison BD, Baulcombe DC (1997) A similarity between viral defense and gene silencing in plants. Science 276(5318):1558–1560

Reddy D, Thirumala-Devi K (2003) Peanuts. In: Virus and virus-like diseases of major crops in developing countries. Springer, Dordrecht, pp 397–423

Rego-Machado CM, Nakasu EYT, Silva JMF, Lucinda N, Nagata T, Inoue-Nagata AK (2020) siRNA biogenesis and advances in topically applied dsRNA for controlling virus infections in tomato plants. Sci Rep 10(1):22277

Robinson KE, Worrall EA, Mitter N (2014) Double stranded RNA expression and its topical application for non-transgenic resistance to plant viruses. J Plant Biotechnol 23:231–237

Rosa C, Kuo YW, Wuriyanghan H, Falk BW (2018) RNA Interference mechanisms and applications in plant pathology. Annu Rev Phytopathol 56:581–610

Roth BM, Pruss GJ, Vance VB (2004) Plant viral suppressors of RNA silencing. Virus Res 102(1):97–108

Šafářová D, Brázda P, Navrátil M (2014) Effect of Artificial dsRNA on Infection of Pea Plants by Pea seedborne mosaic virus.

Sahu PP, Rai NK, Chakraborty S, Singh M, Chandrappa PH, Ramesh B, Chattopadhyay D, Prasad M (2010) Tomato cultivar tolerant to Tomato leaf curl New Delhi virus infection induces virus-specific short interfering RNA accumulation and defence-associated host gene expression. Mol Plant Pathol 11(4):531–544

Sandoval-Mojica AF, Hunter WB, Aishwarya V, Bonilla S, Pelz-Stelinski KS (2021) Antibacterial FANA oligonucleotides as a novel approach for managing the Huanglongbing pathosystem. Sci Rep 11(1):2760

Saxena S, Singh N, Ranade SA, Babu SG (2011) Strategy for a generic resistance to geminiviruses infecting tomato and papaya through in silico siRNA search. Virus Genes 43:409–434

Saxena S, Kesharwani RK, Singh V, Singh S (2013) Designing of putative siRNA against geminiviral suppressors of RNAi to develop geminivirus-resistant papaya crop. Int J Bioinforma Res Appl 9(1):3–12

Sharma L, Dalal M, Verma RK, Kumar SV, Yadav SK, Pushkar S, Chinnusamy V (2018) Auxin protects spikelet fertility and grain yield under drought and heat stresses in rice. Environ Exp Bot 150:9–24

Shen C, Wei C, Li J, Zhang X, Zhong Q, Li Y, Bai B, Wu Y (2020) Barley yellow dwarf virus-GAV-derived vsiRNAs are involved in the production of wheat leaf yellowing symptoms by targeting chlorophyll synthase. Virol J 17:158

Shepherd DN, Martin DP, Van Der Walt E, Dent K, Varsani A, Rybicki EP (2010) Maize streak virus: an old and complex 'emerging' pathogen. Mol Plant Pathol 11(1):1–12

Shimizu T, Nakazono-Nagaoka E, Akita F, Wei T, Sasaya T, Omura T, Uehara-Ichiki T (2012) Hairpin RNA derived from the gene for Pns9, a viroplasm matrix protein of Rice gall dwarf virus, confers strong resistance to virus infection in transgenic rice plants. J Biotechnol 157(3):421–427

Simon-Mateo C, Garcia JA (2011) Antiviral strategies in plants based on RNA silencing. Biochim Biophys Acta 1809:722–731

Singh RK, Cooper TA (2012) Pre-mRNA splicing in disease and therapeutics. Trends Mol Med 18(8):472–482

Singh A, Taneja J, Dasgupta I, Mukherjee SK (2015) Development of plants resistant to tomato geminiviruses using artificial trans-acting small interfering RNA. Mol Plant Pathol 16(7):724–734

Singh K, Wegulo SN, Skoracka A, Kundu JK (2018) Wheat streak mosaic virus: a century old virus with rising importance worldwide. Mol Plant Pathol 19(9):2193–2206

Sohrab SS, Kamal MA, Ilah A, Husen A, Bhattacharya PS, Rana D (2016) Development of Cotton leaf curl virus resistant transgenic cotton using antisense ssC1 gene. Saudi J Biol Sci 23(3):358–362

Song Y, Thomma B (2018) Host-induced gene silencing compromises verticillium wilt in tomato and arabidopsis. Mol Plant Pathol 19(1):77–89

Song L, Fang Y, Chen L, Wang J, Chen X (2021) Role of non-coding RNAs in plant immunity. Plant Commun 2(3):100180

Soyars CL, Peterson BA, Burr CA, Nimchuk ZL (2018) Cutting edge genetics: CRISPR/Cas9 editing of plant genomes. Plant Cell Physiol 59(8):1608–1620

Suarez MM, Rodríguez M, Bejerman N, Laguna IG, Pardina PR (2021) Biological, molecular and phiysiological characterization of four soybean mosaic virus isolates present in Argentine soybean crops. pp. 2021-06

Sun Z-N, Yin G-H, Song Y-Z, An H-L, Zhu C-X, Wen F-J (2010) Bacterially Expressed Double-Stranded RNAs against Hot-Spot Sequences of Tobacco Mosaic Virus or Potato Virus Y Genome Have Different Ability to Protect Tobacco from Viral Infection. Appl Biochem Biotechnol 162(7):1901–1914. https://doi.org/10.1007/s12010-010-8968-2

Sutic DD, Ford RE, Tosic MT (1999) Handbook of plant virus diseases. CRC Press, Boca Raton

Taliansky M, Samarskaya V, Zavriev SK, Fesenko I, Kalinina NO, Love AJ (2021) RNA-based technologies for engineering plant virus resistance. Plants 10:82

Tarn W-Y, Cheng Y, Ko S-H, Huang L-M (2021) Antisense oligonucleotide-based therapy of viral infections. Pharmacology 13(12):2015

Tenllado F, Díaz-Ruíz JR (2001) Double-stranded RNA-mediated interference with plant virus infection. J Virol 75(24):12288–12297

Thomas JE, Gronenborn B, Harding RM, Mandal B, Grigoras I, Randles JW, Sano Y, Timchenko T, Vetten HJ, Yeh HH, Ziebell H, Ictv Report C (2021) ICTV virus taxonomy profile: nanoviridae. J Gen Virol 102(3):001544

Timmons L, Court DL, Fire A (2001) Ingestion of bacterially expressed dsRNAs can produce specific and potent genetic interference in Caenorhabditis elegans. Gene 263:103–112

Tinoco ML, Dias BB, Dall'astta RC, Pamphile JA, Aragão FJ (2010) In vivo trans-specific gene silencing in fungal cells by in planta expression of a double-stranded RNA. BMC Biol 8:27

Tong X, Liu SY, Zou JZ, Zhao JJ, Zhu FF, Chai LX, Wang Y, Han C, Wang XB (2021) A small peptide inhibits siRNA amplification in plants by mediating autophagic degradation of SGS3/RDR6 bodies. EMBO J 40:e108050

Trebicki P (2020) Climate change and plant virus epidemiology. Virus Res 286:198059

Trebicki P, Finlay K (2019) Pests and diseases under climate change; its threat to food security. In: Food security and climate change. Wiley, Hoboken, pp 229–249

Valadkhan S, Mohammadi A, Jaladat Y, Geisler S (2009) Protein-free small nuclear RNAs catalyze a two-step splicing reaction. Proc Natl Acad Sci USA 106(29):11901–11906

Van Der Krol AR, Mur LA, De Lange P, Mol JN, Stuitje AR (1990) Inhibition of flower pigmentation by antisense CHS genes: promoter and minimal sequence requirements for the antisense effect. Plant Mol Biol 14:457–466

Vadlamudi T, Patil BL, Kaldis A, Sai Gopal DVR, Mishra R, Berbati M, Voloudakis A (2020) DsRNAmediated protection against two isolates of Papaya ringspot virus through topical application of dsRNA in papaya. J Virol Methods 275:113750. https://doi.org/10.1016/j.jviromet.2019.113750

Várallyay E, Válóczi A, Agyi A, Burgyán J, Havelda Z (2010) Plant virus-mediated induction of miR168 is associated with repression of ARGONAUTE1 accumulation. EMBO J 29(20):3507–3519

Vergani-Junior CA, Tonon-da-Silva G, Inan MD, Mori MA (2021) DICER: structure, function, and regulation. Biophys Rev 13(6):1081–1090. https://doi.org/10.1007/s12551-021-00902-w

Vetukuri RR, Dubey M, Kalyandurg PB, Carlsson AS, Whisson SC, Ortiz R (2021) Spray-induced gene silencing: an innovative strategy for plant trait improvement and disease control. Crop Breed Appl Biotechnol 21:24

Voinnet O (2009) Origin, biogenesis, and activity of plant microRNAs. Cell 136(4):669–687

Voloudakis AE, Kaldis A, Patil BL (2022) RNA-based vaccination of plants for control of viruses. Annu Rev Virol 9:521–548

Wang MB, Abbott DC, Waterhouse PM (2000) A single copy of a virus-derived transgene encoding hairpin RNA gives immunity to barley yellow dwarf virus. Mol Plant Pathol 1(6):347–356

Wang H, Piatkov KI, Brower CS, Varshavsky A (2009) Glutamine-specific N-terminal amidase, a component of the N-end rule pathway. Mol Cell 34(6):686–695

Wang W, Min L, Qiu X, Wu X, Liu C, Ma J, Zhang D, Zhu L (2021) Biological function of long non-coding RNA (LncRNA) xist. Front Cell Dev Biol 9:645647

Waterhouse PM, Graham MW, Wang MB (1998) Virus resistance and gene silencing in plants can be induced by simultaneous expression of sense and antisense RNA. Proc Natl Acad Sci USA 95(23):13959–13964

Whitten MM, Facey PD, Del Sol R, Fernández-Martínez LT, Evans MC, Mitchell JJ, Bodger OG, Dyson PJ (2016) Symbiont-mediated RNA interference in insects. Proc Biol Sci 283(1825):20160042

Whitehead KA, Langer R, Anderson DG (2009) Knocking down barriers: advances in siRNA delivery. Nat Rev Drug Discov 8(2):129–138. https://doi.org/10.1038/nrd2742

Will CL, Schneider C, Macmillan AM, Katopodis NF, Neubauer G, Wilm M, Lührmann R, Query CC (2001) A novel U2 and U11/U12 snRNP protein that associates with the pre-mRNA branch site. EMBO J 20(16):4536–4546

Worrall EA, Bravo-Cazar A, Nilon AT, Fletcher SJ, Robinson KE, Carr JP, Mitter N (2019) Exogenous application of RNAi-inducing double-stranded RNA inhibits aphid-mediated transmission of a plant virus. Front Plant Sci 10:265

Wu J, Yang Z, Wang Y, Zheng L, Ye R, Ji Y, Zhao S, Ji S, Liu R, Xu L, Zheng H, Zhou Y, Zhang X, Cao X, Xie L, Wu Z, Qi Y, Li Y (2015) Viral-inducible Argonaute18 confers broad-spectrum virus resistance in rice by sequestering a host microRNA. elife 4:e05733

Wu J, Yang R, Yang Z, Yao S, Zhao S, Wang Y, Li P, Song X, Jin L, Zhou T, Lan Y, Xie L, Zhou X, Chu C, Qi Y, Cao X, Li Y (2017) ROS accumulation and antiviral defence control by microRNA528 in rice. Nat Plants 3(1):16203

Yang Z, Li Y (2018) Dissection of RNAi-based antiviral immunity in plants. Curr Opin Virol 32:88–99

Yang Y, Liu T, Shen D, Wang J, Ling X, Hu Z, Chen T, Hu J, Huang J, Yu W, Dou D, Wang MB, Zhang B (2019) Tomato yellow leaf curl virus intergenic siRNAs target a host long noncoding RNA to modulate disease symptoms. PLoS Pathog 15(1):e1007534

Ye J (2013) Roles of regulated intramembrane proteolysis in virus infection and antiviral immunity. Biochim Biophys Acta 1828(12):2926–2932

Yin K, Qiu JL (2019) Genome editing for plant disease resistance: applications and perspectives. Philos Trans R Soc Lond Ser B Biol Sci 374(1767):20180322

Yin C, Jurgenson JE, Hulbert SH (2011) Development of a host-induced RNAi system in the wheat stripe rust fungus Puccinia striiformis f. sp. tritici. Mol Plant-Microbe Interact 24(5):554–561

Zaidi SS, Tashkandi M, Mansoor S, Mahfouz MM (2016) Engineering plant immunity: using CRISPR/Cas9 to generate virus resistance. Front Plant Sci 7:1673

Zhang H, Kolb FA, Jaskiewicz L, Westhof E, Filipowicz W (2004) Single processing center models for human Dicer and bacterial RNase III. Cell 118(1):57–68. https://doi.org/10.1016/j.cell.2004.06.017

Zhang ZY, Yang L, Zhou SF, Wang HG, Li WC, Fu FL (2011) Improvement of resistance to maize dwarf mosaic virus mediated by transgenic RNA interference. J Biotechnol 153(3-4):181–187

Zhang K, Zhang Y, Yang M, Liu S, Li Z, Wang X, Han C, Yu J, Li D (2017) The Barley stripe mosaic virus γb protein promotes chloroplast-targeted replication by enhancing unwinding of RNA duplexes. PLoS Pathog 13(4):e1006319

Zhang C, Chen D, Yang G, Yu X, Wu J (2019) Rice stripe mosaic virus–encoded P4 is a weak suppressor of viral RNA silencing and is required for disease symptom development. Mol Plant-Microb Interact 33(3):412–422

Zhao Y, Yang X, Zhou G, Zhang T (2020) Engineering plant virus resistance: from RNA silencing to genome editing strategies. Plant Biotechnol J 18(2):328–336

Zhou J, Tzanetakis IE (2019) Soybean vein necrosis virus: an emerging virus in North America. Virus Genes 55:12–21

Zvereva AS, Pooggin MM (2012) Silencing and innate immunity in plant defense against viral and non-viral pathogens. Viruses 4(11):2578–2597

RNAi-Based Biopesticide: A Powerful Tool for Crop Protection

12

Rekha Yadav, Sambhavana Chauhan, Hemakshi Sharma, Ruby Tiwari, and Chandra Pal Singh

Abstract

Agriculture production suffers substantially due to insect pest infestation. In the majority of cases, pest management programs involved the use of harmful chemical pesticides, which has serious human health and environmental implications. On the contrary, the use of biopesticides over several years is proven as safe alternative method which supports agriculture in a sustainable manner. Prevalent biopesticides include the bioactive compounds which are majorly expressed in microorganisms such as algae, fungi, bacteria and some plant species. Since past few decades, the double-stranded RNA (dsRNA)-based RNA interference (RNAi) technology has become the upcoming revolutionized approach to deliver insect-resistant plant varieties which acts as biopesticides. RNAi works on the principle of sequence-specific gene silencing. In this phenomenon, dsRNA acts as the key triggering molecule that induces the formation of RNA-induced silencing complex (RISC) and regulates the expression of specific mRNA targets based on complementarity. RNAi has been shown to work effectively without involving the expression of foreign protein in the plant system and therefore reduces the selection pressure on notorious insect-pest to evolve. These properties make it selective and topically applicable approach for crop protection. In this chapter, the basic RNAi technology and its application in insect-pest control as biopesticides are discussed. Moreover, the advantages and risk-associated

Rekha Yadav and Sambhavana Chauhan contributed equally with all other contributors.

R. Yadav · H. Sharma · C. P. Singh (✉)
Department of Botany, University of Rajasthan, Jaipur, Rajasthan, India

S. Chauhan · R. Tiwari
Department of Genetics, University of Delhi, South Campus, New Delhi, India

K. Goswami et al. (eds.), *Non-coding RNAs for Crop Improvement*,
https://doi.org/10.1007/978-981-97-7213-1_12

with this technology are also highlighted. The increasing understanding of RNAi technology and its cost-effective applications, provide a potential method to sustainable agriculture.

Keywords

RNAi · Gene regulation · Biopesticides · Pest-insect control · Agriculture · Crop protection

12.1 Introduction

The expected rise in the current global population of 7.9 billion would be nearly 10 billion in the coming 50 years, which would consequently be responsible for greater consumption of agricultural products (Rask and Rask 2011; Gardner 2013). Additionally, many external factors, including high variations in the environmental conditions, are manifesting a decline in the quality and yield of the agricultural product (Liliane and Charles 2020). The area witnessing a major decline in agricultural yield is associated with pests and pathogens (Shukla et al. 2019; Kaur et al. 2021). Presently, Integrated Pest Management, genetic engineering for producing resistant varieties, and use of chemical pesticides are promising methods for increasing agricultural yield to meet the global food demand (Abrol and Shankar 2014; Anderson et al. 2019). Although efforts are being made to increase the agricultural yield to the required levels, it becomes equally important to critically evaluate the safety measures associated with the crop production which is made available for consumption. However, despite having hazardous impacts on environment and humans, production of chemical pesticides dominates the market because of its strong inhibitory effect against insect pests (Abrol and Shankar 2014; Anderson et al. 2019). These agrochemicals lack in target and non-target organism differentiation. Therefore, it becomes utmost important to adapt sustainable techniques for improving agricultural practices and crop productivity in a safer manner.

Looking forward to the impact of chemical pesticides, the interest shifted towards genetic engineering which provided an alternative for introducing variations within existing plant varieties and developing new variety resilient to pest and pathogens. This plant genetic engineering approach involved genetic manipulation and modification of host plant genome at molecular level which has social and ethical limitations (Joga et al. 2016). It has a potential risk associated with the health implications on consumption of these genetically modified organisms (GMOs) or crop products. To circumvent the use of agrochemicals such as pesticides etc. and genetic modifications of crops, bio-pesticides were considered as a potential approach for pest management. Bio-pesticides are based on living organisms or natural products known to have a profound effect on the yield without affecting the quality of the plant or its products (Kumar and Singh 2015). The active molecules isolated from these organisms have been proven to attack specific target (Tijjani et al. 2016). Hence, the approach of designing RNAi-based biopesticides was adopted. RNA interference technology was used to develop products that aimed to provide protection against specific pests and pathogens (Sharma et al. 2023). This technique is efficient in gene function validation by targeting the cognate mRNAs involved in an important biological function. The technique triggers dsRNA molecule which leads to transcription repression of complementary mRNA in a sequence specific manner (Menezes et al. 2022). RNAi has been reported to provide viral resistance in plants which was further exploited to develop resistance against various insect pests (Ratcliff et al. 1999; Sharma et al. 2023).

RNAi is present in eukaryotes for regulating of RNA translation which emerged as a promising tool for development of plant specific insect pest-management strategies. The RNAi mechanism is conserved throughout eukaryotes which are triggered in a cell by exogenously expressed dsRNA which is synthesized through viral RNA or by endogenous expression of dsRNA which is synthesized from transposons or aberrant transcripts (Kaur et al. 2021). In plants, non-cell autonomous silencing of transcripts by RNAi is reported using grafting experiments showing the movement of silencing signals from silenced stocks to grafted scions (Palauqui et al. 1997). To develop insect resistance, plants silence the insect-pest specific gene by using insect RNAi machinery. As a result, it interferes the insect protein translation which results in arresting the growth and development of insect pests (Sharma et al. 2023). RNAi follows transcription where the dsRNA precursor is processed into short single stranded RNA molecules which further modulate the process of transcription. There are three commonly known pathways of the RNAi which depend upon the class of effector molecule involved i.e. siRNA, piRNA and miRNA. siRNA follows the classical pathways wherein the dsRNA is the trigger molecule for gene silencing. The dsRNA when taken up by the cell gets cleaved by the endonuclease- DICER, into 21–23 nucleotide long siRNA duplexes with a 2-nucleotide overhang at 3′ termini of both ends. The siRNA duplex is then bound with the Argonaute proteins in RNAi Silencing Complex (RISC). The passenger strand in the siRNA duplex undergo degradation, however the guide strand of siRNA duplex remains bound with RISC and forms activated RISC. This activated RISC then screen for complementary mRNA sequence and based on sequence complementarity leads to mRNA degradation and thereby inhibits the protein translation of cognate mRNAs. This ultimately results in RNAi-mediated gene silencing of specific mRNAs in a sequence specific manner (Menezes et al. 2022).

Correspondingly, the delivery approach of the dsRNA into the target organism arises as an important aspect which includes transmembrane channel mediated uptake and the endocytosis mediated uptake (Menezes et al. 2022). Transmembrane-mediated channel uptake was best studied and examined in *Caenorhabditis elegans* more specifically with worms carrying mutant *sid* (systemic RNA interference defective) gene that lacks systemic RNAi capacity, while endocytosis mediated uptake was first studied in *Drosophila melanogaster* but since it had no *sid* gene orthologous, it was later confirmed in *C.elegans.* This pathway is active, specific, and conserved in other animal cells (Menezes et al. 2022). As reported in some organisms, exogenous uptake of dsRNA triggers the RNAi response in an entire animal system, alternatively in some organisms the silencing signal is not even transmitted to the neighboring cell. After the cellular uptake of dsRNA in insects, an efficient dsRNA carrier system would provide stable transmission of silencing signal in the form of dsRNA and enhance RNAi efficacy of the immune system. The gut endonucleases will process or encapsulate the dsRNA. The encapsulation of dsDNA provides additive effects on pest control, and most of them are innocuous, biodegradable, and stable in multiple chemical environments which favors the controlled release of silencing signals using RNAi (Menezes et al. 2022). This provides a potential application of RNAi in plant insect-pest control management. Therefore, RNAi can be effectively used as biopesticides in crop improvement. According to the established studies, factors such as dsRNA concentration, sequence length, effective period of the dsRNA induced silencing, and target selection play key role in influencing the effectiveness of RNAi as biopesticides. To increase the RNAi efficacy, multiple target sequences should be selected for single transcript silencing using in-silico nucleotide basic local alignment search tool (BLAST) from National Center for Biotechnology Information (NCBI) (Kaur et al. 2021; Menezes et al. 2022).

12.2 Overview of RNAi Technology

An ancient mechanism called RNAi is present in the majority of eukaryotic cells, and it regulates gene expression by inhibiting translation and degrading mRNA. The trigger is a molecule of RNA called dsRNA that has two complimentary strands. Nearly a century ago, Wingard discovered the physiological effects of RNAi. He noticed that tobacco infection with the Tobacco ringspot virus (TRSV) in the lower leaves was linked to resistance to secondary infection in the higher leaves (Wingard 1928). The development of genetically modified plants and fungi led to the discovery that the integration of transgenes homologous to endogenous genes occasionally resulted in the suppression of both genes' expression. This phenomenon is known as "co-suppression" in plants (Napoli et al. 1990) and "quelling" in fungi (Romano and Macino 1992). Fire et al. demonstrated that dsRNA was a far more effective suppressor of target transcript production than single-stranded RNA (ssRNA) in *Caenorhabditis elegans* (Fire et al. 1998). Utilizing host-induced gene silencing (HIGS) and spray-induced gene silencing (SIGS), respectively, RNAi is being used to modify endogenous gene expression in plants as well as to target pest and disease genes. To briefly explain the molecular mechanism, siRNA molecules are first produced from dsRNAs that have been converted into 20–30–nt duplexes by the ribonuclease III enzyme, Dicer. The two siRNA strands are then identified by RISC as either passenger or guide strands. The passenger strand is degraded. On the other hand, the guide strand is integrated into the RISC. This is used as a guide to specifically target messenger RNAs (mRNA). RISC cleaves mRNA, which contain amino acid sequences. The RISC can become repeatedly involved in mRNA degradation, preventing the production of new proteins. It is not surprising to use its functioning mechanism in crop protection given the existence of RNAi pathways in pests. A few methods have been tested such as host-induced gene silencing, foliar sprays, recombinant microbes. The first strategy involves spraying crop leaves with pesticides made of synthetic dsRNA or sRNA obtained from pest genes. Pests and diseases on stems, foliage, or fruits can be controlled by spraying RNAi-based products, like dsRNAs, on the foliage. The products are examined similarly to topical pesticides, where the effects on insects are then assessed when the RNA solution is applied to leaves or fed to the intended pests (Andrade and Hunter 2017). This method, also known as spray-induced gene silencing (SIGS), opens the door to the creation of biopesticides that are favorable to the environment. The ability of RNAi technology to manage insect infestations has been established.

Plants can be treated directly with dsRNAs generated in vitro. When insects consume these dsRNA, RNAi-mediated gene silencing is induced (Fig. 12.1). When these plants provide dsRNA to insects, the dsRNA is first converted by DCL in the insect's gut into multiple siRNA molecules before moving on to RISC loading. Generated siRNA molecules target the particular insect gene against which they were initially designed based on the complementary sequence between the siRNA and the target mRNA (Fig. 12.1). Insect growth is slowed as a result of this.

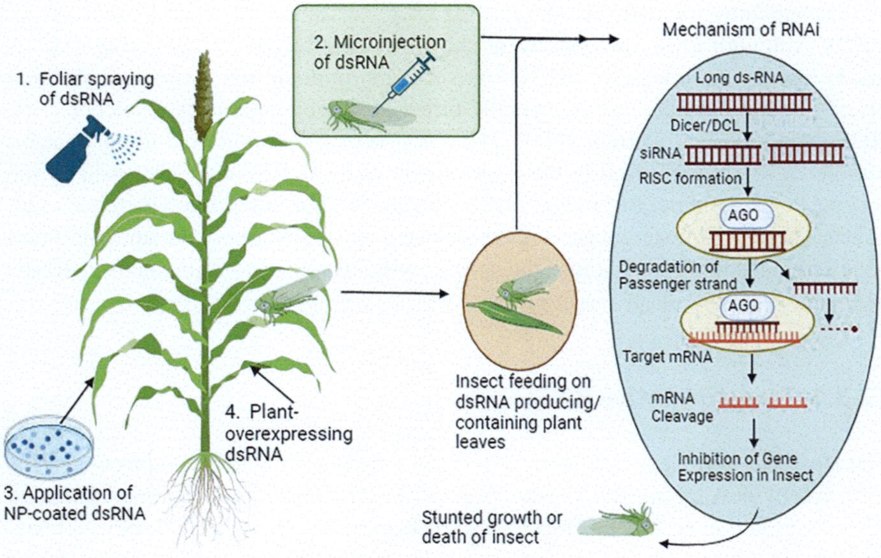

Fig. 12.1 Various modes of dsRNA delivery; (1) Spray of dsRNA on host plants. (2) Microinjection of dsRNA. (3) Application of nanoparticle (NP)-coated dsRNA. (4) Plant overexpressing dsRNA, provide crop protection against pests. The dsRNA would penetrate the insect pest gut upon feeding by insect. DCL processes dsRNA into siRNA molecules, which are small-interfering RNAs (siRNAs, 21–24 nt). In order to achieve RNAi-mediated gene silencing, the RNA-induced silencing complex (RISC) binds to the siRNAs and directs the sequence-dependent destruction or translational inhibition of cognate mRNA targets in insect, and eventually leading to either stunted growth or death of the insect.

12.3 RNAi as Biopesticides

RNAi is a mechanism by which gene expression is regulated. In a cell, dsRNA targets the cognate mRNA and lead to the silencing of targeted transcripts. RNAi phenomenon was exploited to deliver dsRNA within insect pests, to target its endogenously expressing vital genes (Hussain et al. 2019; Zubair et al. 2020; Sharif et al. 2022). Targeting the desired genes resulted in reduced growth and development of insect and thereby gaining resistance in plants against pests (Mamta et al. 2016; Yogindran and Rajam 2020). The successful delivery of dsRNA inside insect gut led to activation of RNAi machinery against insect specific genes to suppress the expression of proteins involved in physiological processes and therefore acts as a biopesticide (Seiber et al. 2014; Fletcher et al. 2020; Zhang et al. 2020). There are only two ways for dsRNA delivery inside the insect system (Cagliari et al. 2019). First, raising a transgenic plant expressing dsRNA against insect specific gene/s, such that when insect feed on the plant it co-ingest the dsRNA along with plant tissue (Mamta et al. 2016; Hussain et al. 2019;Yogindran and Rajam 2020; Kontogiannatos et al. 2021). Second, spraying insect gene specific dsRNA on plants and replenishing the dsRNA time to time on plant surface (Lucena-Leandro et al. 2022). Although these methods were shown to be effective in delivering dsRNA inside insects, the release of dsRNA into the environment imposes potential threat of off-targeting effects on non-specific target gene/s or organisms (Fletcher et al. 2020; Rodrigues and Petrick 2020). The risk is directly proportional to the concentrations of dsRNA released in the environment as the dsRNAs are absorbed by the soil particles or undergo biodegradation certainly after sometime (Fletcher et al. 2020). Therefore, the target gene selection becomes utmost important and minimum gene sequence homology across the species will ensure reduced impact of dsRNA-mediated silencing of off-target genes.

12.3.1 Transformative RNAi

The Host-Induced Gene Silencing (HIGS) of target genes specific to insect pests is achieved by developing RNAi plant transgenics and also known as transformative RNAi (Cagliari et al. 2019). The crop plants expressing these dsRNA showed resistance against notorious insect pests and helped in crop improvement. The US and Chinese regulators approved the commercialization of Bayer synthesiszed and RNAi based Mon87411—"SmartStax Pro" maize against *Diabrotica virgifera* in 2017 and 2021, respectively (Bachman et al. 2016; Zotti et al. 2018). Mon87411 carried the combination of three transgenes—*Bacillus thuringiensis* (*Bt*) specific *Cry3Bt1*, dsRNA against Snf7 of corn rootworm and glyphosphate resistant gene. SmartStax Pro was provided to the farmers for cultivation in US and Canada in 2022 and 2023. Similarly, Bayer has also launched RNAi based soyabean product called "Vistive gold—Mon87705", where a gene involved in fatty acid biosynthesis is targeted (ISAAA 2020). The governments of USA, Canada, Japan and Europe also approved the cultivation and marketing of Mon87705, by keeping health benefits into consideration. Likewise, Australian transgenic regulatory body is expected to release Super-High Olieic (SHO) safflower in 2023, as it targets the enzyme converting oleic acid into polyunsaturated fatty acid (SHO safflower—CSIRO 2015; Wood et al. 2018). This leads to accumulation of oleic acid within seed and does affect the seed viability. Despite delivery promising transgenic crop against several traits, the time constraint is the problem in transformative RNAi. As it is difficult to target the multiple pests by single transgenic crop, and stable RNAi transgenic lines of plants are obtained after several generations which delays the implementation of applications.

12.3.2 Non-Transformative RNAi

RNAi transgenic plants impose a risk of resistance breakdown and evolution of resistant variety of insects (Cagliari et al. 2019; Parker et al. 2019). Also, the investment and labor in terms of input, and huge time is required to bring the transgenic from lab to field conditions as it undergoes several regulatory steps involving Institutional Biosafety Committee (IBSC) and Genetic Engineering Appraisal Committee (GEAC). In this regard, non-transformative RNAi has an advantage because here SIGS approach is utilized to trigger insect specific RNAi machinery to target its own gene. Spraying dsRNA on the plant surface will not lead to any DNA modifications or foreign protein expression which is heritable within plant system (Fletcher et al. 2020; Guo et al. 2023). Therefore, this approach overcomes the resistance breakdown problem and because there is no involvement of genetic manipulation within organisms or genetically modified organism (GMO), it does not have to pass through the regulation process (Cagliari et al. 2019; Fletcher et al. 2020). The main drawback of the technology is plants will keep sprouting new leaves and the continuous spray of dsRNA has to be adapted in regular practice. But this is practically non-economic as the farmers could not afford such high cost for dsRNA synthesis. However, there are reports which states that dsRNA can enter the vascular system and move from one part to another even when sprayed of leaves, injected in trunks or supplemented in soil (Wise et al. 2022). Because of the short half-life of dsRNAs farmers may not be interested to continuously spray it in the field, but due to shorter life the dsRNAs are evacuated from the system within few days unlike GM crops where it last for generations. The success of this application underlies within the potent target gene selection and effective dsRNA concentration to act on insects upon ingestion.

12.3.3 Pest-Insect Control in Different Plants Using RNAi

The transformative as well as non-transformative RNAi methods have been used to control devastation caused by insect pests. Transformative methods covered HIGS driven siRNA and amiRNA-based gene silencing of insect-pests specific genes. For example—*Helicoverpa armigera*, a polyphagous notorious insect pest, cause devastation in various crops such as tomato (Mamta et al. 2016), tobacco (Jin et al. 2015; Mamta et al. 2016), cotton (Mao et al. 2011) and *Arabidopsis* (Liu et al. 2015). The resistance against *H. armigera* was attained after targeting the vital genes involved in physiological processes of insect development like *Acetylcholinesterase—AChE2* involved in nervous system (Kumar et al. 2009), *Argininekinase* involved in metabolism (Ai et al. 2019; Fu et al. 2020; Ma et al. 2022a), *Chitinase* involved in moulting (Mamta et al. 2016; Zhang et al. 2021a; Kottaipalayam-Somasundaram et al. 2022; Mahmood et al. 2022), *Cytochrome P450-CYP6AE14* involved in defense against plant products (Mao et al. 2011; Jin et al. 2015) and, *Ecdysone receptor-EcR* (Yogindran and Rajam 2020) and *HR3* (Ma et al. 2022b) are involved in reproduction, metamorphosis and other developmental stages which on targeting resulted in developmental deformities. Targeting some of these genes helped in enduring resistance against other insect pests such as *Bemisia tabaci*and *Myzus persicae* (Zubair et al. 2020), *H. armigera* and *Leptinotarsadecemlineata* (Hussain et al. 2019; Sharif et al. 2022) and, *N. lugens* (Yu et al. 2014) in tobacco, potato and rice, respectively.

In silico analysis of insect-specific AChE genes, a potential insecticide target showed that mutations in this gene will lead to resistance against insecticides in different insect pests (Yousafi et al. 2021). This was further supported by knockout studies of *acetylcholinesterase receptor, AChERα6* or *Sfα6* which revealed the any kind of mutation (indels) in this gene show resistance in *H. armigera, Spodoptera frugiperda* and *Spodoptera exigua* against insecticides such as spinosyns and spinosad (Wang et al. 2020; Shi et al. 2022; Zuo et al. 2022). This suggested that the selection pressure created by the overdoses of harmful insecticides will lead to evolution of insect pests into a highly resistant species, which cannot be controlled later using chemicals (Yousafi et al. 2021).

Artificial miRNA or dsRNA mediated targeting of *EcR* led to larval deformaties and impaired with oogenesis of *H. armigera* and *Schistocerca gregaria* in the subsequent generation (Yogindran and Rajam 2016; Lenaerts et al. 2019). Further, HIGS of *HaEcR* via tomato transgenics also showed similar results in the T1 generation which further inferred that the targeting of this gene might be helpful in controlling *H. armigera* population in the field (Yogindran and Rajam 2020). *EcR* when targeted in *Henosepilachna vigintioctopunctata* and *Tetranychus cinnabarinus* resulted in transitional blockage from larval to pupal stage (Shen et al. 2019; Wu et al. 2021). When the larvae of *Leptinotarsa decemlineata* and *Sesamia nonagrioides* were fed on diet or leaves coated with dsRNA against *EcR*, it resulted in reduced larval weight and developmental deformaties (Hussain et al. 2019; Kontogiannatos et al. 2021). The dsRNA-mediated gene silencing of *EcR* when corroborated with the ecdysteriod antagonist, Methoxyfenozide showed an adverse lethal phenotype in the subjected larvae of *S. exigua* (Zhang et al. 2021b).

Similarly, insect larval growth retardation and mortality was observed after down-regulating the endogenous gene expression levels of *V-ATPase* (Baum et al. 2007; Thakur et al. 2014; Liu et al. 2019), *serine protease* (Bhatia et al. 2012), *SHP* (Abdellatef et al. 2015), *COPB2* (Shin et al. 2020), *Spo* (He et al. 2019), *Orc* and *Sx1* (Zubair et al. 2020), *dib* (Zheng et al. 2019) using HIGS in different crops (Kaur et al. 2021). The corn rootworm, *Diabrotica virgifera* was controlled by targeting vacuolar sorting gene, *Snf7* using dsRNA technology and further deployed with *Cry3B* gene for improved resistance (Bolognesi et al. 2012). Likewise, targeting *cytochrome P450* genes helped in insect-pest management by causing larval growth defects in several insect pests such as *H. armigera* (Jin et al. 2015), *S. frugiperda* (Zhong et al. 2020), *Scirpophaga incertulas* (Kola et al. 2016), *Ectropis oblique* (Peng et al. 2022), *Leptinotarsa decemlineata* (Moural et al. 2020), *Menduca quinquemaculata* (Poreddy et al. 2017), Conifers (Liu et al. 2022) as mentioned in Table 12.1.

Table 12.1 Potential insect specific genes for HIGS target.

S. No.	Target	Gene name	Insect pest	Host plant	References
1	*Acetylcholinestrase*	*AChE2*	*Helicoperva armigera*	[a]	Kumar et al. (2009)
2	*Acetylcholinestrase receptor*	*ACHeRα6 or Sfα6*	*Helicoperva armigera*	[a]	Wang et al. (2020)
			Spodoptera exigua	[a]	Zuo et al. (2022)
			Spodoptera frugiperda	[a]	Shi et al. (2022)
3	*Arginine kinase*	*AK*	*Helicoverpa armigera*	[a]	Ai et al. (2019)
			Plutella xylostella	*Arabidopsis thaliana*	Fu et al. (2020)
			Tribolium castaneum	–	Zhang et al. (2020)
			Lasioderma serricorne	Tobacco	Ma et al. (2022b)
4	*Chitinase*	*Cht*	*Helicoverpa armigera*	[a]	Mahmood et al. (2022), Mamta et al. (2016)
			Hyphantria cunea	[a]	Zhang et al. (2021a)
			Plutella xylostella	*Arabidopsis thaliana*	Zhu et al. (2019)
			Hyblaea puera	Teak	Kottaipalayam-Somasundaram et al. (2022)
			Agrotis ipsilon	[a]	Li et al. (2022)
5	*Coatomer protein complex subunit 2*	*COPB2*	*Tetranychusurticae*	Cabbage	Shin et al. (2020)
6	*Cuticle protein 19*	*CP19*	*Aphis citricidus, Acyrthosiphon pisum*, and *Myzus persicae*	[a]	Shang et al. (2020)

(continued)

Table 12.1 (continued)

S. No.	Target	Gene name	Insect pest	Host plant	References
7	*Cytochrome P450*	*CYP6BG1*	*Plutella xylostella*	[a]	Bautista et al. (2009)
		CYP6AE14	*Helicoverpa armigera*	Cotton	Jin et al. (2015)
		CY6	*Scirpophaga incertulas*	Rice	Kola et al. (2016)
		CYP6B46	*Menduca quinquemaculata*	Tobacco	Poreddy et al. (2017)
		Cytochrome P450 reductase (CPR)	*Leptinotarsa decemlineata*	Potato	Moural et al. (2020)
		CYP321A8, CYP321A9, CYP321B1	*Spodoptera frugiperda*	[a]	Zhong et al. (2020)
		CYP6CR2 and CYP6DE5	*Dendroctonus armandi*	Conifer	Liu et al. (2022)
		CYP316	*Ectropis oblique*	Tea	Peng et al. (2022)
8	*Ecdysone receptor*	*EcR*	*Spodoptera exigua*	Tobacco	Zhu et al. (2012)
			Leptinotarsa decemlineata	Potato	Hussain et al. (2019)
			Schistocerca gregaria	[a]	Lenaerts et al. (2019)
			Tetranychus cinnabarinus	[a]	Shen et al. (2019)
			Helicoverpa armigera	Tomato	Yogindran and Rajam (2020)
			Sesamia nonagrioides	Corn	Kontogiannatos et al. (2021)
			Henosepilachna vigintioctopunctata	[a]	Wu et al. (2021)
			Spodoptera exigua	[a]	Zhang et al. (2021b)
8	*Ecdysone signalling pathway gene*	*HR3*	*Helicoverpa armigera*	[a]	Xiong et al. (2013)
			Panonychus citri	*Citrus sinensis*	Li et al. (2022)
			Lasioderma serricorne	Tobacco	Ma et al. (2022a)
9	*Ecdysis triggering hormone receptor*	*ETHR*	*Leptinotarsa decemlineata*	Potato	Shen et al. (2020)
10	*Orcokinin and Sex lethal*	*Orc and Sx1*	*Bemisia tabaci*	Tobacco	Zubair et al. (2020)

(continued)

Table 12.1 (continued)

S. No.	Target	Gene name	Insect pest	Host plant	References
11	*Salivary sheath protein*	*SHP*	*Sitobion avenae*	Barley	Abdellatef et al. (2015)
12	*Serine protease*	*SP*	*Myzus persicae*	*Arabidopsis thaliana*	Bhatia et al. (2012)
13	*Spook*	*Spo*	*Chilo suppressalis*	Rice	He et al. (2019)
14	*Ras GTPase gene*	*Ran*	*Leptinotarsa decemlineata*	Potato	Shen et al. (2022)
15	*Vacuolar sorting gene*	*Snf7*	*Diabrotica virgifera vigifera*	Corn	Bolognesi et al. (2012)
16	*Vacuolar type ATPase*	*vATPase*	*Diabrotica virgiferavirgifera*	Maize	Baum et al. (2007)
			Leptinotarsa decemlineata	Potato	Zhu et al. (2011)
			Bemisia tabaci	Tobacco	Thakur et al. (2014)
			Apolyguslucorum	Maize and soybean	Liu et al. (2019)
			Oryctes rhinoceros	Coconut and palm	Watanabe et al. (2020)

[a] Polyphagous insect pest-feed on more than one host plant

These studies showed that the dsRNA *in planta* expression against its cognate mRNA transcript of insect-specific genes resulted in reduction of endogenous gene expression levels which further impaired with developmental of insect pests. These reports suggested that the potent insect-specific gene selection is a crucial step for effective RNAi, which can later be utilized for crop protection using HIGS approach. Although this enhances the plants resistance to sustain in insect pest infestation period, by controlling the growing population of insects, field trial still needs to be conducted to generate the supporting data.

12.4 Modes of dsRNA Delivery in Insect System

12.4.1 Microinjection of dsRNA

In this dsRNA delivery approach, dsRNA in the form of chemically synthesized siRNA duplex or precursor dsRNA is delivered directly in the insect gut (Fig. 12.1) (Gong et al. 2011) or in the plant (Wise et al. 2022) to endure resistance against insect pests. The dsRNA when enters the insect gut, activates insect RNAi machinery and targets insect specific genes and therefore interferes with insect growth and development (Gong et al. 2011; Wise et al. 2022). Although this approach is effective in controlling insect pests, because of involvement of high cost in dsRNA synthesis at large scale it is not economic to be used by poor farmers.

12.4.2 Plant-Mediated dsRNA Delivery in Insect System

In this method the transgenic plants are produced which express dsRNA against insect genes (Fig. 12.1). RNAi proved useful in the downregulation of insect-specific vital genes which resulted in affecting the fast-growing insect-pest population and therefore proven to be an efficient biopesticide. The plant-mediated silencing or downregulation of *H. armigera* specific cytochrome P450 monooxygenase gene *CYP6AE14* resulted in reduced tolerance of insect larvae towards gossypol and thereby imparts insect-resistance in cotton plant (Mao et al. 2007). Similarly, RNAi was used as a forefront for biosafe insecticide technology with specificity towards the target gene based on sequence complementarity. Therefore, RNAi was adopted for controlling other biotic agents such as bacteria (*Agrobacterium*, *Pseudomonas syringae etc.*), fungi (*Fusarium graminearum*, *Botrytis cinerea etc.*) and nematodes (*Meloidogyne incognita*, *Heterodera glycines etc.*) which also contributes significantly to annual crop yield losses (Kaur et al. 2021).

12.4.3 Spray of dsRNA on Host Plants

This dsRNA delivery approach involved the spray-induced gene silencing of insect-specific genes (Fig. 12.1). The dsRNA specific to insect pest are first synthesized chemically and then sprayed on the surface of the host plant. When the insect-pest feed on this plant, the dsRNA enters the insect gut and activates the RNAi machinery of insect against its own endogenous genes. This approach was utilized to target four potential gene targets such as *TREH*, *ATPD*, *ATPE* and *CHS1* provided resistance against *Aphis glycines* in soybean (Yan et al. 2020). The only limitation is involvement of high costs.

12.4.4 Nanoparticle-Mediated dsRNA Delivery

The direct dsRNA delivery methods in the field shows less stability of dsRNA or a reduced number of effective molecules for ingestion by insects and thus reduced RNAi efficiency in field applications. Therefore, there was a strong need to encapsulate this dsRNA to increase the shelf life and also made it available at slow rates for activation of RNAi machinery in insect pests (Fig. 12.1). The dsRNA delivery mediated by nanoparticles such as chitosan, liposomes, cationic dendrimers, silica etc. provides an advantage over traditional dsRNA delivery systems (Das et al. 2015; Yan et al. 2020).

12.5 Risk Associated with RNAi-based Biopesticides

The dsRNA molecules are widespread in eukaryotes, especially in plants and animals. The history shows strong evidence of dsRNA safe consumption by humans and other vertebrates (Dàvalos et al. 2019; OECD 2020). Besides, no scientific evidence exists that suggests that small RNA present in GM crops poses a health risk in conventional foods than the crops having abundant small RNAs which are naturally expressing in the system (Health and Food Safety 2017). Despite the safety of this active molecule, all technologies do carry potential set of associated risks and the careful assessment of these risks is crucial in designing a holistic approach for safe use of adopted practices (Fletcher et al. 2020). There are primarily three routes by which human body may be exposed to dsRNAs applied topically on the plants i.e. by ingestion, inhalation, and through dermal. Nonetheless, a dsRNA should be available in sufficient copies to pass through multiple barriers to enter inside human body and activates gene silencing signal to mediate a biological function (Rodrigues and Petrick 2020). Although dsRNA is considered to impart significantly negligible impact on humans; however, the products used for the formulation of dsRNA with other constituents like chitosan or silica used in nanoparticle mediated dsRNA delivery, could have a potential impact on human health and therefore, should be examined or assessed carefully on case-by-case basis (Fletcher et al. 2020). Moreover, an unintended impact on the environment could take place in species which share sequence homology at the transcriptional level with the target organisms which is often case-specific (Fletcher et al. 2020). Therefore, a thorough understanding is required to develop testable risk hypotheses (Romeis and Widmer 2020). It is known that the shelf life of dsRNA when exposed to environmental conditions, significantly reduced due to microbial nucleases present in soil or leaves, exposure to UV radiations, etc. which limits the availability of dsRNA to pests (Parker et al. 2019; Bachman et al. 2020). This requires stabilizing formulations of these dsRNA for successful application as biopesticides in the fields, which also improves their cellular internalization process and protects the dsRNA against nucleases (Christiaens et al. 2018; Martinez et al. 2021). Thus, these formulations increase the persistence of dsRNA in the environment which is not sustainable and may require risk assessment (Bachman et al. 2020). Although RNAi is highly sequence specific for a target organism, a partial homology could cause potential off-target effects in a non-target group of species (Christiaens et al. 2018). This is due to certain knowledge gaps that impede our bioinformatics predictions which are related to the fundamental insights into the RNAi mechanism (Christiaens et al. 2018; Santos et al. 2019).

Exogenous dsRNA is also known to stimulate innate immune response under experimental conditions which are not relevant to dietary and environmental exposure (Rodrigues and Petrick 2020). However, exposure to a high dsRNA concentration could theoretically saturate the active sites of enzymes involved in core RNAi machinery (Haller et al. 2019). RNAi is a part of the antiviral immunity of cells and such saturation could compromise the antiviral defence mechanism of cells (Piot et al. 2015; Brutscher et al. 2017; Arora et al. 2021). Further, RNAi-based products take a longer time to display efficacy as compared to conventional pesticides so a risk assessment should study the non-lethal phenotypes and life cycle (Romeis and Widmer 2020; Dietz-Pfeilstetter et al. 2021). Therefore, a comprehensive approach with proper risk assessment and alleviation should consider bioinformatics-based design which should be strongly supported by biological data. This data should be corroborated by an in-depth analysis of biological as well as ecological systems in which the dsRNA would be deployed (Fletcher et al. 2020).

12.6 Conclusion

RNAi is still regarded as a promising pest management technique, primarily because of its potential for eco-friendly control. Insect pest management is one of the many uses for which RNAi technology has been put to use, along with tools for crop improvement and functional genomics. With its distinctive insecticidal mode of action, suppression of gene expression, it can work both independently and in conjunction with currently used pest control techniques. The first RNAi-based products, which target insects that are particularly susceptible to dietary dsRNA uptake, will shortly be offered in stores. The breadth and success of RNAi studies that were performed to control pests that affect plants indicate that dsRNA has a bright future in addressing current weaknesses in pest control. However, a number of difficulties still prevent the application from being effective against a variety of insect species. These issues, which are primarily related to oral RNAi in insects' variable RNAi sensitivity, are likely to be resolved by the use of various formulation strategies that increase dsRNA persistence and cellular uptake in these insects.

12.7 Future Perspective

Although some promising proof-of-concept studies in this area have already been published, more work must be done before these RNAi products can compete with the conventional chemical pesticides presently in use against a wide variety of insect species. Understanding the movement of the dsRNA within the target pest is necessary to create effective RNAi strategies that will increase crop output. This will help the pest's penetration and persistence in the environment. Future research will help to advance this technology by examining the role of nucleases, physiological pH, mechanisms of dsRNA uptake and systemic spread, interactions with viruses, and possible resistance mechanisms.

References

Abdellatef E, Will T, Koch A, Imani J, Vilcinskas A, Kogel KH (2015) Silencing the expression of the salivary sheath protein causes transgenerational feeding suppression in the aphid *Sitobion avenae*. Plant Biotechnol J 13:849–857

Abrol DP, Shankar U (2014) Pesticides, food safety and integrated pest management. In: Pimentel D, Peshin R (eds) Integrated pest management, vol 3. Springer, Dordrecht, pp 167–199

Ai X, Ren S, Liu N, Huang L, Liu XN (2019) Transgenic tobacco expressing dsRNA of the *arginine kinase* gene exhibits enhanced resistance against *Helicoverpa armigera*. Bull Insectol 72:115–124

Anderson JA, Ellsworth PC, Faria JC, Head GP, Owen MD, Pilcher CD, Shelton AM, Meissle M (2019) Genetically engineered crops: importance of diversified integrated pest management for agricultural sustainability. Front Bioeng Biotechnol 7:24

Andrade EC, Hunter WB (2017) RNAi feeding bioassay: development of a non-transgenic approach to control Asian citrus psyllid and other hemipterans. Entomol Exp Appl 16:389–396

Arora AK, Chung SH, Douglas AE (2021) Non-target effects of dsRNA molecules in hemipteran insects. Gene 12:407

Bachman PM, Huizinga KM, Jensen PD, Mueller G, Tan J, Uffman JP, Levine SL (2016) Ecological risk assessment for *DvSnf7* RNA: a plant-incorporated protectant with targeted activity against western corn rootworm. Regul Toxicol Pharmacol 81:77–88

Bachman P, Fischer J, Song Z, Urbanczyk-Wochniak E, Watson G (2020) Environmental fate and dissipation of applied dsRNA in soil, aquatic systems, and plants. Front Plant Sci 11:21

Baum JA, Bogaert T, Clinton W, Heck GR, Feldmann P, Ilagan O, Johnson S, Plaetinck G, Munyikwa T, Pleau M, Vaughn T (2007) Control of coleopteran insect pests through RNA interference. Nat Biotechnol 25:1322–1326

Bautista MA, Miyata T, Miura K, Tanaka T (2009) RNA interference-mediated knockdown of a cytochrome P450, *CYP6BG1*, from the diamondback moth, *Plutellaxylostella*, reduces larval resistance to permethrin. Insect Biochem Mol Biol 39:38–46

Bhatia V, Bhattacharya R, Uniyal PL, Singh R, Niranjan RS (2012) Host generated siRNAs attenuate expression of serine protease gene in *Myzus persicae*. PLoS One 7:e46343

Bolognesi R, Ramaseshadri P, Anderson J, Bachman P, Clinton W, Flannagan R, Ilagan O, Lawrence C, Levine S, Moar W, Mueller G (2012) Characterizing the mechanism of action of double stranded RNA activity against Western Corn Rootworm (*Diabrotica virgifera virgifera* LeConte). PLoS One 7:e47534

Brutscher LM, Daughenbaugh KF, Flenniken ML (2017) Virus and dsRNA-triggered transcriptional responses reveal key components of honeybee antiviral defense. Sci Rep 7:1–15

Cagliari D, Dias NP, Galdeano DM, Dos Santos EÁ, Smagghe G, Zotti MJ (2019) Management of pest insects and plant diseases by non-transformative RNAi. Front Plant Sci 10:1319

Christiaens O, Dzhambazova T, Kostov K, Arpaia S, Joga MR, Urru I, Sweet J, Smagghe G (2018) Literature review of baseline information on RNAi to support the environmental risk assessment of RNAi-based GM plants. EFSA Support Publ 15:1424

CSIRO (2015) SHO safflower. https://www.csiro.au/en/Research/AF/Areas/Plant-Science/Bio-based-oils/SHO-safflower. Accessed 12 December 2017

Das S, Debnath N, Cui Y, Unrine J, Palli SR (2015) Chitosan, carbon quantum dot, and silica nanoparticle mediated dsRNA delivery for gene silencing in *Aedes aegypti*: a comparative analysis. ACS Appl Mater Interfaces 7:19530–19535

Dávalos A, Henriques R, Latasa MJ, Laparra M, Coca M (2019) Literature review of baseline information on noncoding RNA (ncRNA) to support the risk assessment of ncRNA-based genetically modified plants for food and feed. EFSA Support Publ 16:1688E

Dietz-Pfeilstetter A, Mendelsohn M, Gathmann A, Klinkenbuß D (2021) Considerations and regulatory approaches in the USA and in the EU for dsRNA-based externally applied pesticides for plant protection. Front Plant Sci 12:682387

Fire A, Xu S, Montgomery MK, Kostas SA, Driver SE, Mello CC (1998) Potent and specific genetic interference by double-stranded RNA in *Caenorhabditis elegans*. Nature 391:806–811

Fletcher SJ, Reeves PT, Hoang BT, Mitter N (2020) A perspective on RNAi-based biopesticides. Front Plant Sci 11:51

Fu S, Liu Z, Chen J, Sun G, Jiang Y, Li M, Xiong L, Chen S, Zhou Y, Asad M, Yang G (2020) Silencing arginine kinase/integrin β1 subunit by transgenic plant expressing dsRNA inhibits the development and survival of *Plutella xylostella*. Pest Manag Sci 76:1761–1771

Gardner B (2013) Global food futures: feeding the world in 2050. Bloomsbury Publishing, London

Gong L, Yang X, Zhang B, Zhong G, Hu M (2011) Silencing of Rieske iron–sulfur protein using chemically synthesised siRNA as a potential biopesticide against *Plutella xylostella*. Pest Manag Sci 67:514–520

Guo H, Liu XZ, Long GJ, Gong LL, Zhang MQ, Ma YF, Hull JJ, Dewer Y, He M, He P (2023) Functional characterization of developmentally critical genes in the white-backed planthopper: efficacy of nanoparticle-based dsRNA sprays for pest control. Pest Manag Sci 79:1048–1061

Haller S, Widmer F, Siegfried BD, Zhuo X, Romeis J (2019) Responses of two ladybird beetle species (Coleoptera: Coccinellidae) to dietary RNAi. Pest Manag Sci 75:2652–2662

He K, Xiao H, Sun Y, Ding S, Situ G, Li F (2019) Transgenic micro-RNA-14 rice shows high resistance to rice stem borer. Plant Biotechnol J 17:461–471

Health and Food Safety (2017) Genetically modified organisms register. https://webgate.ec.europa.eu/dyna/gm_register/gm_register_auth.cfm?pr_id=92. Accessed 18 November 2021

Hussain T, Aksoy EC, Alıskan ME, Bakhsh A (2019) Transgenic potato lines expressing hairpin RNAi construct of moltingassociated EcR gene exhibit enhanced resistance against Colorado potato beetle (*Leptinotarsa decemlineata, Say*). Transgenic Res 28:151–164

ISAAA (2020) GM approval database – Mon87705. https://www.isaaa.org/gmapprovaldatabase/event/default.asp?EventID=177. Accessed 18 November 18 2021

Jin S, Singh ND, Li L, Zhang X, Daniell H (2015) Engineered chloroplast dsRNA silences cytochrome p450 monooxygenase, V-ATPase and chitin synthase genes in the insect gut and disrupts *Helicoverpa armigera* larval development and pupation. Plant Biotechnol J 13:435–446

Joga MR, Zotti MJ, Smagghe G, Christiaens O (2016) RNAi efficiency, systemic properties, and novel delivery methods for pest insect control: what we know so far. Front Physiol 7:553. https://doi.org/10.3389/fphys.2016.00553

Kaur R, Choudhury A, Chauhan S, Ghosh A, Tiwari R, Rajam MV (2021) RNA interference and crop protection against biotic stresses. PMBP 27:2357–2377

Kola VS, Renuka P, Padmakumari AP, Mangrauthia SK, Balachandran SM, Ravindra BV, Madhav MS (2016) Silencing of *CYP6* and *APN* genes affects the growth and development of rice yellow stem borer, *Scirpophagaincertulas*. Front Physiol 7:20

Kontogiannatos D, Swevers L, Kourti A (2021) Assessment of *Sesamia nonagrioides* (Lepidoptera: *Noctuidae) EcR* and *USP* genes as targets for exogenous non-persistent RNAi. Diversity 13:677

Kottaipalayam-Somasundaram SR, Jacob JP, Aiyar B, Merzendorfer H, Nambiar-Veetil M (2022) Chitin metabolism as a potential target for RNAi-based control of the forestry pest *Hyblaea puera Cramer* (Lepidoptera: *Hyblaeidae*). Pest Manag Sci 78:296–303

Kumar S, Singh A (2015) Biopesticides: present status and the future prospects. J Fertil Pestic 6:1–2

Kumar M, Gupta GP, Rajam MV (2009) Silencing of *acetylcholinesterase* gene of *Helicoverpaarmigera* by siRNA affects larval growth and its life cycle. J Insect Physiol 55:273–278

Lenaerts C, Marchal E, Peeters P, Vanden Broeck J (2019) The ecdysone receptor complex is essential for the reproductive success in the female desert locust, *Schistocerca gregaria*. Sci Rep 9:15

Li D, Zhang J, Yang Y, Liu J, Lu J, Ren M, Abbas M, Zhu KY, Zhang J (2022) Identification and RNAi-based functional analysis of chitinase family genes in *Agrotis ipsilon*. Pest Manag Sci 78:4278–4287

Liliane TN, Charles MS (2020) Factors affecting yield of crops. In: Anamulah A (ed) Agronomy-climate change & food security. IntechOpen, London, pp 9–24

Liu F, Wang XD, Zhao YY, Li YJ, Liu YC, Sun J (2015) Silencing the *HaAK* gene by transgenic plant-mediated RNAi impairs larval growth of *Helicoverpa armigera*. Int J Biol Sci 11:67–74

Liu F, Yang B, Zhang A, Ding D, Wang G (2019) Plant-mediated RNAi for controlling *Apolygus lucorum*. Front Plant Sci 10:64

Liu B, Fu D, Ning H, Tang M, Chen H (2022) Knockdown of *CYP6CR2* and *CYP6DE5* reduces tolerance to host plant allelochemicals in the Chinese white pine beetle *Dendroctonusarmandi*. Pestic Biochem Physiol 187:105180

Lucena-Leandro VS, Abreu EF, Vidal LA, Torres CR, Junqueira CI, Dantas J, Albuquerque ÉV (2022) Current scenario of exogenously induced RNAi for lepidopteran agricultural pest control: from dsRNA design to topical application. Int J Mol Sci 23:15836

Ma YM, Chen N, Tan JY, Li MY, Liu S (2022a) Molecular characterization of an arginine kinase from *Lasiodermaserricorne* and the effect of gene silencing on larval survival. J Stored Prod Res 99:102032

Ma LX, He RT, Yan SY, Yang WJ (2022b) RNAi suppression of hormone receptor *HR3* blocks larval molting and metamorphosis in the cigarette beetle, *Lasioderma serricorne*. Agriculture 12:1257

Mahmood S, Kumari P, Kisku AV, Mahapatro GK, Banerjee N, Sarin NB (2022) Ectopic expression of *Xenorhabdus Nematophila Chitinase*in tobacco confers resistance against *Helicoverpaarmigera*. PCTOC 151:593–604

Mamta, Reddy KRK, Rajam MV (2016) Targeting chitinase gene of *Helicoverpaarmigera* by host-induced RNA interference confers insect resistance in tobacco and tomato. Plant Mol Biol 90:281–292

Mao YB, Cai WJ, Wang JW, Hong GJ, Tao XY, Wang LJ, Huang YP, Chen XY (2007) Silencing a cotton bollworm P450 monooxygenase gene by plant-mediated RNAi impairs larval tolerance of gossypol. Nat Biotechnol 25:1307–1313

Mao YB, Tao XY, Xue XY, Wang LJ, Chen XY (2011) Cotton plants expressing *CYP6AE14* double-stranded RNA show enhanced resistance to bollworms. Transgenic Res 20:665–673

Martinez Z, De Schutter K, Van Damme EJ, Vogel E, Wynant N, Broeck JV, Christiaens O, Smagghe G (2021) Accelerated delivery of dsRNA in lepidopteran midgut cells by a Galanthus nivalis lectin (GNA)-dsRNA-binding domain fusion protein. Pestic Biochem Physiol 175:104853

Menezes PS, Yan Y, Yang Y, Mitter N, Mahony TJ, Mody KT (2022) RNAi-Based biocontrol of pests to improve the productivity and welfare of livestock production. Appl Biosci 1:229–243

Moural TW, Ban L, Hernandez JA, Wu M, Zhao C, Palli SR, Alyokhin A, Zhu F (2020) Silencing NADPH-Cytochrome P450 reductase affects imidacloprid susceptibility, fecundity, and embryonic development in *Leptinotarsa decemlineata*. *bioRxiv*

Napoli C, Lemieux C, Jorgensen R (1990) Introduction of a chimeric chalcone synthase gene into petunia results in reversible co-suppression of homologous genes in trans. Plant Cell 2:279–289

OECD (2020) Considerations for the environmental risk assessment of the application of sprayed or externally applied dsRNA-based pesticides. OECD, Paris, p 26

Palauqui JC, Elmayan T, Pollien JM, Vaucheret H (1997) Systemic acquired silencing: transgene-specific post-transcriptional silencing is transmitted by grafting from silenced stocks to non-silenced scions. EMBO J 16:4738–4745

Parker KM, Barragán Borrero V, van Leeuwen DM, Lever MA, Mateescu B, Sander M (2019) Environmental fate of RNA interference pesticides: adsorption and degradation of double-stranded RNA molecules in agricultural soils. Environ Sci Technol 53:3027–3036

Peng C, Yin H, Liu Y, Mao XF, Liu ZY (2022) RNAi mediated gene silencing of detoxification related genes in the ectropis oblique. Gene 13:1141

Piot N, Snoeck S, Vanlede M, Smagghe G, Meeus I (2015) The effect of oral administration of dsRNA on viral replication and mortality in *Bombus terrestris*. Viruses 7:3172–3185

Poreddy S, Li J, Baldwin IT (2017) Plant-mediated RNAi silences midgut-expressed genes in congeneric lepidopteran insects in nature. BMC Plant Biol 17:199

Rask KJ, Rask N (2011) Economic development and food production–consumption balance: a growing global challenge. Food Policy 36:186–196

Ratcliff FG, Baulcombe DC, MacFarlane SA (1999) Gene silencing without DNA RNA-mediated cross-protection between viruses. Plant Cell 11:1207–1216

Rodrigues TB, Petrick JS (2020) Safety considerations for humans and other vertebrates regarding agricultural uses of externally applied RNA molecules. Front Plant Sci 11:407

Romano N, Macino G (1992) Quelling: transient inactivation of gene expression in *Neurospora crassa* by transformation with homologous sequences. Mol Microbiol 6:3343–3353

Romeis J, Widmer F (2020) Assessing the risks of topically applied dsRNA-based products to non-target arthropods. Front Plant Sci 11:679

Santos D, Mingels L, Vogel E, Wang L, Christiaens O, Cappelle K, Wynant N, Gansemans Y, Van Nieuwerburg HF, Smagghe G, Swevers L (2019) Generation of virus-and dsRNA-derived siRNAs with species-dependent length in insects. Virus 11:738

Seiber JN, Coats J, Duke SO, Gross AD (2014) Biopesticides: state of the art and future opportunities. J Agric Food Chem 62:11613–11619

Shang F, Ding BY, Ye C, Yang L, Chang TY, Xie J, Tang LD, Niu J, Wang JJ (2020) Evaluation of a cuticle protein gene as a potential RNAi target in aphids. Pest Manag Sci 76:134–140

Sharif MN, Iqbal MS, Alam R, Awan MF, Tariq M, Ali Q, Nasir IA (2022) Silencing of multiple target genes via ingestion of dsRNA and PMRi affects development and survival in *Helicoverpaarmigera*. Sci Rep 12:10405

Sharma Y, Padha S, Dhar A, Baweja V, Singh IK (2023) RNAi-based biopesticides: a review of recent studies in lepidopteran insects. In: Proceedings of the zoological society. Springer, New Delhi, pp 1–9

Shen GM, Chen W, Li CZ, Ou SY, He L (2019) RNAi targeting ecdysone receptor blocks the larva to adult development of *Tetranychuscinnabarinus*. Pestic Biochem Physiol 159:85–90

Shen CH, Xu QY, Fu KY, Guo WC, Jin L, Li., GQ. (2020) Two splice isoforms of *Leptinotarsa* ecdysis triggering hormone receptor have distinct roles in larva-pupa transition. Front Physiol 11:593962

Shen CH, Jin L, Fu KY, Guo WC, Li GQ (2022) RNA interference targeting *RasGTPase* gene *Ran* causes larval and adult lethality in *Leptinotarsadecemlineata*. Pest Manag Sci 78:3849–3858

Shi T, Tang P, Wang X, Yang Y, Wu Y (2022) CRISPR-mediated knockout of nicotinic *acetylcholine ereceptor* (*nAChR*) *α6* subunit confers high levels of resistance to spinosyns in *Spodopterafrugiperda*. Pestic Biochem Physiol 187:105191

Shin YH, Lee SH, Park Y (2020) Development of mite (*Tetranychus urticae*)-resistant transgenic Chinese cabbage using plant-mediated RNA interference. Hortic Environ Biotechnol 61:305–315

Shukla N, Singh EA, Kabadwa BC, Sharma RK (2019) Present status and future prospects of bioagents in agriculture. Int J Curr Microbiol App Sci 8:2138–2153

Thakur N, Upadhyay SK, Verma PC, Chandrashekar K, Tuli R, Singh PK (2014) Enhanced whitefly resistance in transgenic tobacco plants expressing double stranded RNA of v-ATPase A gene. PLoS One 9:e87235

Tijjani A, Bashir KA, Mohammed I, Muhammad A, Gambo A, Musa H (2016) Biopesticides for pests control: a review. J Biopesticide Agric 3:6–13

Wang J, Ma H, Zuo Y, Yang Y, Wu Y (2020) CRISPR-mediated gene knockout reveals nicotinic *acetylcholinereceptor* (*nAChR*) subunit *α6* as a target of spinosyns in *Helicoverpaarmigera*. Pest Manag Sci 76:2925–2931

Watanabe S, Adams BL, Kong A, Masang JN, Vowell T, Melzer M (2020) Identification of genes that result in high mortality of *Oryctes rhinoceros* (Scarabaeidae: *Coleoptera*) when targeted using an RNA interference approach: implications for large invasive insects. Ann Entomol Soc Am 113:310–317

Wingard SA (1928) Hosts and symptoms of ring spot, a virus disease of plants. J Agric Res 37:127–153

Wise JC, Wise AG, Rakotondravelo M, Vandervoort C, Seeve C, Fabbri B (2022) Trunk injection delivery of dsRNA for RNAi-based pest control in apple trees. Pest Manag Sci 78:3528–3533

Wood CC, Okada S, Taylor MC, Menon A, Mathew A, Cullerne D, Stephen SJ, Allen RS, Zhou XR, Liu Q, Oakeshott JG (2018) Seed-specific RNAi in safflower generates a superhigh oleic oil with extended oxidative stability. Plant Biotechnol J 16:1788–1796

Wu JJ, Mu LL, Kang WN, Ze LJ, Shen CH, Jin L, Anjum AA, Li GQ (2021) RNA interference targeting ecdysone receptor blocks the larval–pupal transition in *Henosepilachna vigintioctopunctata*. Insect Sci 28:419–429

Xiong Y, Zeng H, Zhang Y, Xu D, Qiu D (2013) Silencing the *HaHR3* gene by transgenic plant-mediated RNAi to disrupt *Helicoverpa armigera* development. Int J Biol Sci 9:370

Yan S, Qian J, Cai C, Ma Z, Li J, Yin M, Ren B, Shen J (2020) Spray method application of transdermal dsRNA delivery system for efficient gene silencing and pest control on soybean aphid Aphis glycines. J Pest Sci 93:449–459

Yogindran S, Rajam MV (2016) Artificial miRNA-mediated silencing of ecdysone receptor (*EcR*) affects larval development and oogenesis in *Helicoverpa armigera*. Insect Biochem Mol Biol 77:21–30

Yogindran S, Rajam MV (2020) Host-derived artificial miRNA-mediated silencing of ecdysone receptor gene provides enhanced resistance to *Helicoverpa armigera* in tomato. Genomics 7543:31963–31967

Yousafi Q, Sarfaraz A, Khan MS, Saleem S, Shahzad U, Khan AA, Sadiq M, Abid AD, Shahzad MS, Ul Hassan N (2021) In silico annotation of unreviewed *acetylcholinesterase* (AChE) in some lepidopteran insect pest species reveals the causes of insecticide resistance. Saudi J Biol Sci 28:2197–2209

Yu R, Xu X, Liang Y, Tian H, Pan Z, Jin S, Wang N, Zhang W (2014) The insect ecdysone receptor is a good potential target for RNAi-based pest control. Int J Biol Sci 10:1171–1180

Zhang N, Jiang H, Meng X, Qian K, Liu Y, Song Q, Stanley D, Wu J, Park Y, Wang J (2020) Broad-complex transcription factor mediates opposing hormonal regulation of two phylogenetically distant *arginine kinase* genes in *Tribolium castaneum*. Commun Biol 3:631

Zhang X, Wang Y, Zhang S, Kong X, Liu F, Zhang Z (2021a) RNAi-mediated silencing of the chitinase 5 gene for fall webworm (*Hyphantria cunea*) can inhibit larval molting depending on the timing of dsRNA injection. Insect 12:406

Zhang Z, Ma Y, Ma X, Hu H, Wang D, Song X, Ren X, Ma Y (2021b) Combined transcriptomic analysis and RNA interference reveal the effects of methoxyfenozide on ecdysone signaling pathway of *Spodopteraexigua*. Int J Mol Sci 22:9080

Zheng Y, Hu Y, Yan S, Zhou H, Song D, Yin M, Shen J (2019) A polymer/detergent formulation improves dsRNA penetration through the body wall and RNAi-induced mortality in the soybean aphid Aphis glycines. Pest Manag Sci 75:1993–1999

Zhong ZB, Xu S, Ai ZC, Yang LL, She LY, Xing G, Mei CD, Zhang P, Wang SM, Ling CX (2020) Silencing of cytochrome P450 in *Spodopterafrugiperda* (Lepidoptera: *Noctuidae*) by RNA interference enhances susceptibility to chlorantraniliprole. J Insect Sci 20:12

Zhu F, Xu J, Palli R, Ferguson J, Palli SR (2011) Ingested RNA interference for managing the populations of the Colorado potato beetle, *Leptinotarsadecemlineata*. Pest Manag Sci 67:175–182

Zhu JQ, Liu S, Ma Y, Zhang JQ, Qi HS, Wei ZJ, Yao Q, Zhang WQ, Li S (2012) Improvement of pest resistance in transgenic tobacco plants expressing dsRNA of an insect associated gene *EcR*. PLoS One 7:e38572

Zhu B, Shan J, Li R, Liang P, Gao X (2019) Identification and RNAi-based function analysis of chitinase family genes in diamondback moth, *Plutella xylostella*. Pest Manag Sci 75:1951–1961

Zotti M, Dos Santos EA, Cagliari D, Christiaens O, Taning CN, Smagghe G (2018) RNA interference technology in crop protection against arthropod pests, pathogens and nematodes. Pest Manag Sci 74:1239–1250

Zubair M, Khan MZ, Rauf I, Amin I, Mansoor S (2020) Artificial microRNA (amiRNA)-mediated resistance against whitefly (*Bemisia tabaci*) targeting three genes. Crop Prot 137:105308

Zuo YY, Xue YX, Wang ZY, Ren X, Aioub AA, Wu YD, Yang YH, Hu ZN (2022) Knockin of the G275E mutation of the nicotinic *acetylcholinereceptor* (*nAChR*) α6 confers high levels of resistance to spinosyns in *Spodopteraexigua*. Insect Sci 29:478–486

Potential Management of a Serious Pest of Cocoa Plantation, Cocoa Pod Borer, Using RNAi Technology: A Case Study

13

Mohd Firdaus Abdullah, M. G. K. Jones, and C. L. Tan

Abstract

Cocoa pod borer is the most serious and damaging pest of cocoa in Southeast Asia. Annually, crop production losses due to cocoa pod borer (CPB) damages are about 5–20%. Occasionally, total loss in production can be due to CPB damages. Chemical insecticide spraying is the usual method employed in the plantations. This resulted in chemical residues that are not biodegradable found on the plants and cocoa beans. Therefore, alternative approaches to managing CPB need to be explored. This chapter describes the use of RNA interference (RNAi) technology to manage CPB. Several genes involved in vital metabolic processes of the insect have been identified via bioinformatics. Feeding the insects with the dsRNA causes increased insect mortality. Analysis using quantitative PCR confirmed the suppression of the target genes.

Keywords

RNA interference · *Theobroma cacao* · Cocoa pod borer

A. F. Abdullah · C. L. Tan (✉)
Malaysian Cocoa Board, Wisma SEDCO, Kota Kinabalu, Sabah, Malaysia
e-mail: tancl@koko.gov.my

M. G. K. Jones
State Agricultural Biotechnology Centre, Murdoch University, Perth, WA, Australia
e-mail: m.jones@murdoch.edu.au

K. Goswami et al. (eds.), *Non-coding RNAs for Crop Improvement*,
https://doi.org/10.1007/978-981-97-7213-1_13

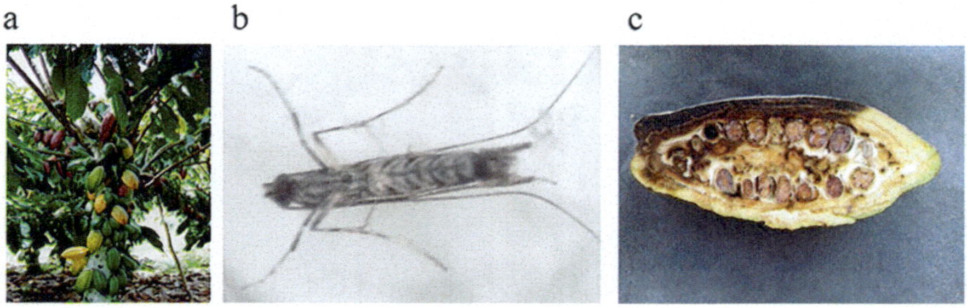

Fig. 13.1 (**a**) Cocoa (*Theobroma cacao*) tree, (**b**) Cocoa pod borer insect, (**c**) Cocoa pod damaged by CPB

13.1 Introduction

Theobroma cacao L., or commonly called cocoa (Fig. 13.1a), is a major tropical commodity crop. Its seed contains rich oil and flavour. The unique source of cocoa solids and cocoa butter. Cocoa is mainly used in manufacturing of chocolate, cocoa confectionery, cocoa butter, and powder (Wood and Lass 2001). Other uses include the pharmaceutical and health industries. The cocoa bean is obtained after the fresh seeds were fermented and dried. It is the most economical part of the plant. This constitutes the raw material for manufacturing of semi-finished products, for example, cocoa mass and cocoa butter, or finished end-products such as chocolates and confectioneries that are used for direct consumption by the customers. Cocoa mass is used for manufacturing chocolate, biscuits, and confectionery. Cocoa butter is used for making sweets, chocolate, and perfume and even in pharmaceutical and health industries. Other parts of the cocoa plants that are useful include husks, fats and organic chemicals that are extracted from the husks, which can be used to feed cattle, manufacture fertilizers, pharmaceutical products, and soap.

The major pest in cocoa plantation in South-east Asia is cocoa pod borer, scientifically known as *Conopomorpha cramerella* (Snellen) (Fig. 13.1b). It is of the order Lepidoptera and family Gracillariidae (Bradley 1986). The origin is from South Asia (Posada and Vega 2005). The insect is mainly found in Malaysia, Indonesia, Thailand, Vietnam, Papua New Guinea and surrounding countries. The larvae eat the cocoa pulp; the material that surrounds the individual seeds. The larval feeding resulted in the clumping of the cocoa seeds, which produces smaller than normal seeds and poor-quality cocoa beans (Fig. 13.1c). The losses in monetary terms due to this insect can be up to 80% in some geographical areas (Posada et al. 2011).

13.1.1 How RNAi Works in the Cell of Insect

RNAi has been used widely to identify functional genes accumulated from many sequencing projects (Abdurakhmonov et al. 2016; Vogel et al. 2018). Among the components involved in the RNAi mechanism is the Dicer, which cuts the long dsRNA into short interfering RNAs. Subsequently, the RISC guide the attack on mRNA with sequence complimentary to siRNA. These mechanisms are evolutionarily conserved across all eukaryotic taxa (Agrawal et al. 2003). This implies that the mechanism also works in insects and the possibility of using this technology to manage insect pests.

Several studies (Christiaens et al. 2020; Nitnavare et al. 2021) have shown that by customizing the dsRNA to certain essential genes that are vital to insect growth and metabolism; it could be custom-made into species-specific pesticides. In addition, the current chemical pesticides are generally broad spectrum and can pose harmful effects to non-target insects and human beings. Also, insect pests are developing resistance to the current reservoir of chemical pesticides.

13.1.2 RNAi and Insect Control

One of the essential criteria for utilizing RNAi as an insect management tool is that the target insect must be able to ingest the dsRNA through feeding, digesting, and absorption into their cells (Christiaens et al. 2020; He et al. 2022). Therefore, plants can be protected by these novel, environmentally friendly RNA-based pesticides. Besides, they are very specific in their target insects and do not kill non-target organisms such as useful pollinators (Wynant et al. 2014).

One early example of the usage of RNAi technology on insect pests is research studies on Western corn rootworm (WCRW, *Diabrotica virgifera*) and cotton ballworm (CBW, *Helicoverpa armigera*) (Fishilevich et al. 2016; Darlington et al. 2022). The WCRW larva was fed with 14 different dsRNAs of the essential genes, which resulted in reducing the performance of the larva. For the research on cotton ballworm (CBW), cytochrome P450 was the target gene. The gene, CYP6AE14, detoxifies gossypol, a common metabolite in cotton plants. Genetically transformed cotton plants expressing CYP6AE dsRNA greatly reduced the growth CBW larvae. As a result, the transformed plants suffered less damage as compared to the control (Mao et al. 2011).

In another case study, scientists used transgenic tobacco plants expressing hairpin RNA to decrease mRNA levels of moult-regulating transcription factor in larval *H. armigera*, causing retarded growth and subsequent death of the insect larva (Kim et al. 2015). Another study conducted on nicotine showed that the tobacco hornworm *Manduca sexta* can withstand high concentration of nicotine. In *M. sexta*, nicotine induces the cytochrome P450 gene, CYP6B46. Genetically transformed tobacco plants expressing dsRNA targeting CYP6B46 were developed. Tobacco hornworm larvae fed on the transformed tobacco leaves showed more susceptibility to predation by spider due to less nicotine been exhaled through their spiracles (Kumar et al. 2014). These results confirmed that RNAi is useful in controlling agriculture insect pests.

13.2 Identification of Insect Vital Genes

Vital genes necessary for lepidopteran insects' growth and metabolism with their function (Table 13.1) were collected by reviewing the literature. The sequences of these mRNA genes were retrieved from the NCBI database for different lepidopteran insects (*Epiphyas postvittana, Spodoptera frugiperda, Manduca sexta, Plutella xylostella* and *Helicoverpa armigera*). These gene sequences were aligned using DNA* software (MegAlign). Consensus sequences were obtained from aligning the various gene sequences. Appropriate primers were designed using Primer 3 software (Table 13.2).

Table 13.1 Functions of target genes identified

Gene	Function	References
Thioredoxin (Tdxn)	It is involved in the mechanisms of defence preventing oxidative damage caused by ROS. Examples of ROS are hydroxyl radical, superoxide anion and hydrogen peroxide. They are noted for their increased reactivity and damage to DNA, proteins, and lipid membranes	Huang et al. (2019)
Calreticulin (Crt)	It is involved in the system of defence towards potential parasites and pathogens. It functions by recognising the molecular structures found in pathogen-specific features, such as microorganisms' cell wall and causes phagocytosis. Some of the possible functions are in insect olfactory system and neuronal activities	Jia et al. (2009)
Aminopeptidase (Amp)	These are the known putative receptors of *Bacillus thuringiensis* toxins	Wang et al. (2021)
Chitinase (Ctns)	It helps to dissolve chitin into soluble and insoluble oligosaccharides. It plays a vital role during insect post-embryonic moulting processes	Zhang et al. (2017)
Cathepsin L (Cpl)	Cathepsin is involved in the moulting process of larvae by producing proteolytic enzymes essential for embryonic development	Hashmi et al. (2002)
Gluthathione peroxidase (Gp)	It is involved in the removal of cytotoxic peroxides formed in aerobic metabolism, by helping to reduce the damage by free radical to mitochondria and enhance the intrinsic radio resistance of lepidopteran insects	Suman et al. (2009)
Annexin (Axn)	Possible functions of Annexin are in hemolymph coagulation, and physiological mechanisms of the insect salivary gland	Schmid et al. (2019)

Table 13.2 Degenerate PCR primers used for PCR and qRT-PCR

Primer	Forward (from 5′ to 3′)	Reverse (from 5′ to 3′)
Cpl	ATACTCGAGTGGAAGGACAACATTTC	ATAGGTACCTACGTGCCACCATTAT
Crt	ATACTCGAGGGTGAAACTCCGTATGA	GTAGGTACCCCCATACTCCCTTGTAAT
Gp	ATACTCGAGGTAACACAGCGAACAA	AGTGGTACCTTCTCCAAACCCAAAG
Axn	TACCTCGAGGCAAGGATATCGTAGAGA	TAGGGTACCCGCATCATAGCTAAAGAG
Tdxn	ATACTCGAGTCAAGACGGTAACTTAGAACCA	ATAGGTACCTCTTTGCCATCTTTGGGATTT C
Amp	ATACTCGAGATTTGCCTTGGAAGTTG	ATAGGTACCGATGGCCTACAGGTATT
Ctns	AATCTCGAGGGTTACGCTGGTATTGA	ATAGGTACCTTACGTGTGGGAGAATG
Qcpl	CATGCATGCTTTGGCTACTG	CACCATGGTCCAGTTCTGTG
Qamp	TCTGGTGCTATGGAAAATTGG	AACCATTGGTGAGCCAACTC
Qact	GAGAAGATCTGGCATCACACC	TGGGTCATCTTCTCCCTGTT
Qcrt	TGATGACCCCAATGATGAGA	ATCATTGCTGGTTCCCATTC
Actin	AGGTATCCTCACCCTGAAGT	CTTCATGAGGTAGTCGGTCAAG

Note: Those primers prefix with Q are primers for qPCR

13.3 Isolation of CPB RNA and Amplification of the Identified Vital Genes

Total RNA was extracted from different stages of mixed populations of *C. cramerella* with TRIzol reagent (Life Technologies) and ethanol precipitation. The insect's quantity used depended on the purpose of the experiment. Initially, the insects were macerated together with liquid nitrogen in a 1.5 mL RNAse-free centrifuge tube with 800 mL of TRIzol reagent and subsequently vortexed for 5 min. The insect RNA was extracted with 200 mL of chloroform. The aqueous phase was ethanol-precipitated for 30 min at 12,000 g and resuspended in DEPC-treated water and added with DNase I (Qiagen, Australia). The total RNA readings for 260/280 and 260/230 were 1.97 and 2.27 respectively (Fig. 13.2). Preparations of cDNA were made using the High-Capacity cDNA Reverse Transcription kit (Applied Biosystems, Australia) following the manufacturer's manual. PCRs of the vital genes were carried out by amplifying from 300 ng of cDNA with MyTaq DNA Polymerase (Bioline, Australia) according to the manufacturer's instructions in a 20 mL reaction tubes. The PCR reaction was carried out at this temperature profile: 95 °C for 3 min, subsequently 35 cycles of 95 °C for 15 s, then 55 °C for 15 s and 72 °C for 30 s and finally at 72 °C for 10 min in a Veriti 96-Well Thermal Cycler (Applied Biosystems).

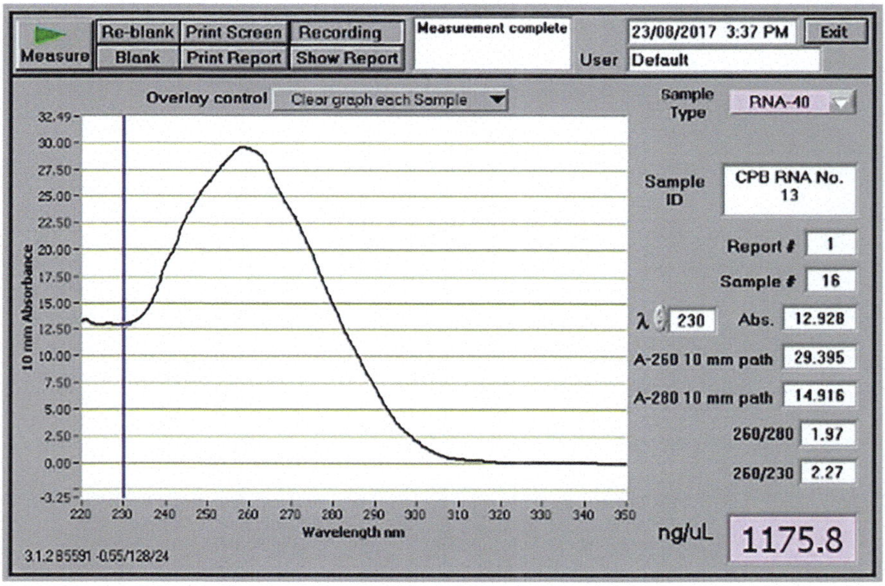

Fig. 13.2 CPB total RNA

13.3.1 Isolation of Genomic DNA of Cocoa Pod Borer

CPB adult moth genomic DNA was isolated using the caesium chloride ultracentrifugation method (Fig. 13.3a). Genomic DNA was ultra-centrifuged in caesium chloride gradient to ensure the purity of the DNA (Fig. 13.3a). Nanodrop® 2000 was used to quantify the DNA. The readings of the DNA at 260/280 and 260/230 were 1.92 and 2.12 respectively (Fig. 13.3b).

13.3.2 Amplification of Vital Genes from CPB cDNA

Vital genes from cDNA were amplified using degenerate PCR primers (Table 13.2). The amplicons were eluted, purified, and then sequenced. From BLAST analysis, the genes were found to be more than 90% similar to target genes in other insect species. Some examples are shown in Fig. 13.4.

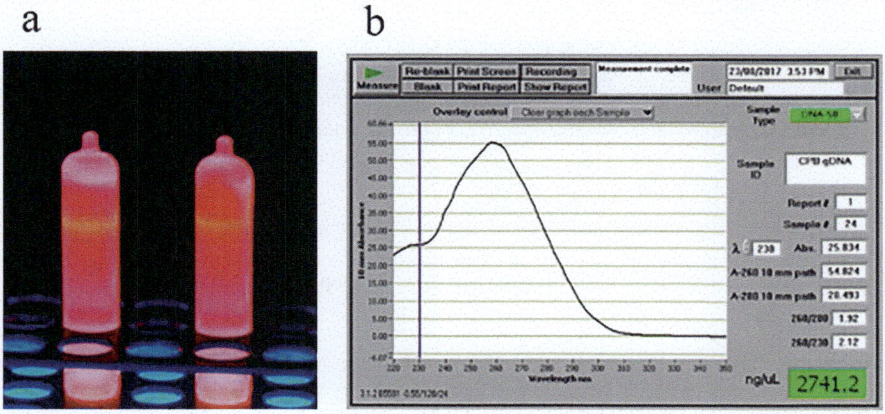

Fig. 13.3 Caesium chloride ultracentrifugation (**a**) and Nanodrop measurement (**b**) of CPB gDNA. For RNA interference studies, these cDNA were used to isolate CPB vital genes

a

Accession	Description	Identity
XM_022311334.1	*Myzus persicae calreticulin-like mRNA*	99%
XM_003240040.3	*Acyrthosiphon pisum calreticulin mRNA*	94%
XM015524006.1	*Diuraphis noxia calreticulin (LOC107173458), mRNA*	92%

b

Accession	Description	Identity
XM_022305806.1	*Myzus persicae puromycin-sensitive aminopeptidase (LOC111027432), transcript variant X5, mRNA*	97%
XM_015524093.1	*Diuraphis noxia puromycin-sensitive aminopeptidase (LOC107173520), transcript variant X3, mRNA*	94%
XM016808083.1	*Acyrthosiphon pisum puromycin-sensitive aminopeptidase (LOC100162520), transcript variant X5, mRNA*	94%

Fig. 13.4 (**a**) Blast results of CPB Crt gene and (**b**) CPB Amp gene

13.4 Cloning Target Genes into Vector to Generate Double-stranded RNA

13.4.1 DsRNA Vector Cloning

The verified genes via sequencing and BLAST analysis were cloned into the dsRNA vector, pDoubler (Fig. 13.5a). The vector and genes were excised with restriction enzymes KpnI and XhoI overnight at 37 °C. The following day, the vector and the corresponding genes were ligated overnight at 4 °C. After ligation was completed, the products were transformed into *E. coli* JM109 and selected on LB media with 25 mg ml^{-1} kanamycin antibiotics (Fig. 13.5b).

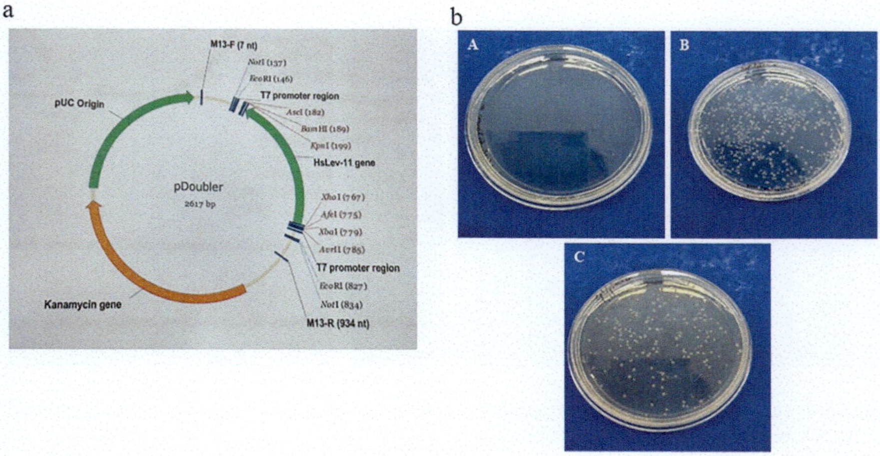

Fig. 13.5 (**a**) dsRNA vector, pDoubler, (**b**) identifying clones of pDoubler containing CPB genes transformed in to bacteria JM109. (A) Control; (B) pDoubler containing Crt gene; (C) pDoubler containing Tdxn gene

13.4.2 Double-Stranded (ds) RNA Generation

The control gene used was a 745-nucleotide long dsRNA of the *Aequorea Victoria gfp* gene. Double-stranded RNAs (dsRNAs) were synthesized with MEGAscript RNAi Kit following the manufacturer's manual (Ambion, USA). The insect genes DNA templates were generated using primer pairs, each with a T7 promoter sequence upstream of the gene-specific portion so that in vitro transcription with the T7 RNA polymerase promoter could be carried out. The primers used to amplify the *gfp* gene were T7GFP-F and T7GFP-R. Briefly, DNA (2 mg) was incubated with the T7 enzyme mix and each of the ribonucleotides (75 mM) The temperature profile of the reaction was; 16 h at 37 °C, subsequently followed by 1 h of DNase-1 treatment at 37 °C. Finally, an extra annealing step at 75 °C for 5 min was carried out so that longer dsRNA of the *gfp* gene was completely annealed before nuclease treatment. Double-stranded RNAs (dsRNAs) were purified and screened for integrity on a non-denaturing agarose gel (1.5%) with 1X TBE.

13.4.3 Positive Identification of Clones by Colony PCR and dsRNA Vector Restriction Digest

Positive clones containing CPB genes were identified by colony PCR (Fig. 13.6a). Confirmation was carried out with restriction digests of the dsRNA vector (Fig. 13.6b).

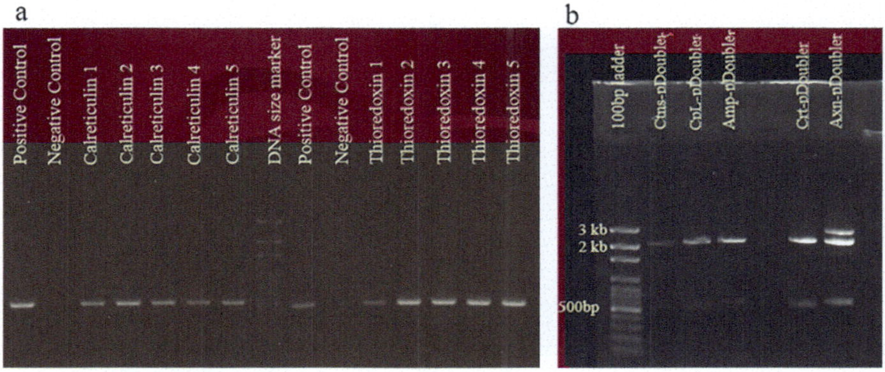

Fig. 13.6 (**a**) Colony PCR to identify positive clones containing CPB genes, (**b**) restriction digest of dsRNA vector to validate insert presence

13.5 Feeding Bioassays of CPB Moths with Double-Stranded RNA

CPB-infested pods were collected from farmers' fields in Sabah, Malaysia. These pods were wrapped in paper and stored for 2 weeks in the dark. During this time, the pods were frequently checked for pupae. The pupae collected and were kept in Petri dishes with aeration holes until the eclosion of the adult moth. Subsequently, the CPB insects were treated with control (negative and test control) and seven different in vitro synthesized dsRNA (30 μg/ml). Each type of control and treatment was experimented with a minimum of 4 insects each and repeated thrice. Feeding bioassays were carried out for a duration of seven days. After 4 days of feeding with *Amp* dsRNA, complete mortality of the insects was observed. Feeding with *Crt* and *Cpl* dsRNA resulted in 60% mortality of the insects (Fig. 13.7).

13.6 Analysis of CPB Insects Using Real-time Quantitative Reverse Transcriptase PCR

After feeding for 3 days with dsRNA, one live CPB insect (in triplicates) was removed for total RNA extraction using Trizol reagent (Invitrogen). Any residual DNA from the total RNA was removed using RNase free DNase I added at 37 °C. Synthesis of the first strand cDNA was carried out using oligo-dT primer (Fermentas). The newly synthesized cDNA was used as a template for qRT-PCR. An internal control gene encoding actin (Forward: 5′-AGGTATCCTCACCCTGAAGT-3′ and Reverse 5′-CTTCATGAGGTAGTCGGTCAAG-3′) was used.

Gene expression in cocoa pod borer insects were analyzed with the GoTaq qPCR Master Mix (Promega Corporation, USA) in a Corbett Quantitative Thermal Cycler (Qiagen Pty Ltd., Australia). Briefly, 1 μL of the cDNA was converted from 500 ng of RNA of experimental and control samples. Subsequently, it was treated with 1x

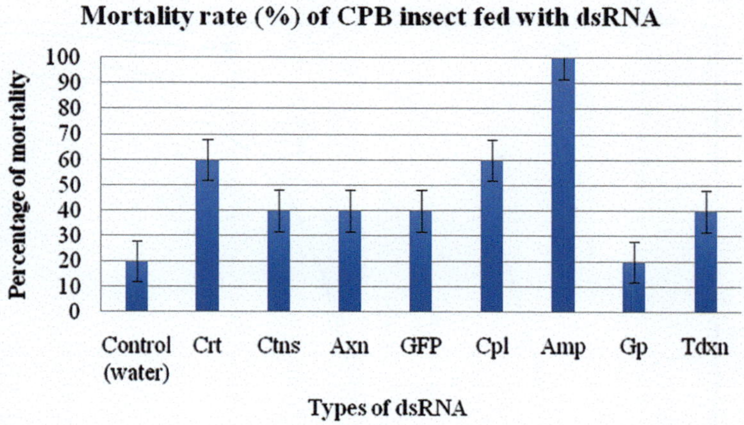

Fig. 13.7 Feeding bioassays of CPB insects with dsRNA

GoTaq qPCR master mix, then 5 μM of each of the gene specific primer pair with sterile deionized water were added to make the reaction mixture to 20 μL. All PCRs were carried out in triplicate and mean of Ct values were determined. Expression of Crt, Cpl and Amp was quantified with primer pairs qCrt-F and qCrt-R, qCpl-F and qCpl-R and qAmp-F and qAmp-R, which amplified at 128, 123 and 134 bp respectively with CPB insect cDNA (Table 13.2). The PCR reactions were carried out at 95 °C for 10 min, then for 45 cycles at 95 °C for 15 s and 55 °C for 60 s. Normalization of gene expression was carried out using the actin gene with primer pairs of qActin-F and qActin-R that amplified 116 bp (Table 13.2). Relative gene expression was analyzed using the $\Delta\Delta$Ct method as according to the ABI PRISM 7700 Sequence Detection System Bulletin 2 (Applied Biosystems). Among the three genes studied, *Amp* gene was most effective in causing mortality in CPB moths in the bioassays (Fig. 13.8).

Feeding dsRNA of the *Cpl* gene caused a reduction of 3-million-fold in its transcripts versus the control; dsRNA feeding of *Crt* gene caused a reduction of 58,000-fold in transcript whereas feeding dsRNA of the *Amp* gene caused a 10,000-fold reduction in transcripts (Table 13.3).

13.7 Conclusion and Future Perspective

The highest mortality of CPB insects were those fed with Amp dsRNA, followed by Cpl and Crt dsRNAs. This experiment confirmed that gene silencing occurred in CPB as demonstrated by the high-fold-reduction in transcripts compared to those fed without dsRNA and mortality of the insects. The study also showed that RNA interference is a promising tool for developing an environmentally friendly biopesticide to control CPB infestations in cocoa cultivation.

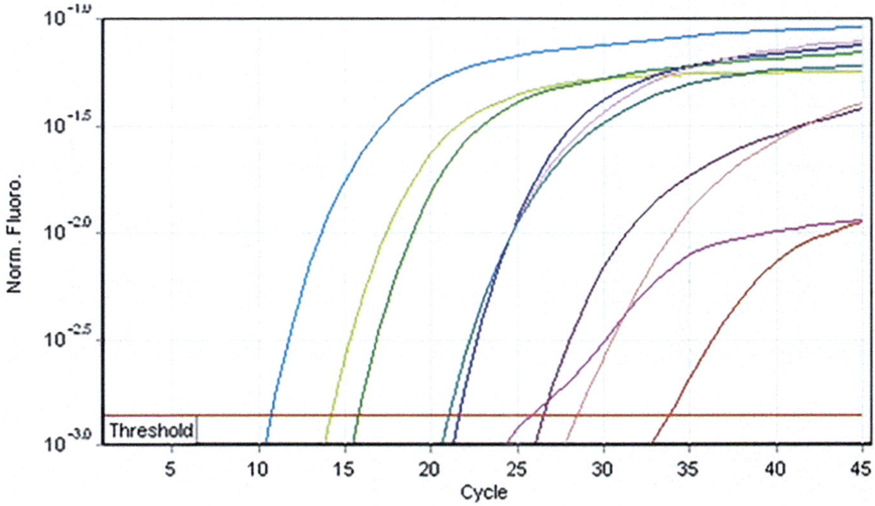

Fig. 13.8 Quantitative RT-PCR analyses of CPB target genes

Table 13.3 Quantification of decrease in gene expression after feeding with dsRNA

Vital gene	dsRNA concentration (μg/ml)	No. of fold reduction in transcript level
Cpl	30	23,562,675
Crt	30	58,656
Amp	30	10,960

RNAi offers a promising avenue for controlling this destructive pest and enhance cocoa plantation management. It could revolutionize pest control by providing a targeted, environmentally friendly solution. By silencing essential genes in the cocoa pod borer, RNAi can significantly reduce pest populations without harming beneficial insects or the surrounding ecosystem. The specificity of RNAi reduces the likelihood of resistance development, a common issue with conventional pesticides. Advances in genetic engineering and molecular biology will enable the identification and validation of critical genes involved in the pest's growth, reproduction, and survival. The integration of these RNAi constructs into cocoa plants can provide continuous protection, reducing the need for chemical interventions. Embracing this innovative approach could lead to a more resilient and productive future for cocoa plantations globally.

Acknowledgement We are grateful to the Australian Government for providing a Research Fellowship to enable the main author to carry out part of the research at the WA State Agricultural Biotechnology Centre (SABC), Murdoch University, Perth, Western Australia. The project was funded under the 11th Malaysian Development Fund.

References

Abdurakhmonov IY, Ayubov MS, Ubaydullaeva KA, Buriev ZT, Shermatov SE, Ruziboev HS, Shapulatov UM, Saha S, Ulloa M, Yu JZ, Percy RG, Devor EJ, Sharma GC, Sripathi VR, Kumpatla SP, van der Krol A, Kater HD, Khamidov K, Salikhov SI, Jenkins JN, Abdukarimov A, Pepper AE (2016) RNA interference for functional genomics and improvement of cotton (Gossypium sp.). Front Plant Sci 7:202. https://doi.org/10.3389/fpls.2016.00202

Agrawal N, Dasaradhi PV, Mohmmed A, Malhotra P, Bhatnagar RK, Mukherjee SK (2003) RNA interference: biology, mechanism, and applications. Microbiol Mol Biol Rev 67(4):657–685. https://doi.org/10.1128/MMBR.67.4.657-685.2003

Bradley JD (1986) Identity of the South East Asian cocoa moth, *Conopomorpha cramerella* (Snellen) (Lepidoptera: Gracillariidae), with descriptions of three allied new species. Bull Entomol Res 76(1):41–51

Christiaens O, Whyard S, Vélez AM, Smagghe G (2020) Double-stranded RNA technology to control insect pests: current status and challenges. Front Plant Sci 11:451. https://doi.org/10.3389/fpls.2020.00451

Darlington M, Reinders JD, Sethi A, Lu AL, Ramaseshadri P, Fischer JR, Boeckman CJ, Petrick JS, Roper JM, Narva KE, Vélez AM (2022) RNAi for western corn rootworm management: lessons learned, challenges, and future directions. Insects 13(1):57. https://doi.org/10.3390/insects13010057

Fishilevich E, Vélez AM, Storer NP, Li H, Bowling AJ, Rangasamy M, Worden SE, Narva KE, Siegfried BD (2016) RNAi as a management tool for the western corn rootworm, *Diabrotica virgifera virgifera*. Pest Manag Sci 72(9):1652–1663. https://doi.org/10.1002/ps.4324

Hashmi S, Britton C, Liu J, Guiliano DB, Oksov Y, Lustigman S (2002) Cathepsin L is essential for embryogenesis and development of *Caenorhabditis elegans*. J Biol Chem 277(5):3477–3486. https://doi.org/10.1074/jbc.M106117200

He L, Huang Y, Tang X (2022) RNAi-based pest control: production, application and the fate of dsRNA. Front Bioeng Biotechnol 10:1080576. https://doi.org/10.3389/fbioe.2022.1080576

Huang H, Ullah F, Zhou D-X, Yi M, Zhao Y (2019) Mechanisms of ROS regulation of plant development and stress responses. Front Plant Sci 10:800. https://doi.org/10.3389/fpls.2019.00800

Jia XY, He LH, Jing RL, Li RZ (2009) Calreticulin: conserved protein and diverse functions in plants. Physiol Plant 136(2):127–138. https://doi.org/10.1111/j.1399-3054.2009.1223.x

Kim YH, Soumaila IM, Cooper AM, Zhu KY (2015) RNA interference: applications and advances in insect toxicology and insect pest management. Pestic Biochem Physiol 120:109–117. https://doi.org/10.1016/j.pestbp.2015.01.002

Kumar P, Pandit SS, Steppuhn A, Baldwin IT (2014) Natural history-driven, plant-mediated RNAi-based study reveals CYP6B46's role in a nicotine-mediated antipredator herbivore defense. Proc Natl Acad Sci USA 111:1245–1252. https://doi.org/10.1073/pnas.1314848111

Mao YB, Tao XY, Xue XY, Wang LJ, Chen XY (2011) Cotton plants expressing CYP6AE14 double-stranded RNA show enhanced resistance to bollworms. Transgenic Res 20:665–673

Nitnavare RB, Bhattacharya J, Singh S, Kour A, Hawkesford MJ, Arora N (2021) Next generation dsRNA-based insect control: success so far and challenges. Front Plant Sci 12:673576. https://doi.org/10.3389/fpls.2021.673576

Posada F, Vega FE (2005) Establishment of the fungal entomopathogen *Beauveria bassiana* (Ascomycota: Hypocreales) as an endophyte in cocoa seedlings (*Theobroma cacao*). Mycologia 97(6):1195–1200

Posada FJ, Virdiana I, Navies M, Pava-Ripoll M, Hebbar P (2011) Sexual dimorphism of pupae and adults of the cocoa pod borer, *Conopomorpha cramerella*. J Insect Sci 11:52

Schmid MR, Dziedziech A, Arefin B, Kienzle T, Wang Z, Akhter M, Berka J, Theopold U (2019) Insect hemolymph coagulation: kinetics of classically and non-classically secreted clotting factors. Insect Biochem Mol Biol 109:63–71. https://doi.org/10.1016/j.ibmb.2019.04.007

Suman S, Seth RK, Chandna S (2009) Mitochondrial antioxidant defence in radio-resistant Lepidopteran insect cells. Bioinformation 4(1):19–23. https://doi.org/10.6026/97320630004019

Vogel E, Santos D, Mingels L, Verdonckt TW, Broeck JV (2018) RNA interference in insects: protecting beneficials and controlling pests. Front Physiol 9:1912. https://doi.org/10.3389/fphys.2018.01912

Wang J, Yang X, He H, Chen J, Liu Y, Huang W, Ou L, Yang Z, Guan X, Zhang L, Wu S (2021) Knockout of two cry-binding aminopeptidase N isoforms does not change susceptibility of Aedes aegypti Larvae to Bacillus thuringiensis subsp. israelensis Cry4Ba and Cry11Aa toxins. Insects 12(3):223. https://doi.org/10.3390/insects12030223

Wood GA, Lass RA (2001) Cocoa, 4th edn. Wiley-Blackwell, Hoboken

Wynant N, Santos D, Wielendaele PV, Broeck JV (2014) Scavenger receptor-mediated endocytosis facilitates RNA interference in the desert locust, *Schistocerca gregaria*. Insect Mol Biol 23(3):320–329. https://doi.org/10.1111/imb.12083

Zhang J, Khan SA, Heckel DG, Bock R (2017) Next-generation insect-resistant plants: 492 RNAi-mediated crop protection. Trends Biotechnol 35(9):871–882. https://doi.org/10.1016/j.tibtech.2017.04.009

microRNAs as Cogent Buttons for Mediating Drought Stress Tolerance in Plants

14

Temesgen Assefa Gelaw and Neeti Sanan-Mishra

Abstract

Drought stress hampers plant's entire life and over the past several years, numerous efforts have been made towards the improvement of crops, mostly using traditional or modern breeding approaches. Over the last few decades, the domain of molecular biology has flourished and made it possible to understand the various genes required to achieve desirable plant traits under different environmental conditions. Noncoding RNAs are one of the most important regulators of gene expression discovered in recent times and playing a gear shifting role in plant growth, development, and stress response. Expression profiling of drought responsive microRNAs have been reported in different plants and functional characterization of microRNAs and targets showed the significant role in drought stress mitigation. Therefore, in this chapter, we have summarized and presented the role of plant microRNAs in response to drought stress in different crops. We have also highlighted some of the alternative promising strategies for drought stress tolerance.

Keywords

Drought · Noncoding RNAs · microRNAs · Plants · Stress

T. A. Gelaw
Plant RNAi Biology Group, International Centre for Genetic Engineering and Biotechnology, New Delhi, India

Department of Biotechnology, College of Agriculture and Natural Resource Sciences, Debre Birhan University, Debre Birhan, Ethiopia

N. Sanan-Mishra (✉)
Plant RNAi Biology Group, International Centre for Genetic Engineering and Biotechnology, New Delhi, India
e-mail: neeti@icgeb.res.in

K. Goswami et al. (eds.), *Non-coding RNAs for Crop Improvement*,
https://doi.org/10.1007/978-981-97-7213-1_14

14.1 Introduction

Drought is one of the most devastating environmental stresses hampering plant growth, development, and yield across the globe (Ferdous et al. 2015). Exploiting yield potential and maintaining the stability of crops in water-limited environments are urgent and just-to-do tasks to guarantee the availability of food supply for the increasing world population (Yang et al. 2010). Trait improvements in plants will increase agricultural productivity even under stressful environments (Zhang et al. 2017). Over the past years, numerous efforts have been made towards the improvement of crops, mostly using traditional or modern breeding approaches. In the last few decades, the domain of molecular biology has flourished in plants and made it possible to understand different genes required to achieve desirable traits under different conditions (Ansari et al. 2020) and to find new targets for genetic improvement. Since drought stress response is a quantitatively controlled trait (Anupama et al. 2018) and plants are sessile in nature, they evolved their own mechanisms to maintain optimal growth and dévelopment under such stressful conditions. There have been tremendous efforts devoted to identifying key regulators in plant drought stress response through wet lab and computational approaches. However, the identified potential candidate genes constitute only a small portion of these regulators (Yang et al. 2010). Noncoding RNAs (ncRNAs) are one of the most important regulators of gene expression discovered in recent times and they play a gear shifting role in plant growth, development, and stress response (Patil et al. 2021). MicroRNAs (miRNAs) are one class of noncoding endogenous smallRNAs (20–24 nucleotides in size) that regulate gene expression at post-transcriptional level via cleavage or/and translational inhibition of target messenger RNA (mRNA) (Zhang et al. 2022). miRNAs can bind to the 3′UTR (untranslated region) of the mRNA in a sequence complementarity manner. Several miRNAs can regulate a particular mRNA and/or single miRNA can regulate a combination of several mRNAs depending on the necessity of the response pathway activation or inhibition (Ma and Hu 2023). It is reported that miRNA expression is altered in response to stress in several plants such as in *Arabidopsis* (Liu et al. 2008; Ma et al. 2015), maize (Wang et al. 2014), rice (Chung et al. 2016; Zhang et al. 2016, 2018; Fard et al. 2017), chickpea (Hajyzadeh et al. 2015), sorghum (Katiyar et al. 2015), wheat (Akdogan et al. 2016; Giusti et al. 2017; Hua et al. 2019), potato (Zhang et al. 2014), switch grass (Hivrale et al. 2015), creeping bent grass (Zhou et al. 2013), *Dendrobium huoshanense* (Wang et al. 2022), common bean (Wu et al. 2017), sugarcane (Gentile et al. 2015), and others. These findings help sheds light on the growing body of drought response mechanisms research which can potentially be used in the development of new drought-tolerant crops and plants at large (Ferdous et al. 2015). Therefore, in this chapter, we highlighted the fate of agriculture and drought stress and the role of miRNAs in assisting sustainable agricultural production in water-limited conditions.

14.2 How Do Plants Sense and Respond to Drought Stress?

Plants must make numerous decisions in their lifetime and such decisions are based on the input from signals of environmental cues. Extreme weather events such as unfavourable water availability, suboptimal temperatures, and increased soil ion content, result in different stresses and alternations in plant growth and development (Lamers et al. 2020). Abiotic stresses are perceived by primary sensory mechanisms which translate the physio-chemical signals into a biological signal (Zhang et al. 2023). Although, the question of how plants perceive stress signals and adapt to harsh conditions is still enigmatic, different stresses exert various signals and plants consider those cues and respond in different manners. For instance, drought causes a primary signal called hyperosmotic stress, which is often named osmotic stress. But when it comes to salt stress, it has both osmotic and ionic effects. Moreover, the secondary effects of both drought and salt stresses might cause oxidative damage to cellular components like proteins, nucleic acids, membrane lipids, and dysfunction of metabolic pathways (Zhu 2016). Thus, both primary and secondary signals possibly give rise to cellular responses indicating the unique overlapping signal could be shared between drought and salt stress where both recruit abscisic acid (ABA) accumulation to elicit adaptive responses (Zhu 2002).

Since the discovery of SOS (large family of protein kinases) pathway in plants in 2000s as the first abiotic stress-signaling pathway (Zhu et al. 1998), several stress signaling pathways have been investigated in plants. It is well documented that plants have large family of MAP kinase pathways (Zhu 2016). The rapid activation of multiple MAPKs has been reported during salt, drought, heat, cold, and wounding response (de Zelicourt et al. 2016). During drought, salinity, and osmotic stress exposure, SnRK2 family of protein kinases also gets activated. For instance, Boudsocq et al. (2004) showed in *Arabidopsis*, SnRK2 proteins are activated by osmotic stress, and protein kinases such as SnRK2.2/3/6/7/8 are activated by ABA. The decuple *Arabidopsis* SnRK2 mutant plants are impaired in osmotic stress regulation of genes and in the accumulation of ABA, proline, and IP3, but they are unaffected in the osmotic stress-induced accumulation of ROS indicating SnRK2 activation is upstream of ABA accumulation and such activation controls osmotic adjustment and adaptation (Fujii et al. 2011). But how osmotic stress activates the kinases needs further understanding. Also, osmotic and heat stress cause the activity of many lipid signals such as phosphatidic acid, sphingolipids, phosphoinositides, oxylipins, lysophospholipids, and N-acylethanolamines (Sharma et al. 2023). Although, the mechanism how stress regulates the biosynthesis of lipid signals and effects of lipid signaling targets in plants is not well understood, but membrane lipids are crucial for maintaining the chloroplast and membrane stability under stressful conditions (Hou et al. 2016; Sharma et al. 2023).

Water deficit in plants results in osmotic stress where the water availability is first sensed as reduction in osmotic potential and it induces stomatal closure, reduced turgor pressure, reduced photosynthesis rates, altered leaf gas exchange parameters, and crop yield (Lamers et al. 2020). The adaptive mechanism of plants under osmotic stress are mainly through increasing solute concentration such as organic solutes (proline, betaine, soluble sugars, and polyols), and inorganic solutes (K^+, Na^+, and Ca^{2+}), thereby reducing the osmotic potential and enhancing water retention which are collectively called osmoregulation (Zhang et al. 2023). As soil moisture content gradually decreases, the proline content usually showed significant accumulation, but the more severe stress brings it down. Also, during drought stress the soluble sugar content continues to increase (Kishor et al. 2005). Plants use a universal process to overcome and/or respond drought stress (Fig. 14.1). This involves regulating stomatal movement, osmotic adjustments, and accumulation of dehydration-induced proteins, restricting plant growth, regulating the movement of ions and substances, regulating the onset of senescence and fine-tuning of phytohormones, modulate the redox pathway and balancing the production of ROS scavenging enzymes (Gelaw and Sanan-Mishra 2021), all guided by complex molecular regulatory networks.

Whatever the morphological, physiological, biochemical, and molecular coordinated responses takes place in plants, the ultimate target falls to one of the two possibilities named as; drought escape, and drought avoidance or tolerance. Some plants utilize a drought escape mechanism involving shortening of the life cycle by employing self-reproduction, remobilization of photosynthates, elasticity development, seasonal growth, and early flowering before the beginning of the driest part of the year (Seleiman et al. 2021). The other group of plants employ drought avoidance and tolerance mechanism by reducing stomatal transpiration, increasing of water uptake using extended root system, maintaining water potential and leaf temperature via hairy leaves and cuticles (Dobra et al. 2010; Boulard et al. 2017).

Since drought stress is a very complex process, studying the stress response mechanism in plants is a complex task as well. Therefore, understanding the response mechanism requires lot of research with techniques as well as methods (Osmolovskaya et al. 2018; Marchin et al. 2020; Ogbaga et al. 2020; Kröel-Dulay et al. 2022; Juenger and Verslues 2023). For simplicity, morphological, physiological, biochemical, and molecular responses from germination to harvest could be considered. For this reason, little is known about drought, though it affects the global crop production largely.

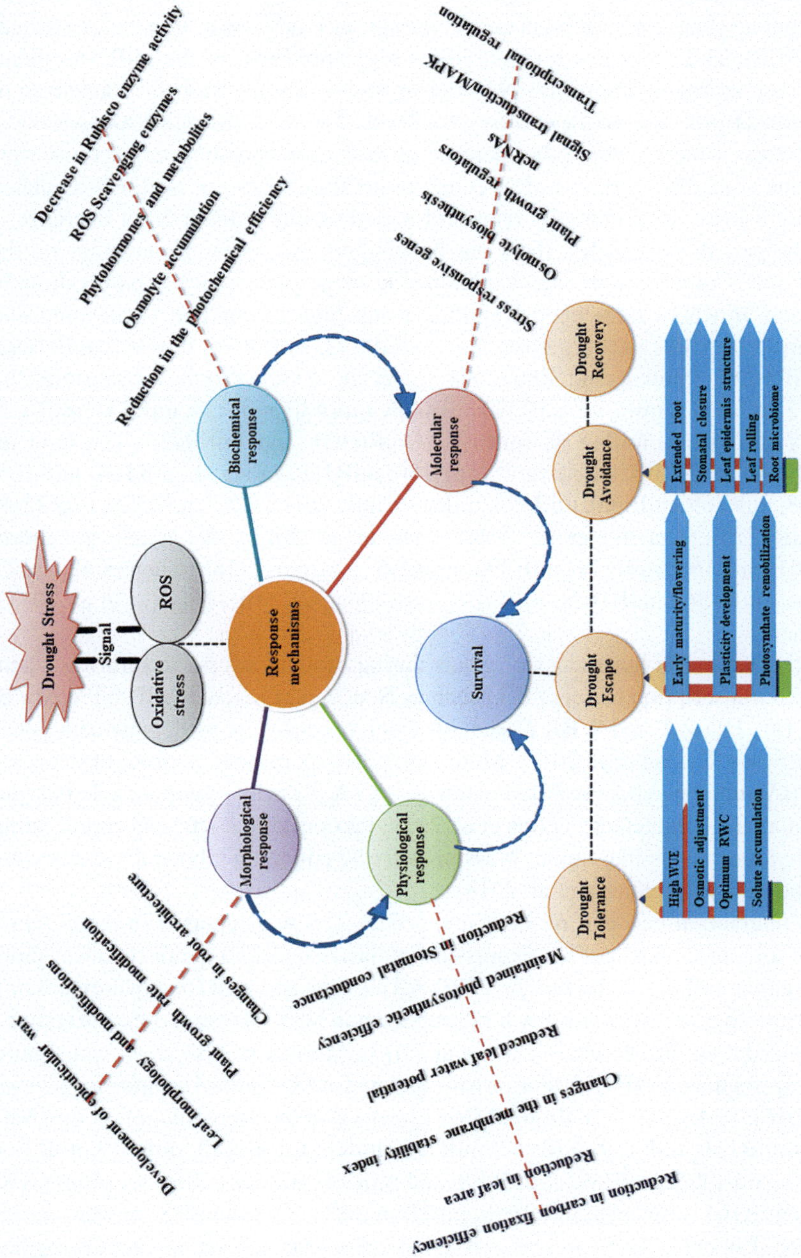

Fig. 14.1 Generalized drought stress response mechanism in plants

14.3 Role of miRNAs in Plant Drought Stress Tolerance

miRNAs are the effectual gene regulators (Ferdous et al. 2015). During stress, miRNAs regulate gene expression via target cleavage or inhibition of translation (Kumar et al. 2018). The up or down-regulation is also applicable to the miRNAs themselves. The up-regulation of miRNAs during stress can fine-tune the abundance of target transcript(s) and on the contrary, the level of some of the miRNAs goes down during stress which leads to the increase of their corresponding target transcripts (Singroha et al. 2021). Both miRNAs and target transcripts up- or down-regulated by drought stress are potentially relevant for engineering plant drought tolerance.

In the past three decades, there has been significant research outcomes on the role of miRNAs in drought stress tolerance in many plant species. Novel drought responsive miRNAs have been identified using high throughput sequencing and microarray in plants such as *Arabidopsis* (miR393, miR397b, miR402, miR169a, and miR169c) (Sunkar and Zhu 2004; Li et al. 2008), *Oryza sativa* (miR170, miR172, miR8175, miR395, miR397, miR408, miR440, miR529, miR531, miR827, miR977, miR1861, miR1876, miR3982, miR6300, and miR9773) (Zhou et al. 2010; Singh et al. 2020), *Zea mays* (miR156, miR159, miR160, miR164, miR166, miR168, miR169, miR390, miR393, miR395, miR396,miR399, miR529, miR2218, and miR2275) (Aravind et al. 2017; Tang et al. 2022), *Medicago truncatula* (miR164, miR169, miR171, miR396, miR399, and miR2118) (Chen et al. 2012), Populus (miR156a, miR162a, miR319a, miR396a, and miR166n) (Li et al. 2011), and others.

miRNAs target a huge class of genes during drought stress, and most of them belong to transcription factors (TFs) such as NAC/NAM, AREB/ABF, NFY, MYB, TAF, CUC, HD-ZIP and WRKY, each of which regulates many downstream genes and pathways (Samad et al. 2017). Moreover, miRNAs respond to drought stress via both ABA and non-ABA mediated pathways. ABA plays an important role in regulating stomatal conductance (Zhou et al. 2019; Farooq et al. 2020) and coordinating with phytohormones like auxin, brassinosteroids, ethylene, gibberellins, jasmonic acid, and salicylic acid (Ku et al. 2018).

The interaction between miR169 and NFYA TFs was reported in *Arabidopsis* drought tolerance response. Co-expression studies revealed that miR169a was more effective than miR169c. *Arabidopsis* NFYA5 mutants and miR169a overexpressing plants showed more susceptibility for drought while NFYA5 overexpression resulted in enhanced drought tolerance (Ding et al. 2013). Moreover, ABA treatment significantly up-regulated NFYA5 and down-regulated miR169 in *Arabidopsis* (Cheng et al. 2016). In *Arabidopsis*, drought stress resulted in the up-regulation of miRNAs such as miR156, miR159, miR167, miR168, miR171, miR172, miR319, miR393, miR394a, miR395c, miR395e, miR396 and miR397 whereas other set of miRNAs such as miR161, miR168a, miR168b, miR169, miR171a and miR319c were down-regulated (Liu et al. 2008). In research reports, it has been observed that the expression level of miRNAs and targets depend on plant species. For instance, miR1510 was down and up-regulated in *Medicago truncatula* and *Glycine max* respectively, during drought (Mantri et al. 2013). miR156 showed up-regulation in *Arabidopsis*, barley, peach, and wheat while it was down-regulated in rice and maize (Sunkar and Zhu 2004; Kantar et al. 2010; Eldem et al. 2012). Overexpression of miR169a resulted in more sensitivity to drought stress via increased leaf transpiration (Li et al. 2008) while in tomato contrasting results were observed, like decreased transpiration rate, reduced stomatal opening, decreased leaf water loss (Zhang et al. 2011). In *Brassica napus* L, miR169n acted as a negative regulator of drought. The short tandem target mimic (STTM) knock-down of miR169n resulted in enhanced tolerance to osmotic stress at the post-germination stage and drought stress at the seedling stage, reflected by higher relative water content and proline content, lower relative electrolyte leakage/membrane damage, and showed higher antioxidant enzyme activities (Li et al. 2021). In *Arabidopsis*, overexpression of gma-miR394a resulted in and enhanced drought tolerance via lowered leaf water loss and reduced the expression of F-box gene (*At1g27340*) (Ni et al. 2012). In tobacco plants, overexpression of Sp-miR396a-5p resulted in increased tolerance to drought, salt, and cold stress but it makes the plant susceptible to *Phytophthora nicotianae* infection (Chen et al. 2015). The transgenic plants exhibited reduced ROS production and enhanced osmoregulation by down-regulating target transcripts, *NtGRF1*, *NtGRF3*, *NtGRF7* and *NtGRF8*. In chickpea, overexpression of one of the conserved miRNAs (miR408), resulted in better drought adaptation leading to repression of plantacyanin transcripts, DREB and other drought responsive genes (Hajyzadeh et al. 2015). It was earlier reported that, transgenic *Arabidopsis* plants overexpressing DREB1A and DREB2A showed improved dehydration tolerance (Maruyama et al. 2009). It was noted that DREB2A binds to protein2a during dehydration response proving DREB1A and DREB2A TFs are involved in drought tolerance to the plants. In another study, overexpression of miR408 in *Arabidopsis*, was found to improve tolerance to salinity, cold and oxidative stress, but the transgenic plants were found to be sensitive to drought and osmotic stress. With elevated expression of miR408, reduced ROS accumulation and increased expression of genes coding for antioxidant enzymes was reported (Ma et al. 2015). There are many reports supporting the role of miRNAs in plants drought stress tolerance (reviewed in Ferdous et al. 2015; Gelaw and Sanan-Mishra 2021) and in Table 14.1 we have presented some of the transgenic based experimental results for drought stress tolerance in plants.

During drought stress, plants inhibit the lateral root growth for deeper root production (Esfahani and Sonnewald 2024). Under stressful conditions, usually ABA takes advantage of the lateral root inhibition. This phenomenon is guided by miR393. Chen et al. (2012a, b) reported that in ABA hypersensitive *Arabidopsis* mutants, the increased level of miR393 had resulted in the inhibition of lateral root formation via direct cleavage of two auxin receptors namely TIR1 and AFB2. In contrary, *Arabidopsis* seedlings expressing miR393-resistant TIR1 or AFB2 showed no inhibition to lateral root growth during osmotic stress or ABA treatment. Leaf water transpiration is also the most important physiological process during drought stress. Some plants use leaf senescence to avoid drought or reduce transpiration and canopy size. In *Arabidopsis*, miR164 has been reported to be involved in leaf senescence (Kim et al. 2009). Thus, the involvement of miR164 in leaf senescence sheds light on aiding the gene regulatory role during drought response.

Table 14.1 miRNA transgenics developed for drought stress tolerance experiments.

S. No.	miRNA	Method	Plants	Features observed	References
1.	miR156	Overexpression	*Nicotiana tabacum* cv. Xanthi	Better growth, biomass, and higher antioxidant activity	Kang et al. (2020)
			Medicago sativa L.	Decreased plant height and increased branches and leaves with improved drought and salt tolerance	Wang et al. (2021)
			Solanum lycopersicum	Strigolactones enhanced the opening of stomata after drought post trauma and enhanced plant recovery after drought stress	Visentin et al. (2020)
2.	miR164	Overexpression	*Arabidopsis*	Leaf longevity via ethylene signaling/ drought tolerance	Kim et al. (2009), Wang et al. (2016)
3.	miR165/166	Knock-down	*Arabidopsis*	Cold and drought tolerance	Yan et al. (2016)
4.	miR168	Overexpression	*Arabidopsis*	ABA hypersensitivity and drought tolerance	Li et al. (2012)
5.	miR169c	Overexpression	Tomato	Reduced stomatal conductance and transpiration rate/ enhanced drought tolerance	Li et al. (2008), Zhang et al. (2011)
		Overexpression	*Arabidopsis*	Increased sensitivity to drought stress with reduced survival, increased leaf water loss, and shorter root length	Yu et al. (2019)
6.	miR169i/l	Overexpression	*Arabidopsis*	Drought tolerance	Du et al. (2017)
7.	miR169o	Overexpression	*Populus trichocarpa*	Significantly improved photosynthetic rate and reduced t leave water loss under drought stress	Jiao et al. (2021)
8.	miR172c	Overexpression	*Arabidopsis*	Increased water deficit and salinity tolerance	Li et al. (2016)

(continued)

Table 14.1 (continued)

S. No.	miRNA	Method	Plants	Features observed	References
9.	Osa-miR319	Constitutive expression	Creeping bentgrass	Increased leaf wax content and water retention capacity	Zhou et al. (2013)
10.	miR393	Overexpression	Rice	Increased tillering, early flowering and less tolerance to salt and drought, auxin hyposensitivity	Xia et al. (2012)
		Overexpression	*Nicotiana tabacum* and *Agrostis stolonifera*	Resulted in fewer and longer tillers, enhanced tolerance due to reduced stomatal density and denser cuticles	Zhao et al. (2019)
11.	Gma-miR394a	Overexpression	*Arabidopsis*	Increased drought recovery aided by F-box proteins in stress-response pathway	Ni et al. (2012)
12.	miR396a	Overexpression	Tobacco	Increased tolerance to salt, drought, cold stress	Chen et al. (2015)
13.	miR399f	Overexpression	*Arabidopsis*	Hypersensitivity to drought	Baek et al. (2016)
14.	miR408	Overexpression	*Arabidopsis*	Drought sensitivity	Ma et al. (2015)
		Overexpression	*Cicer arietinum*	Drought tolerance through indirect regulation of DREB and other drought responsive genes	Hajyzadeh et al. (2015)
			Lolium Perenne L.	Increased drought tolerance via higher leaf relative water content, lower electrolyte leakage and less lipid peroxidation	Hang et al. (2021)
15.	miR535	Knock-down	*Oryza sativa*	Increased stress tolerance by negatively regulating the tolerance response via OsSPL7/12/16	Yue et al. (2020)

(continued)

Table 14.1 (continued)

S. No.	miRNA	Method	Plants	Features observed	References
16.	miR827	Overexpression	*Hordeum vulgare* L.	Enhanced leaf stomatal conductance and photosynthetic assimilation via SPX, NBS-LRR domain and APT1	Ferdous et al. (2017)
17.	miR1508a	Overexpression	*Glycine max* L.	Dwarf phenotypes, thick cell walls, lower survival rates and greater leaf water loss during drought	Sun et al. (2020)
18.	miR1916	Overexpression	*Solanum lycopersicum*	Negative effect on osmoregulation and leads to increased ROS accumulation	Chen et al. (2019)
19.	miR2118	Overexpression	*Tobacco*	Induced drought tolerance	Wu et al. (2015)

The increased number of research on the roles of miRNAs in orchestrating plant agronomic traits under normal and stressed conditions such as drought is indicative of the huge potential behind miRNA making it a promising candidate for molecular breeding of crops (Shah and Ullah 2023). Overall, miRNA-target mediated drought stress response involves diverse molecular, biochemical, and physiological processes.

14.4 Other Feasible Strategies for Drought Stress Tolerance

14.4.1 Seed Priming for Drought Tolerance

One of the phenomena which has gained more attention in improving the plants' response to stress and increase productivity in recent years, is molecular priming (Khan et al. 2022). Stress priming is an approach to induce short- or long-term stress memory (memory of stress) which enables plants to remember the stress when it encounters again (Chinnusamy and Zhu 2009; Migicovsky et al. 2014; Liu et al. 2022). Stress leaves an imprint on plants and stress memory depends on imprints left on the physical structure. When the stress comes again, plants can better respond to stress, or the retain abundance of regulators controlling functional gene expression by promote or suppress the role of transcription factors, signaling molecules, and stress-response proteins (Hu et al. 2016). It is reported that the memory of stress and stress-tolerance can happen in the current or subsequent generations via the junction of synergistic signaling pathways (Wang et al. 2020; Thakur et al. 2020).

Several reports have showed seed priming as a feasible and low-cost technique to enhance tolerance to stress (Marthandan et al. 2020; Liu et al. 2022). The seed treatment is mainly given during the pre-germination time and it helps in boosting the percentage of germination, seedling emergence, and stress tolerance via activation of a panel of enzymes, increase antioxidants and metabolite production (Işeri et al. 2014; Hussain et al. 2016, 2017). Seed priming can be performed in different methods such as osmo-priming (Kurek et al. 2019), chemical priming (Jisha et al. 2013), bio-priming (Mahmood et al. 2016), hormonal priming (Umezawa et al. 2010; Wei et al. 2017) and others.

Also, priming with signaling molecules such as H_2O_2 and osmolytes such as proline were reported to ameliorate drought and other stress tolerance in plants. Seed priming with H_2O_2 at different concentrations (0, 1, 5, 10, and 15 mM) resulted that seedlings that were showing enhanced antioxidant enzyme activities including superoxide dismutase, catalase, ascorbate peroxidase, and guaiacol peroxidase. Priming with 2–10 mM H_2O_2 was reported beneficial for enhanced drought tolerance in rice seedlings (Jira-Anunkul and Pattanagul 2020). Reports indicated that priming with H_2O_2 results in the accumulation of stress tolerance signaling networks such as transcription factors, and activation of antioxidant enzymes, and ROS scavenging networks (Savvides et al. 2016). Another study by Ashraf et al. (2015) in maize showed that higher germination rates were recorded when maize seeds were primed with 140 mM H_2O_2 but a marked decline in photosynthesis, and increase in the concentration of proteins, H_2O_2 accumulation, malondialdehyde content, and ascorbic acid were recorded during drought stress. In maize, individual and combined priming with 10- and 15-mM proline and H_2O_2 resulted in improved tolerance to drought in a genotype-dependent manner (Gelaw and Sanan-Mishra 2024). In Chinese cabbage, priming with potassium nitrate and urea (200 $mmol^{-1}$ each) resulted in enhanced drought tolerance conferred by increased superoxide dismutase, peroxidase, and catalase enzyme activities and levels as well as soluble sugar and proline accumulation (Yan 2015). Moreover, exogenously applied proline was reported to significantly enhance growth, photosynthetic efficiency, plant water status, osmoprotectants, and bulb yield (Semida et al. 2020). There are emerging reports indicating the beneficial roles of rhizospheric and plant-associated microbiomes in drought stress mitigation. The role of priming with Plant Growth Promoting Rhizobacteria (PGPR) in drought stress tolerance is reviewed in Barnawal et al. (2019).

14.4.2 Nanotechnology for Drought Stress Tolerance

Nanoparticles are materials of 1–100 nm in size, at least in one dimension that are bioactive in nature and can exhibit multiple quantum roles (Arya et al. 2019; Gelaw and Sanan-Mishra 2022). Their small size and bioactive nature allow them to traverse the cellular membranes and play a role in modulating the plant machinery. Nanoparticles have opened a new avenue in agricultural sector (Gelaw and Sanan-Mishra 2022). Reports have shown that nanoparticles can be used as herbicides (Siddiqui et al. 2015), pesticides (Alejandro and Rubiales 2009), plant growth regulators (Singh et al. 2021), nano-fertilizers (Derosa et al. 2010; Zulfiqar et al. 2019), post-harvest management tools (González-Estrada et al. 2020), and tools for stress tolerance (Elsakhawy et al. 2018; Almutairi 2019; Das and Das 2019; Van Nguyen et al. 2022) in the agricultural sector. Nanoparticles are important effecters of plant agronomic traits to maximize crop productivity (Duhan et al. 2017; Verma et al. 2019; Ali and Khan 2021) through enhancing seed germination (Gopinath et al. 2014), biomass, plant height, and yield (Mahmood 2015), as well as metabolite content (Kole et al. 2013), and so on.

There are many reports on the use of different nanoparticles for alleviating various stresses in a variety of plants (reviewed in Gelaw and Sanan-Mishra 2022). During drought stress, Zerovalent Iron nanoparticles in *Arabidopsis thaliana* activate plasma membrane H^+-ATPase, promote stomatal opening, increase chlorophyll content and plant biomass, and increase CO_2 assimilation (Kim et al. 2015). Metal-based (Fe, Cu, Co, and Zn oxide) nanoparticles in *Glycine max* also improved drought tolerance index, relative water content and biomass reduction rate by activating the expression of drought responsive genes (Linh et al. 2020). Similar result was reported by Sedghi et al. (2013) where zinc oxide nanoparticles (0.5 and 1 g L^{-1}) increased seed germination and improved drought stress tolerance. Iron oxide and hydrogel nanoparticles (0–100 mg kg^{-1}) also ameliorate drought stress and cadmium toxicity in rice (Ahmed et al. 2021a, b). Moreover, chitosan NPs (30, 60 and 90 ppm) showed significant increase in the leaf area, SPAD, number of grains per spike, the grain yield, and the harvest index in barley. In *Sorghum bicolor*, Cerium oxide (10 mg L^{-1}) (Djanaguiraman et al. 2018), and Zinc oxide nanoparticles (5 mg kg^{-1}) (Dimkpa et al. 2019) showed improved drought stress tolerance by increasing leaf carbon assimilation rates, pollen germination, yield, and nutrient translocation and acquisition. In *Zea mays*, ZnO nanoparticles (100 mg L^{-1}) (Sun et al. 2021) and potassium nano-silica (100–200 ppm) (Aqaei et al. 2020) resulted in improved drought tolerance through alleviating degradation of photosynthetic pigments, maintaining higher net photosynthetic rate, and enhanced water use efficiency. Moreover, extended studies in *Triticum aestivum* also reported the potential roles of different nanoparticles in mitigating drought stress tolerance such as titanium dioxide (Jaberzadeh et al. 2013; Faraji and Sepehri 2019), silicon oxide (Behboudi et al. 2018), selenium (Ikram et al. 2020), Iron (Adrees et al. 2020), and copper nanoparticles (Ahmed et al. 2021a, b). Under stressful conditions, nanoparticles help the plants by triggering the antioxidant defence systems, osmolyte accumulation, maintaining ROS homeostasis, and increasing nutrient uptake efficiency. However, the procedures for the use of nanoparticles should be carefully considered since it might create cell toxicity.

14.4.3 Plant-Microbiome Interaction for Drought Stress Tolerance

Plants and microbes or generally life and microbes cannot be kept separable. The interaction between plants and microbes constitutes a diversified ecosystem. More specifically, PGPRs and arbuscular mycorrhizal fungi plays an indispensable role in improving biotic and abiotic stress tolerance and assisting plant growth development in a variety of ways. It modulates a complex mechanism in the plant cell and the residing microbial flora inhibits pathogens and serves as part of innate defence system (Munir et al. 2022). Drought stress not only impacts the plant growth and development but also it affects microbial population structure of the root zone (Naylor et al. 2017). Drought affects the plant operational mechanism, such as metabolism and root exudates, which in turn influences the plant rhizosphere microbiome (Ait-El-Mokhtar et al. 2023). Plant exudates, such as amino acids, vitamins, sugars, flavonoids, carboxylic acids, organic acids, fatty acids, benzoxazinoids, and ethylene, play a central role in recruiting rhizospheric microbiome (Vives-Peris et al. 2020) putting selective pressure on to them to endure harsh conditions. For instance, glycerol-3-phosphate synthesis induction during drought favors rhizospheric actinobacteria in the root. Actinobacteria promotes plant fitness and health under drought stress condition (Xu et al. 2018). Since drought reduces the availability of iron and siderophore in rhizosphere, actinobacteria will get favourable condition to colonize and promote plant performance (Xu et al. 2021). There are emerging reports indicating the beneficial roles of rhizospheric and plant associated microbiome in drought stress mitigation (Hestrin et al. 2022; Ali et al. 2022; Fadiji et al. 2023) in different plants such as rice (Oubaha et al. 2024), maize (Huang et al. 2017), sorghum (Grover et al. 2014), wheat (Viscardi et al. 2016), garden pea (Saikia et al. 2018) green gram (Fukami et al. 2018), potato (Ngumbi and Kloepper 2016), cucumber (Wang et al. 2012), lettuce (Sahin et al. 2015), foxtail millet (Niu et al. 2018) and others.

14.5 Conclusion

The increased global climate change created various factors for reduction of crop productivity, and drought is one of the most devastating environmental stresses hampering plant growth, development, and yield. Therefore, exploiting and increasing the yield potential and maintaining yield stability of crops in water-limited environments are urgent and just-to-do tasks to guarantee the availability of food supply for the increasing world population. Since the birth of modern biotechnology, various advancements in crop breeding and genetic modifications for better drought adaptation has been done. miRNAs are one of the potential candidates for drought stress tolerance where it binds to the mRNA in a sequence complementarity manner to regulates gene expression at post-transcriptional level via cleavage or/and translational inhibition of target mRNA. Several miRNAs can regulate the same mRNA and/or miRNA scan regulate a combination of several mRNAs depending on plant species and the requirement of the response pathway activation or inhibition. Beyond the expression profiling, functional analysis studies in miRNAs have shown the huge potential of those ncRNAs in mitigating drought and other stresses. Briefly, miRNAs control the regulatory response of plant growth, development, and stress response. Even if the advances and sophistication of molecular techniques supported by bioinformatics tools, the cost is still high and raising transgenics costs more including the regulatory concerns. Therefore, other promising strategies such as priming, nanotechnology, as well as plant microbiome interaction studies could assist the sustainable agriculture under normal and stressed conditions.

14.6 Future Perspectives of Agriculture and miRNAs

The world population is speculated to reach around 10 billion by the year 2050 which imposes a huge burden on the agriculture. However, global climate changes are resulting in extreme environmental stresses which could lead to significant losses in agricultural production. The continuous increase in environmental temperatures has increased the probability of occurrence, duration, and severity of drought, thereby making it challenging to meet the capacity to feed the ever-increasing world population.

miRNAs stand out as potent molecular regulators within plants, holding substantial promise in comprehending and alleviating the repercussions of drought stress on agriculture. As we look toward the future of agriculture, the integration of miRNA research promises to revolutionize crop resilience and productivity in the face of increasingly unpredictable climate patterns. One of the key future perspectives lies in the application of miRNAs in precision agriculture. By understanding the regulatory roles of miRNAs in drought response pathways, scientists can develop precise tools for monitoring plant stress levels and implementing targeted interventions. This could involve the development of miRNA-based biosensors that provide real-time information about plant health and stress levels, enabling farmers to take proactive measures to optimize water usage and mitigate drought damage. Another exciting avenue is the use of miRNAs in crop improvement programs. By manipulating the expression of specific miRNAs associated with drought tolerance, researchers can breed or engineer crops with enhanced resilience to water scarcity. This could involve the development of genetically modified crops with optimized miRNA profiles or the use of genome editing techniques such as CRISPR to precisely modulate miRNA activity. These efforts have the potential to significantly increase crop yields and ensure food security in regions vulnerable to drought.

The future of miRNA research in agriculture also involves advanced bioinformatics and systems biology approaches. By integrating high-throughput sequencing technologies with computational modeling, scientists can unravel complex miRNA regulatory networks underlying drought stress response in plants. This holistic understanding will enable the identification of key regulatory nodes and the development of predictive models for optimizing crop performance under varying environmental conditions. Furthermore, miRNAs also hold promise in environmental monitoring and climate resilience efforts. By analyzing miRNA expression patterns in response to environmental stressors, researchers can gain insights into ecosystem health and resilience. This could involve the development of miRNA-based biomarkers for assessing environmental stress levels and predicting ecosystem responses to climate change. Such tools could inform conservation strategies and facilitate the sustainable management of natural resources in the face of escalating climate challenges.

References

Adrees M, Khan ZS, Ali S, Hafeez M, Khalid S, Ur Rehman MZ, Hussain A, Hussain K, Shahid Chatha SA, Rizwan M (2020) Simultaneous mitigation of cadmium and drought stress in wheat by soil application of iron nanoparticles. Chemosphere 238:124681. https://doi.org/10.1016/j.chemosphere.2019.124681

Ahmed F, Javed B, Razzaq A, Mashwani ZUR (2021a) Applications of copper and silver nanoparticles on wheat plants to induce drought tolerance and increase yield. IET Nanobiotechnol 15(1):68–78. https://doi.org/10.1049/nbt2.12002

Ahmed T, Noman M, Manzoor N, Shahid M, Abdullah M, Ali L, Wang G, Hashem A, Al-Arjani ABF, Alqarawi AA, Abd Allah EF, Li B (2021b) Nanoparticle-based amelioration of drought stress and cadmium toxicity in rice via triggering the stress responsive genetic mechanisms and nutrient acquisition. Ecotoxicol Environ Saf 209:111829. https://doi.org/10.1016/j.ecoenv.2020.111829

Ait-El-Mokhtar M, Meddich A, Baslam M (2023) Plant-microbiome interactions under drought-insights from the molecular machinist's toolbox. Front Sust Food Syst 7:1253735. https://doi.org/10.3389/fsufs.2023.1253735

Akdogan G, Tufekci ED, Uranbey S, Unver T (2016) miRNA-based drought regulation in wheat. Funct Integr Genomics 16(3):221–233. https://doi.org/10.1007/s10142-015-0452-1

Alejandro PDL, Rubiales D (2009) Nanotechnology for parasitic plant control. Pest Manag Sci 65(5):540–545. https://doi.org/10.1002/ps.1732

Ali SMA, Khan N (2021) Uptake, translocation, and consequences of nanomaterials on plant growth and stress adaptation. J Nanomater 2021:6677616. https://doi.org/10.1155/2021/6677616

Ali S, Tyagi A, Park S, Mir RA, Mushtaq M, Bhat B, Mahmoudi H, Bae H (2022) Deciphering the plant microbiome to improve drought tolerance: mechanisms and perspectives. Environ Exp Bot 201:104933. https://doi.org/10.1016/j.envexpbot.2022.104933

Almutairi ZM (2019) Plant molecular defense mechanisms promoted by nanoparticles against environmental stresses. Int J Agric Biol 21(2):259–270. https://doi.org/10.17957/IJAB/15.0890

Ansari WA, Chandanshive SU, Bhatt V, Nadaf AB, Vats S, Katara JL, Sonah H, Deshmukh R (2020) Genome editing in cereals: approaches, applications and challenges. Int J Mol Sci 21:11. https://doi.org/10.3390/ijms21114040

Anupama A, Bhugra S, Lall B, Chaudhury S, Chugh A (2018) Assessing the correlation of genotypic and phenotypic responses of indica rice varieties under drought stress. Plant Physiol Biochem 127:343–354. https://doi.org/10.1016/j.plaphy.2018.04.001

Aqaei P, Weisany W, Diyanat M, Razmi J, Struik PC (2020) Response of maize (Zea mays L.) to potassium nano-silica application under drought stress. J Plant Nutr 43(9):1205–1216. https://doi.org/10.1080/01904167.2020.1727508

Aravind J, Rinku S, Pooja B, Shikha M, Kaliyugam S, Mallikarjuna MG, Kumar A, Rao AR, Nepolean T (2017) Identification, characterization, and functional validation of drought-responsive microRNAs in subtropical maize inbreds. Front Plant Sci 8:941. https://doi.org/10.3389/fpls.2017.00941

Arya A, Mishra V, Chundawat TS (2019) Green synthesis of silver nanoparticles from green algae (*Botryococcus braunii*) and its catalytic behavior for the synthesis of benzimidazoles. Chem Data Collect 20:1–7. https://doi.org/10.1016/j.cdc.2019.100190

Ashraf MA, Rasheed R, Hussain I, Iqbal M, Haider MZ, Parveen S, Sajid MA (2015) Hydrogen peroxide modulates antioxidant system and nutrient relation in maize (Zea mays L.) under water-deficit conditions. Arch Agron Soil Sci 61(4):507–523. https://doi.org/10.1080/03650340.2014.938644

Baek D, Chun HJ, Kang S, Shin G, Park SJ, Hong H, Kim C, Kim DH, Lee SY, Kim MC, Yun DJ (2016) A role for *Arabidopsis* miR399f in salt, drought, and ABA signaling. Mol Cells 39(2):111–118. https://doi.org/10.14348/molcells.2016.2188

Barnawal D, Singh R, Singh RP (2019) Role of plant growth promoting rhizobacteria in drought tolerance. In: PGPR amelioration in sustainable agriculture. Elsevier, Amsterdam. https://doi.org/10.1016/b978-0-12-815879-1.00006-9

Behboudi F, Tahmasebi Sarvestani Z, Zaman Kassaee M, Modares Sanavi SAM, Sorooshzadeh A (2018) Improving growth and yield of wheat under drought stress via application of SiO_2 nanoparticles. J Agric Sci Technol 20(7):1479–1492

Boudsocq M, Barbier-Brygoo H, Laurière C (2004) Identification of nine sucrose nonfermenting 1-related protein kinases 2 activated by hyperosmotic and saline stresses in *Arabidopsisthaliana*. J Biol Chem 279(40):41758–41766. https://doi.org/10.1074/jbc.M405259200

Boulard T, Roy JC, Pouillard JB, Fatnassi H, Grisey A (2017) Modelling of micrometeorology, canopy transpiration and photosynthesis in a closed greenhouse using computational fluid dynamics. Biosyst Eng 158:110–133. https://doi.org/10.1016/j.biosystemseng.2017.04.001

Chen H, Li Z, Xiong L (2012a) A plant microRNA regulates the adaptation of roots to drought stress. FEBS Lett 586(12):1742–1747. https://doi.org/10.1016/j.febslet.2012.05.013

Chen L, Wang T, Zhao M, Tian Q, Zhang WH (2012b) Identification of aluminum-responsive microRNAs in *Medicago truncatula* by genome-wide high-throughput sequencing. Planta 235(2):375–386. https://doi.org/10.1007/s00425-011-1514-9

Chen L, Luan Y, Zhai J (2015) Sp-miR396a-5p acts as a stress-responsive genes regulator by conferring tolerance to abiotic stresses and susceptibility to Phytophthora nicotianae infection in transgenic tobacco. Plant Cell Rep 34(12):2013–2025. https://doi.org/10.1007/s00299-015-1847-0

Chen L, Meng J, Luan Y (2019) miR1916 plays a role as a negative regulator in drought stress resistance in tomato and tobacco. Biochem Biophys Res Commun 508(2):597–602. https://doi.org/10.1016/j.bbrc.2018.11.165

Cheng HY, Wang Y, Tao X, Fan YF, Dai Y, Yang H, Ma XR (2016) Genomic profiling of exogenous abscisic acid-responsive microRNAs in tomato (*Solanum lycopersicum*). BMC Genomics 17(1):1–13. https://doi.org/10.1186/s12864-016-2591-8

Chinnusamy V, Zhu JK (2009) Epigenetic regulation of stress responses in plants. Curr Opin Plant Biol 12(2):133–139. https://doi.org/10.1016/j.pbi.2008.12.006

Chung PJ, Jung H, Jeong DH, Ha SH, Do Choi Y, Kim JK (2016) Transcriptome profiling of drought responsive noncoding RNAs and their target genes in rice. BMC Genomics 17(1):563. https://doi.org/10.1186/s12864-016-2997-3

Das A, Das B (2019) Nanotechnology a potential tool to mitigate abiotic stress in crop plants: abiotic and biotic stress in plants. IntechOpen, London, pp 1–13. https://doi.org/10.5772/intechopen.83562

de Zelicourt A, Colcombet J, Hirt H (2016) The role of MAPK modules and ABA during abiotic stress signaling. Trends Plant Sci 21(8):677–685. https://doi.org/10.1016/j.tplants.2016.04.004

Derosa MC, Monreal C, Schnitzer M, Walsh R, Sultan Y (2010) Nanotechnology in fertilizers. Nat Nanotechnol 5(2):91. https://doi.org/10.1038/nnano.2010.2

Dimkpa CO, Singh U, Bindraban PS, Elmer WH, Gardea-Torresdey JL, White JC (2019) Zinc oxide nanoparticles alleviate drought-induced alterations in sorghum performance, nutrient acquisition, and grain fortification. Sci Total Environ 688:926–934. https://doi.org/10.1016/j.scitotenv.2019.06.392

Ding Y, Tao Y, Zhu C (2013) Emerging roles of microRNAs in the mediation of drought stress response in plants. J Exp Bot 64(11):3077–3086. https://doi.org/10.1093/jxb/ert164

Djanaguiraman M, Nair R, Giraldo JP, Prasad PVV (2018) Cerium oxide nanoparticles decrease drought-induced oxidative damage in sorghum leading to higher photosynthesis and grain yield. ACS Omega 3(10):14406–14416. https://doi.org/10.1021/acsomega.8b01894

Dobra J, Motyka V, Dobrev P, Malbeck J, Prasil IT, Haisel D, Gaudinova A, Havlova M, Gubis J, Vankova R (2010) Comparison of hormonal responses to heat, drought and combined stress in tobacco plants with elevated proline content. J Plant Physiol 167(16):1360–1370. https://doi.org/10.1016/j.jplph.2010.05.013

Du Q, Zhao M, Gao W, Sun S, Li WX (2017) microRNA/microRNA* complementarity is important for the regulation pattern of NFYA5 by miR169 under dehydration shock in *Arabidopsis*. Plant J 91(1):22–33. https://doi.org/10.1111/tpj.13540

Duhan JS, Kumar R, Kumar N, Kaur P, Nehra K, Duhan S (2017) Nanotechnology: the new perspective in precision agriculture. Biotechnol Rep 15:11–23. https://doi.org/10.1016/j.btre.2017.03.002

Eldem V, Çelikkol Akçay U, Ozhuner E, Bakir Y, Uranbey S, Unver T (2012) Genome-wide identification of miRNAs responsive to drought in peach (Prunus persica) by high-throughput deep sequencing. PLoS ONE 7(12):e50298. https://doi.org/10.1371/journal.pone.0050298

Elsakhawy T, Omara AE-D, Alshaal T, El-Ramady H, Ghazi A, El-Nahrawy S, Elhawat N (2018) Nanomaterials and plant abiotic stress in agroecosystems. Environ Biodiver Soil Secur 2(1):50–55. https://doi.org/10.21608/jenvbs.2018.3897.1030

Esfahani NM, Sonnewald U (2024) Unlocking dynamic root phenotypes for simultaneous enhancement of water and phosphorus uptake. Plant Physiol Biochem 207:108386. https://doi.org/10.1016/j.plaphy.2024.108386

Fadiji AE, Yadav AN, Santoyo G, Babalola OO (2023) Understanding the plant-microbe interactions in environments exposed to abiotic stresses: an overview. Microbiol Res 271:127368. https://doi.org/10.1016/j.micres.2023.127368

Faraji J, Sepehri A (2019) Ameliorative effects of TiO_2 nanoparticles and sodium nitroprusside on seed germination and seedling growth of wheat under peg-stimulated drought stress1. J Seed Sci 41(3):309–317. https://doi.org/10.1590/2317-1545v41n3213139

Fard EM, Ghabooli M, Mehri N, Bakhshi B (2017) Regulation of mir159 and mir396 mediated by Piriformospora indica confer drought tolerance in rice. J Plant Mol Breed 5(1):10–18. https://doi.org/10.22058/jpmb.2017.60864.1129

Farooq A, Bukhari SA, Akram NA, Ashraf M, Wijaya L, Alyemeni MN, Ahmad P (2020) Exogenously applied ascorbic acid-mediated changes in osmoprotection and oxidative defense system enhanced water stress tolerance in different cultivars of safflower (*Carthamus tinctorious* L.). Plants 9(1):1–15. https://doi.org/10.3390/plants9010104

Ferdous J, Hussain SS, Shi BJ (2015) Role of microRNAs in plant drought tolerance. Plant Biotechnol J 13(3):293–305. https://doi.org/10.1111/pbi.12318

Ferdous J, Whitford R, Nguyen M, Brien C, Langridge P, Tricker PJ (2017) Drought-inducible expression of Hv-miR827 enhances drought tolerance in transgenic barley. Funct Integr Genomics 17(2-3):279–292. https://doi.org/10.1007/s10142-016-0526-8

Fujii H, Verslues PE, Zhu JK (2011) *Arabidopsis* decuple mutant reveals the importance of SnRK2 kinases in osmotic stress responses in vivo. Proc Natl Acad Sci USA 108(4):1717–1722. https://doi.org/10.1073/pnas.1018367108

Fukami J, Cerezini P, Hungria M (2018) Azospirillum: benefits that go far beyond biological nitrogen fixation. AMB Express 8(1):1–12. https://doi.org/10.1186/s13568-018-0608-1

Gelaw TA, Sanan-Mishra N (2021) Non-coding RNAs in response to drought stress. Int J Mol Sci 22(22):1–37. https://doi.org/10.3390/ijms222212519

Gelaw TA, Sanan-Mishra N (2022) Nanomaterials coupled with microRNAs for alleviating plant stress: a new opening towards sustainable agriculture. Physiol Mol Biol Plants 28(4):791–818. https://doi.org/10.1007/S12298-022-01163-X

Gelaw TA, Sanan-Mishra N (2024) Molecular priming with H2O2 and proline triggers antioxidant enzyme signals in maize seedlings during drought stress. Biochimica et Biophysica Acta (BBA) - General Subjects 1868(7):130633. https://doi.org/10.1016/j.bbagen.2024.130633

Gentile A, Dias LI, Mattos RS, Ferreira TH, Menossi M (2015) MicroRNAs and drought responses in sugarcane. Front Plant Sci 6:1–13. https://doi.org/10.3389/fpls.2015.00058

Giusti L, Mica E, Bertolini E, De Leonardis AM, Faccioli P, Cattivelli L, Crosatti C (2017) microRNAs differentially modulated in response to heat and drought stress in durum wheat cultivars with contrasting water use efficiency. Funct Integr Genomics 17(2-3):293–309. https://doi.org/10.1007/s10142-016-0527-7

González-Estrada RR, Blancas-Benitez FJ, Montaño-Leyva B, Zambrano-Zaragoza ML, Aguirre-Güitrón L, Moreno-Hernández C, Cortés-Rivera HJ, Fonseca-Cantabrana A, Herrera-González JA, Rayón-Díaz E, Gutierrez-Martinez P (2020) Zinc nanomaterials: a safe tool for postharvest disease management. In: Zinc-based nanostructures for environmental and agricultural applications. Elsevier, Amsterdam, pp 243–265. https://doi.org/10.1016/B978-0-12-822836-4.00018-5

Gopinath K, Gowri S, Karthika V, Arumugam A (2014) Green synthesis of gold nanoparticles from fruit extract of Terminalia arjuna, for the enhanced seed germination activity of Gloriosa superba. J Nanostructure Chem 4(3):115. https://doi.org/10.1007/s40097-014-0115-0

Grover M, Madhubala R, Ali SZ, Yadav SK, Venkateswarlu B (2014) Influence of Bacillus spp. strains on seedling growth and physiological parameters of sorghum under moisture stress conditions. J Basic Microbiol 54(9):951–961. https://doi.org/10.1002/jobm.201300250

Hajyzadeh M, Turktas M, Khawar KM, Unver T (2015) MiR408 overexpression causes increased drought tolerance in chickpea. Gene 555(2):186–193. https://doi.org/10.1016/j.gene.2014.11.002

Hang N, Shi T, Liu Y, Ye W, Taier G, Sun Y, Wang K, Zhang W (2021) Overexpression of Os-microRNA408 enhances drought tolerance in perennial ryegrass. Physiol Plant 172(2):733–747. https://doi.org/10.1111/ppl.13276

Hestrin R, Kan M, Lafler M, Wollard J, Kimbrel JA, Ray P, Blazewicz SJ, Stuart R, Craven K, Firestone M, Nuccio EE, Pett-Ridge J (2022) Plant-associated fungi support bacterial resilience following water limitation. ISME J 16(12):2752–2762. https://doi.org/10.1038/s41396-022-01308-6

Hivrale V, Zheng Y, Puli COR, Jagadeeswaran G, Gowdu K, Kakani VG, Barakat A, Sunkar R (2015) Characterization of drought- and heat-responsive microRNAs in switchgrass. Plant Sci 242:214–223. https://doi.org/10.1016/j.plantsci.2015.07.018

Hou Q, Ufer G, Bartels D (2016) Lipid signalling in plant responses to abiotic stress. Plant Cell Environ 39(5):1029–1048. https://doi.org/10.1111/pce.12666

Hu T, Jin Y, Li H, Amombo E, Fu J (2016) Stress memory induced transcriptional and metabolic changes of perennial ryegrass (Lolium perenne) in response to salt stress. Physiol Plant 156(1):54–69. https://doi.org/10.1111/ppl.12342

Hua Y, Zhang C, Shi W, Chen H (2019) High-throughput sequencing reveals microRNAs and their targets in response to drought stress in wheat (*Triticum aestivum* L.). Biotechnol Biotechnol Equip 33(1):465–471. https://doi.org/10.1080/13102818.2019.1586586

Huang XF, Zhou D, Lapsansky ER, Reardon KF, Guo J, Andales MJ, Vivanco JM, Manter DK (2017) Mitsuaria sp. and Burkholderia sp. from *Arabidopsis* rhizosphere enhance drought tolerance in *Arabidopsis* thaliana and maize (*Zea mays* L.). Plant Soil 419(1-2):523–539. https://doi.org/10.1007/s11104-017-3360-4

Hussain S, Khan F, Hussain HA, Nie L (2016) Physiological and biochemical mechanisms of seed priming-induced chilling tolerance in rice cultivars. Front Plant Sci 7:1–14. https://doi.org/10.3389/fpls.2016.00116

Hussain M, Farooq M, Lee DJ (2017) Evaluating the role of seed priming in improving drought tolerance of pigmented and non-pigmented rice. J Agron Crop Sci 203(4):269–276. https://doi.org/10.1111/jac.12195

Ikram M, Raja NI, Javed B, Mashwani Z, Hussain M, Hussain M, Ehsan M, Rafique N, Malik K, Sultana T, Akram A (2020) Foliar applications of bio-fabricated selenium nanoparticles to improve the growth of wheat plants under drought stress. Green Process Synth 9(1):706–714. https://doi.org/10.1515/gps-2020-0067

Işeri ÖD, Sahin FI, Haberal M (2014) Sodium chloride priming improves salinity response of tomato at seedling stage. J Plant Nutr 37(3):374–392. https://doi.org/10.1080/01904167.2013.859699

Jaberzadeh A, Moaveni P, Tohidi Moghadam HR, Zahedi H (2013) Influence of bulk and nanoparticles titanium foliar application on some agronomic traits, seed gluten and starch contents of wheat subjected to water deficit stress. Not Bot Horti Agrobot Cluj-Napoca 41(1):201–207. https://doi.org/10.15835/nbha4119093

Jiao Z, Lian C, Han S, Huang M, Shen C, Li Q, Niu MX, Yu X, Yin W, Xia X (2021) PtmiR169o plays a positive role in regulating drought tolerance and growth by targeting the PtNF-YA6 gene in poplar. Environ Exp Bot 189:104549. https://doi.org/10.1016/j.envexpbot.2021.104549

Jira-Anunkul W, Pattanagul W (2020) Seed priming with hydrogen peroxide alleviates the effects of drought stress in rice (*Oryza sativa* L.) seedlings. Not Bot Horti Agrobot Cluj-Napoca 48(1):273–283. https://doi.org/10.15835/NBHA48111829

Jisha KC, Vijayakumari K, Puthur JT (2013) Seed priming for abiotic stress tolerance: an overview. Acta Physiol Plant 35(5):1381–1396. https://doi.org/10.1007/s11738-012-1186-5

Juenger TE, Verslues PE (2023) Time for a drought experiment: do you know your plants' water status? Plant Cell 35(1):10–23. https://doi.org/10.1093/plcell/koac324

Kang T, Yu CY, Liu Y, Song WM, Bao Y, Guo XT, Li B, Zhang HX (2020) Subtly manipulated expression of ZmmiR156 in tobacco improves drought and salt tolerance without changing the architecture of transgenic plants. Front Plant Sci 10:1–15. https://doi.org/10.3389/fpls.2019.01664

Kantar M, Unver T, Budak H (2010) Regulation of barley miRNAs upon dehydration stress correlated with target gene expression. Funct Integr Genomics 10(4):493–507. https://doi.org/10.1007/s10142-010-0181-4

Katiyar A, Smita S, Muthusamy SK, Chinnusamy V, Pandey DM, Bansal KC (2015) Identification of novel drought-responsive microRNAs and trans-acting siRNAs from Sorghum bicolor (L.) Moench by high-throughput sequencing analysis. Front Plant Sci 6:506. https://doi.org/10.3389/fpls.2015.00506

Khan A, Khan V, Pandey K, Sopory SK, Sanan-Mishra N (2022) Thermo-priming mediated cellular networks for abiotic stress management in plants. Front Plant Sci 13:1–29. https://doi.org/10.3389/fpls.2022.866409

Kim JH, Woo HR, Kim J, Lim PO, Lee IC, Choi SH, Hwang D, Nam HG (2009) Trifurcate feed-forward regulation of age-dependent cell death involving miR164 in *Arabidopsis*. Science 323:1053–1057

Kim JH, Oh Y, Yoon H, Hwang I, Chang YS (2015) Iron nanoparticle-induced activation of plasma membrane H^+-ATPase promotes stomatal opening in *Arabidopsis* thaliana. Environ Sci Technol 49(2):1113–1119. https://doi.org/10.1021/es504375t

Kishor KPB, Sangam S, Amrutha RN, Sri Laxmi P, Naidu KR, Rao KRSS, Rao S, Reddy KJ, Theriappan P, Sreenivasulu N (2005) Regulation of proline biosynthesis, degradation, uptake and transport in higher plants: its implications in plant growth and abiotic stress tolerance. Curr Sci 88(3):424–438

Kole C, Kole P, Randunu KM, Choudhary P, Podila R, Ke PC, Rao AM, Marcus RK (2013) Nanobiotechnology can boost crop production and quality: first evidence from increased plant biomass, fruit yield and phytomedicine content in bitter melon (*Momordica charantia*). BMC Biotechnol 13:37. https://doi.org/10.1186/1472-6750-13-37

Kröel-Dulay G, Mojzes A, Szitár K, Bahn M, Batáry P, Beier C, Bilton M, De Boeck HJ, Dukes JS, Estiarte et al (2022) Field experiments underestimate aboveground biomass response to drought. Nat Ecol Evol 6(5):540–545. https://doi.org/10.1038/s41559-022-01685-3

Ku YS, Sintaha M, Cheung MY, Lam HM (2018) Plant hormone signaling crosstalks between biotic and abiotic stress responses. Int J Mol Sci 19:10. https://doi.org/10.3390/ijms19103206

Kumar V, Khare T, Shriram V, Wani SH (2018) Plant small RNAs: the essential epigenetic regulators of gene expression for salt-stress responses and tolerance. Plant Cell Rep 37(1):61–75. https://doi.org/10.1007/s00299-017-2210-4

Kurek K, Plitta-Michalak B, Ratajczak E (2019) Reactive oxygen species as potential drivers of the seed aging process. Plants 8(6):1–13. https://doi.org/10.3390/plants8060174

Lamers J, Van Der Meer T, Testerink C (2020) How plants sense and respond to stressful environments. Plant Physiol 182(4):1624–1635. https://doi.org/10.1104/PP.19.01464

Li WX, Oono Y, Zhu J, He XJ, Wu JM, Iida K, Lu XY, Cui X, Jin H, Zhu JK (2008) The *Arabidopsis* NFYA5 transcription factor is regulated transcriptionally and posttranscriptionally to promote drought resistance. Plant Cell 20(8):2238–2251. https://doi.org/10.1105/tpc.108.059444

Li B, Qin Y, Duan H, Yin W, Xia X (2011) Genome-wide characterization of new and drought stress responsive microRNAs in Populus euphratica. J Exp Bot 62(11):3765–3779. https://doi.org/10.1093/jxb/err051

Li W, Cui X, Meng Z, Huang X, Xie Q, Wu H, Jin H, Zhang D, Liang W (2012) Transcriptional regulation of *Arabidopsis* MIR168a and ARGONAUTE1 homeostasis in abscisic acid and abiotic stress responses. Plant Physiol 158(3):1279–1292. https://doi.org/10.1104/pp.111.188789

Li W, Wang T, Zhang Y, Li Y (2016) Overexpression of soybean miR172c confers tolerance to water deficit and salt stress, but increases ABA sensitivity in transgenic *Arabidopsisthaliana*. J Exp Bot 67(1):175–194. https://doi.org/10.1093/jxb/erv450

Li J, Duan Y, Sun N, Wang L, Feng S, Fang Y, Wang Y (2021) The miR169n-NF-YA8 regulation module involved in drought resistance in *Brassica napus* L. Plant Sci 313:111062. https://doi.org/10.1016/j.plantsci.2021.111062

Linh TM, Mai NC, Hoe PT, Lien LQ, Ban NK, Hien LTT, Chau NH, Van NT (2020) Metal-based nanoparticles enhance drought tolerance in soybean. J Nanomater 2020:4056563. https://doi.org/10.1155/2020/4056563

Liu HH, Tian X, Li YJ, Wu CA, Zheng CC (2008) Microarray-based analysis of stress-regulated microRNAs in *Arabidopsisthaliana*. RNA 14(5):836–843. https://doi.org/10.1261/rna.895308

Liu H, Able AJ, Able JA (2022) Priming crops for the future: rewiring stress memory. Trends Plant Sci 27(7):699–716. https://doi.org/10.1016/j.tplants.2021.11.015

Ma Z, Hu L (2023) MicroRNA: a dynamic player from signalling to abiotic tolerance in plants. Int J Mol Sci 24(14):11364. https://doi.org/10.3390/ijms241411364

Ma C, Burd S, Lers A (2015) MiR408 is involved in abiotic stress responses in *Arabidopsis*. Plant J 84(1):169–187. https://doi.org/10.1111/tpj.12999

Mahmood T (2015) Potential of copper nanoparticles to increase growth and yield of wheat. J Nanosci Adv Technol 1(1):6–11. https://doi.org/10.24218/jnat.2015.02

Mahmood A, Turgay OC, Farooq M, Hayat R (2016) Seed biopriming with plant growth promoting rhizobacteria: a review. FEMS Microbiol Ecol 92(8):1–14. https://doi.org/10.1093/femsec/fiw112

Mantri N, Basker N, Ford R, Pang E, Pardeshi V (2013) The role of micro-ribonucleic acids in legumes with a focus on abiotic stress response. Plant Genome 6(3):1–14. https://doi.org/10.3835/plantgenome2013.05.0013

Marchin RM, Ossola A, Leishman MR, Ellsworth DS (2020) A simple method for simulating drought effects on plants. Front Plant Sci 10(January):1–14. https://doi.org/10.3389/fpls.2019.01715

Marthandan V, Geetha R, Kumutha K, Renganathan VG, Karthikeyan A, Ramalingam J (2020) Seed priming: a feasible strategy to enhance drought tolerance in crop plants. Int J Mol Sci 21(21):1–23. https://doi.org/10.3390/ijms21218258

Maruyama K, Takeda M, Kidokoro S, Yamada K, Sakuma Y, Urano K, Fujita M, Yoshiwara K, Matsukura S, Morishita Y, Sasaki R, Suzuki H, Saito K, Shibata D, Shinozaki K, Yamaguchi-Shinozaki K (2009) Metabolic pathways involved in cold acclimation identified by integrated analysis of metabolites and transcripts regulated by DREB1A and DREB2A. Plant Physiol 150(4):1972–1980. https://doi.org/10.1104/pp.109.135327

Migicovsky Z, Yao Y, Kovalchuk I (2014) Transgenerational phenotypic and epigenetic changes in response to heat stress in *Arabidopsisthaliana*. Plant Signal Behav 9:1–11. https://doi.org/10.4161/psb.27971

Munir N, Hanif M, Abideen Z, Sohail M, El-Keblawy A, Radicetti E, Mancinelli R, Haider G (2022) Mechanisms and strategies of plant microbiome interactions to mitigate abiotic stresses. Agronomy 12(9):1–33. https://doi.org/10.3390/agronomy12092069

Naylor D, Degraaf S, Purdom E, Coleman-Derr D (2017) Drought and host selection influence bacterial community dynamics in the grass root microbiome. ISME J 11(12):2691–2704. https://doi.org/10.1038/ismej.2017.118

Ngumbi E, Kloepper J (2016) Bacterial-mediated drought tolerance: current and future prospects. Appl Soil Ecol 105:109–125. https://doi.org/10.1016/j.apsoil.2016.04.009

Ni Z, Hu Z, Jiang Q, Zhang H (2012) Overexpression of gma-MIR394a confers tolerance to drought in transgenic *Arabidopsisthaliana*. Biochem Biophys Res Commun 427(2):330–335. https://doi.org/10.1016/j.bbrc.2012.09.055

Niu X, Song L, Xiao Y, Ge W (2018) Drought-tolerant plant growth-promoting rhizobacteria associated with foxtail millet in a semi-arid and their potential in alleviating drought stress. Front Microbiol 8:1–11. https://doi.org/10.3389/fmicb.2017.02580

Ogbaga CC, Athar H, Amir M, Bano H, Chater CCC, Jellason NP (2020) Clarity on frequently asked questions about drought measurements in plant physiology. Sci Afr 8:e00405. https://doi.org/10.1016/j.sciaf.2020.e00405

Osmolovskaya N, Shumilina J, Kim A, Didio A, Grishina T, Bilova T, Keltsieva OA, Zhukov V, Tikhonovich I, Tarakhovskaya E, Frolov A, Wessjohann LA (2018) Methodology of drought stress research: experimental setup and physiological characterization. Int J Mol Sci 19(12):4089. https://doi.org/10.3390/ijms19124089

Oubaha B, Rathore RS, Bagri J, Singhal NK, Mazumdar K, Rishi V, Pareek A, Singla-Pareek SL (2024) *Bacillus siamensis* strain BW enhances rice growth and salinity tolerance through redox equilibrium and hormone modulation. Curr Plant Biol 37:100321. https://doi.org/10.1016/j.cpb.2024.100321

Patil S, Joshi S, Jamla M, Zhou X, Taherzadeh MJ, Suprasanna P, Kumar V (2021) MicroRNA-mediated bioengineering for climate-resilience in crops. Bioengineered 12(2):10430–10456. https://doi.org/10.1080/21655979.2021.1997244

Sahin U, Ekinci M, Kiziloglu FM, Yildirim E, Turan M, Kotan R, Ors S (2015) Ameliorative effects of plant growth promoting bacteria on water-yield relationships, growth, and nutrient uptake of lettuce plants under different irrigation levels. Hortic Sci 50(9):1379–1386. https://doi.org/10.21273/hortsci.50.9.1379

Saikia J, Sarma RK, Dhandia R, Yadav A, Bharali R, Gupta VK, Saikia R (2018) Alleviation of drought stress in pulse crops with ACC deaminase producing rhizobacteria isolated from acidic soil of Northeast India. Sci Rep 8(1):1–16. https://doi.org/10.1038/s41598-018-21921-w

Samad AFA, Sajad M, Nazaruddin N, Fauzi IA, Murad AMA, Zainal Z, Ismail I (2017) MicroRNA and transcription factor: Key players in plant regulatory network. Front Plant Sci 8:1–18. https://doi.org/10.3389/fpls.2017.00565

Savvides A, Ali S, Tester M, Fotopoulos V (2016) Chemical priming of plants against multiple abiotic stresses: mission possible? Trends Plant Sci 21(4):329–340. https://doi.org/10.1016/j.tplants.2015.11.003

Sedghi M, Hadi M, Toluie SG (2013) Effect of nano zinc oxide on the germination parameters of soybean seeds under drought stress. Ser Biol 16(2):73–78

Seleiman MF, Al-Suhaibani N, Ali N, Akmal M, Alotaibi M, Refay Y, Dindaroglu T, Abdul-Wajid HH, Battaglia ML (2021) Drought stress impacts on plants and different approaches to alleviate its adverse effects. Plants 10(2):1–25. https://doi.org/10.3390/plants10020259

Semida WM, Abdelkhalik A, Rady MOA, Marey RA, Abd El-Mageed TA (2020) Exogenously applied proline enhances growth and productivity of drought stressed onion by improving photosynthetic efficiency, water use efficiency and up-regulating osmoprotectants. Sci Hortic 272:109580. https://doi.org/10.1016/j.scienta.2020.109580

Shah SMS, Ullah F (2023) A comprehensive overview of miRNA targeting drought stress resistance in plants. Braz J Biol 83:1–6. https://doi.org/10.1590/1519-6984.242708

Sharma P, Lakra N, Goyal A, Ahlawat YK, Zaid A, Siddique KHM (2023) Drought and heat stress mediated activation of lipid signaling in plants: a critical review. Front Plant Sci 14:1–19. https://doi.org/10.3389/fpls.2023.1216835

Siddiqui MH, Al-Whaibi MH, Firoz M, Al-Khaishany MY (2015) Role of nanoparticles in plants. In: Siddiqui M, Al-Whaibi M, Mohammad F (eds) Nanotechnology and plant sciences. Springer, Cham. https://doi.org/10.1007/978-3-319-14502-0_2

Singh S, Kumar A, Panda D, Modi MK, Sen P (2020) Identification and characterization of drought responsive miRNAs from a drought tolerant rice genotype of Assam. Plant Gene 21:100213. https://doi.org/10.1016/j.plgene.2019.100213

Singh A, Tiwari S, Pandey J, Lata C, Singh IK (2021) Role of nanoparticles in crop improvement and abiotic stress management. J Biotechnol 337:57–70. https://doi.org/10.1016/j.jbiotec.2021.06.022

Singroha G, Sharma P, Sunkur R (2021) Current status of microRNA-mediated regulation of drought stress responses in cereals. Physiol Plant 172(3):1808–1821. https://doi.org/10.1111/ppl.13451

Sun M, Jing Y, Wang X, Zhang Y, Zhang Y, Ai J, Li J, Jin L, Li W, Li Y (2020) Gma-miR1508a confers dwarfing, cold tolerance, and drought sensitivity in soybean. Mol Breed 40(4):44. https://doi.org/10.1007/s11032-020-01116-w

Sun L, Song F, Zhu X, Liu S, Liu F, Wang Y, Li X (2021) Nano-ZnO alleviates drought stress via modulating the plant water use and carbohydrate metabolism in maize. Arch Agron Soil Sci 67(2):245–259. https://doi.org/10.1080/03650340.2020.1723003

Sunkar R, Zhu JK (2004) Novel and stress regulated microRNAs and other small RNAs from *Arabidopsis* w inside box sign. Plant Cell 16(8):2001–2019. https://doi.org/10.1105/tpc.104.022830

Tang Q, Lv H, Li Q, Zhang X, Li L, Xu J, Wu F, Wang Q, Feng X, Lu Y (2022) Characteristics of microRNAs and target genes in maize root under drought stress. Int J Mol Sci 23(9):4968. https://doi.org/10.3390/ijms23094968

Thakur A, Sharma KD, Siddique KHM, Nayyar H (2020) Cold priming the chickpea seeds imparts reproductive cold tolerance by reprogramming the turnover of carbohydrates, osmoprotectants and redox components in leaves. Sci Hortic 261:108929. https://doi.org/10.1016/j.scienta.2019.108929

Umezawa T, Nakashima K, Miyakawa T, Kuromori T, Tanokura M, Shinozaki K, Yamaguchi-Shinozaki K (2010) Molecular basis of the core regulatory network in ABA responses: sensing, signaling and transport. Plant Cell Physiol 51(11):1821–1839. https://doi.org/10.1093/pcp/pcq156

Van Nguyen D, Nguyen HM, Le NT, Nguyen KH, Nguyen HT, Le HM, Nguyen AT, Dinh NTT, Hoang SA, Van Ha C (2022) Copper nanoparticle application enhances plant growth and grain yield in maize under drought stress conditions. J Plant Growth Regul 41:364–375. https://doi.org/10.1007/s00344-021-10301-w

Verma ML, Kumar P, Sharma D, Verma AD, Jana AK (2019) Advances in nanobiotechnology with special reference to plant systems. In: Plant nanobionics, nanotechnology in the life sciences. Springer, Cham, pp 371–387. https://doi.org/10.1007/978-3-030-12496-0_13

Viscardi S, Ventorino V, Duran P, Maggio A, De Pascale S, Mora ML, Pepe O (2016) Assessment of plant growth promoting activities and abiotic stress tolerance of Azotobacter chroococcum strains for a potential use in sustainable agriculture. J Soil Sci Plant Nutr 16(3):848–863. https://doi.org/10.4067/s0718-95162016005000060

Visentin I, Pagliarani C, Deva E, Caracci A, Turečková V, Novák O, Lovisolo C, Schubert A, Cardinale F (2020) A novel strigolactone-miR156 module controls stomatal behaviour during drought recovery. Plant Cell Environ 43(7):1613–1624. https://doi.org/10.1111/pce.13758

Vives-Peris V, de Ollas C, Gómez-Cadenas A, Pérez-Clemente RM (2020) Root exudates: from plant to rhizosphere and beyond. Plant Cell Rep 39(1):3–17. https://doi.org/10.1007/s00299-019-02447-5

Wang CJ, Yang W, Wang C, Gu C, Niu DD, Liu HX, Wang YP, Guo JH (2012) Induction of drought tolerance in cucumber plants by a consortium of three plant growth-promoting rhizobacterium strains. PLoS ONE 7(12):e52565. https://doi.org/10.1371/journal.pone.0052565

Wang YG, An M, Zhou SF, She YH, Li WC, Fu FL (2014) Expression profile of maize micrornas corresponding to their target genes under drought stress. Biochem Genet 52(11-12):474–493. https://doi.org/10.1007/s10528-014-9661-x

Wang L, Zhao H, Chen D, Li L, Sun H, Lou Y, Gao Z (2016) Characterization and primary functional analysis of a bamboo NAC gene targeted by miR164b. Plant Cell Rep 35(6):1371–1383. https://doi.org/10.1007/s00299-016-1970-6

Wang W, Wang X, Zhang J, Huang M, Cai J, Zhou Q, Dai T, Jiang D (2020) Salicylic acid and cold priming induce late-spring freezing tolerance by maintaining cellular redox homeostasis and protecting photosynthetic apparatus in wheat. Plant Growth Regul 90(1):109–121. https://doi.org/10.1007/s10725-019-00553-8

Wang K, Liu Y, Teng F, Cen H, Yan J, Lin S, Li D, Zhang W (2021) Heterogeneous expression of Osa-MIR156bc increases abiotic stress resistance and forage quality of alfalfa. Crop J 9(5):1135–1144. https://doi.org/10.1016/j.cj.2020.11.009

Wang Y, Dai J, Chen R, Song C, Wei P, Cai Y, Wang Y, Han B (2022) miRNA-based drought regulation in the important medicinal plant *Dendrobium huoshanense*. J Plant Growth Regul 41:1099–1108. https://doi.org/10.1007/s00344-021-10366-7

Wei LX, Lv BS, Li XW, Wang MM, Ma HY, Yang HY, Yang RF, Piao ZZ, Wang ZH, Lou JH, Jiang CJ, Liang ZW (2017) Priming of rice (*Oryza sativa* L.) seedlings with abscisic acid enhances seedling survival, plant growth, and grain yield in saline-alkaline paddy fields. Field Crop Res 203:86–93. https://doi.org/10.1016/j.fcr.2016.12.024

Wu BF, Li WF, Xu HY, Qi LW, Han SY (2015) Role of cin-miR2118 in drought stress responses in *Caragana intermedia* and tobacco. Gene 574(1):34–40. https://doi.org/10.1016/j.gene.2015.07.072

Wu J, Wang L, Wang S (2017) MicroRNAs associated with drought response in the pulse crop common bean (*Phaseolus vulgaris* L.). Gene 628:78–86. https://doi.org/10.1016/j.gene.2017.07.038

Xia K, Wang R, Ou X, Fang Z, Tian C, Duan J, Wang Y, Zhang M (2012) OsTIR1 and OsAFB2 downregulation via OsmiR393 overexpression leads to more tillers, early flowering and less tolerance to salt and drought in rice. PLoS ONE 7(1):e30039. https://doi.org/10.1371/journal.pone.0030039

Xu L, Naylor D, Dong Z, Simmons T, Pierroz G, Hixson KK, Kim YM, Zink EM, Engbrecht KM, Wang Y, Gao C, DeGraaf S, Madera MA, Sievert JA, Hollingsworth J, Birdseye D, Scheller HV, Hutmacher R, Dahlberg J, Coleman-Derr D (2018) Drought delays development of the sorghum root microbiome and enriches for monoderm bacteria. Proc Natl Acad Sci USA 115(18):E4284–E4293. https://doi.org/10.1073/pnas.1717308115

Xu L, Dong Z, Chiniquy D, Pierroz G, Deng S, Gao C, Diamond S, Simmons T, Wipf HML, Caddell D, Varoquaux N, Madera MA, Hutmacher R, Deutschbauer A, Dahlberg JA, Guerinot ML, Purdom E, Banfield JF, Taylor JW, Coleman-Derr D (2021) Genome-resolved metagenomics reveals role of iron metabolism in drought-induced rhizosphere microbiome dynamics. Nat Commun 12(1):3209. https://doi.org/10.1038/s41467-021-23553-7

Yan M (2015) Seed priming stimulate germination and early seedling growth of Chinese cabbage under drought stress. S Afr J Bot 99:88–92. https://doi.org/10.1016/j.sajb.2015.03.195

Yan J, Zhao C, Zhou J, Yang Y, Wang P, Zhu X, Tang G, Bressan RA, Zhu JK (2016) The miR165/166 mediated regulatory module plays critical roles in ABA homeostasis and response in *Arabidopsisthaliana*. PLoS Genet 12(11):1006416. https://doi.org/10.1371/journal.pgen.1006416

Yang S, Vanderbeld B, Wan J, Huang Y (2010) Narrowing down the targets: towards successful genetic engineering of drought-tolerant crops. Mol Plant 3(3):469–490. https://doi.org/10.1093/mp/ssq016

Yu Y, Ni Z, Wang Y, Wan H, Hu Z, Jiang Q, Sun X, Zhang H (2019) Overexpression of soybean miR169c confers increased drought stress sensitivity in transgenic *Arabidopsisthaliana*. Plant Sci 285:68–78. https://doi.org/10.1016/j.plantsci.2019.05.003

Yue E, Cao H, Liu B (2020) Osmir535, a potential genetic editing target for drought and salinity stress tolerance in *Oryza sativa*. Plants 9(10):1–10. https://doi.org/10.3390/plants9101337

Zhang X, Zou Z, Gong P, Zhang J, Ziaf K, Li H, Xiao F, Ye Z (2011) Over-expression of microRNA169 confers enhanced drought tolerance to tomato. Biotechnol Lett 33(2):403–409. https://doi.org/10.1007/s10529-010-0436-0

Zhang N, Yang J, Wang Z, Wen Y, Wang J, He W, Liu B, Si H, Wang D (2014) Identification of novel and conserved microRNAs related to drought stress in potato by deep sequencing. PLoS ONE 9(4):e95489. https://doi.org/10.1371/journal.pone.0095489

Zhang F, Luo X, Zhou Y, Xie J (2016) Genome-wide identification of conserved microRNA and their response to drought stress in Dongxiang wild rice (*Oryza rufipogon* Griff.). Biotechnol Lett 38(4):711–721. https://doi.org/10.1007/s10529-015-2012-0

Zhang H, Zhang J, Yan J, Gou F, Mao Y, Tang G, Botella JR, Zhu JK (2017) Short tandem target mimic rice lines uncover functions of miRNAs in regulating important agronomic traits. Proc Natl Acad Sci USA 114(20):5277–5282. https://doi.org/10.1073/pnas.1703752114

Zhang J, Zhang H, Srivastava AK, Pan Y, Bai J, Fang J, Shi H, Zhu JK (2018) Knockdown of rice microRNA166 confers drought resistance by causing leaf rolling and altering stem xylem development. Plant Physiol 176(3):2082–2094. https://doi.org/10.1104/pp.17.01432

Zhang F, Yang J, Zhang N, Wu J, Si H (2022) Roles of microRNAs in abiotic stress response and characteristics regulation of plant. Front Plant Sci 13:919243. https://doi.org/10.3389/fpls.2022.919243

Zhang Y, Xu J, Li R, Ge Y, Li Y, Li R (2023) Plants' response to abiotic stress: mechanisms and strategies. Int J Mol Sci 24(13):1–17. https://doi.org/10.3390/ijms241310915

Zhao J, Yuan S, Zhou M, Yuan N, Li Z, Hu Q, Bethea FG, Liu H, Li S, Luo H (2019) Transgenic creeping bentgrass overexpressing Osa-miR393a exhibits altered plant development and improved multiple stress tolerance. Plant Biotechnol J 17(1):233–251. https://doi.org/10.1111/pbi.12960

Zhou L, Liu Y, Liu Z, Kong D, Duan M, Luo L (2010) Genome-wide identification and analysis of drought-responsive microRNAs in *Oryza sativa*. J Exp Bot 61(15):4157–4168. https://doi.org/10.1093/jxb/erq237

Zhou M, Li D, Li Z, Hu Q, Yang C, Zhu L, Luo H (2013) Constitutive expression of a miR319 gene alters plant development and enhances salt and drought tolerance in transgenic creeping bentgrass. Plant Physiol 161(3):1375–1391. https://doi.org/10.1104/pp.112.208702

Zhou R, Kong L, Yu X, Ottosen CO, Zhao T, Jiang F, Wu Z (2019) Oxidative damage and antioxidant mechanism in tomatoes responding to drought and heat stress. Acta Physiol Plant 41(2):1–11. https://doi.org/10.1007/s11738-019-2805-1

Zhu JK (2002) Salt and drought stress signal transduction in plants. Annu Rev Plant Biol 53:247–273. https://doi.org/10.1146/annurev.arplant.53.091401.143329

Zhu JK (2016) Abiotic stress signaling and responses in plants. Cell 167(2):313–324. https://doi.org/10.1016/j.cell.2016.08.029

Zhu JK, Liu J, Xiong L (1998) Genetic analysis of salt tolerance in *Arabidopsis*: evidence for a critical role of potassium nutrition. Plant Cell 10(7):1181–1191. https://doi.org/10.1105/tpc.10.7.1181

Zulfiqar F, Navarro M, Ashraf M, Akram NA, Munné-Bosch S (2019) Nanofertilizer use for sustainable agriculture: advantages and limitations. Plant Sci 289:110270. https://doi.org/10.1016/j.plantsci.2019.110270

miRNAs: A Low-Cost Solution to Enhance Rice Yield Under Salinity Stress

15

Sudhir Kumar, Vaishali Saxena, Sudhir K. Sopory, and Neeti Sanan-Mishra

Abstract

Soil salinity is one of the primary concerns for arable land. At low to moderate levels, salt stress impacts plant growth and yields, but extreme salt stress can be devastating to plant development. Global warming is causing saltwater intrusion in to agricultural zones which is causing the salinized areas to grow quickly. To ensure food security, there is an increasing interest in enhancing agricultural productivity, but environmental stresses such as salinity are the major bottlenecks. Plants have developed a variety of defense mechanisms to withstand salt stress. One of the main strategies is the use of miRNAs, which can influence post-transcriptional gene regulation in a variety of contexts. The major goal of this chapter is to explain the miRNA-regulated molecular and physiological mechanisms and biochemical pathways involved in salinity stress tolerance in rice. This chapter emphasizes the multifaceted roles of miRNAs in linking plant growth and responses to salinity stress as well as rice grain yield.

Keywords

Crop improvement · miRNA · Rice · Yield · Salt stress

Sudhir Kumar and Vaishali Saxena contributed equally.

S. Kumar · V. Saxena · N. Sanan-Mishra (✉)
Plant RNAi Biology Group, International Centre for Genetic Engineering and Biotechnology, New Delhi, India
e-mail: Sopory@icgeb.res.in; neeti@icgeb.res.in

S. K. Sopory
International Centre for Genetic Engineering and Biotechnology, New Delhi, India

K. Goswami et al. (eds.), *Non-coding RNAs for Crop Improvement*,
https://doi.org/10.1007/978-981-97-7213-1_15

15.1 Introduction

Plants are facing a variety of environmental challenges and abiotic stresses such as cold, heat, salt, drought, and nutrient inadequacy have been shown to significantly reduce crop yield. For several of the world's most important crops, the yield reduction induced by these stresses might be extremely high, reaching even more than 50% (Qin et al. 2011). One of the most devastating abiotic stressors is salinity as globally, salinity stress has impacted the state of around 20% of total farmed and 33% of irrigated agricultural land (Qadir et al. 2014; Shrivastava and Kumar 2015; Abobatta 2020). Thus far, salinity has impacted about 1.0 billion acres, or more than 5% of the world's total land area (Hopmans et al. 2021). Additionally, sea levels are expected to rise by 12–22 centimeters (cm) worldwide during the twentieth century. Sea levels have been rising continuously, at a pace of roughly 0.34 cm/year, as seen in satellite pictures, which has a significant impact on the water and land quality close to the sea (Zheng et al. 2023), leading to soil salinization. Over 400 million people worldwide, including those in Asia, Africa, and South America, suffer from chronic hunger, necessitating a 38% increase in food production over the course of the following 30 years (Zheng et al. 2023). Rice (*Oryza sativa* L.) is a staple food for more than half of the global population and at the same time, it is relatively sensitive to salt stress. Excessive soil salinity, exceeding the critical threshold of 3.0 dS/m, adversely affects various grain yield factors. These include overall grain yield, seed set rate, 1000-seed weight, number of effective panicles, panicle length, and the quantity of grains per panicle (Mumtaz et al. 2018). For instance, Djaman et al. (2020) reported that rice grain yield declined by 50% when the soil's electrical conductivity (EC) was 6.0 dS/m.

Being sessile in nature plants, have developed a range of defense mechanisms for salinity stress, including exclusion of sodium ions (Na^+), ABA-induced stomatal movement, accumulation of osmolytes, and ROS scavenging systems (Begum 2022). Numerous researches revealed that miRNAs play a critical role in salinity stress response in plants, as crucial post-transcriptional regulators (Sun et al. 2019). For instance, in *Arabidopsis*, miR397 overexpression could suppress the expression of the target gene *LACCASE* (*LAC*), particularly producing a considerable decrease in *LAC2*, *LAC4*, and *LAC17*, which resulted in lower lignin concentration and increased susceptibility to salt stress (Nguyen et al. 2020). *Laccases* are glycoproteins with a lot of copper atoms that are found in plants. High NaCl concentrations cause the expression of the *laccase* gene in the roots of tomato, maize, and *Arabidopsis* (Cai et al. 2006). When salt stress is applied to artichoke roots, miR397a expression gets downregulated, which may cause *laccase* expression to be increased (De Paola et al. 2012). *Arabidopsis*'s constitutive expression of *SvMIR397* resulted in downregulation of the target gene *LAC2/4/17*. miR397 transformants displayed decreased rosette fresh weight and salt sensitivity compared to wild-type plants after being subjected to 150 mM NaCl for 7 days (Nguyen et al. 2020). Therefore, this chapter focuses on providing concise details about how plant miRNAs function in salinity stress response. It delves into the miRNA-related regulatory networks that drive molecular processes during salinity stress in plants.

15.2 The Mechanisms of Salt Perception

Stress sensing and signalling is the first and foremost step in plant's stress response, which results in an effective adaptive strategy. It is critical to comprehend the mechanisms behind salt input, which causes excessive ions to accumulate in plants (Isayenkov and Maathuis 2019). Ion uptake can take place via the symplastic or apoplastic pathways in plants. Several transporters mediate the net intake of salt through these pathways. Na^+ enters the cell through the root cytoplasm and is transported to the vacuole by transporters such as the Na^+/H^+ antiporter. The Na^+/H^+ antiporter known as Salt Overly Sensitive 1 (SOS1) is thought to be involved in Na^+ sensing because of its lengthy hydrophilic cytoplasmic tail (Ponce et al. 2021). The vacuolar-type H^+-ATPase (V-ATPase) and the vacuolar pyrophosphatase (V-PPase) are two types of H^+ pumps involved in the transport of ions across the vacuolar membrane (Dietz et al. 2001). Transmembrane protein receptors, including receptor-like kinases (RLKs) and histidine kinases, are crucial in detecting osmotic stress in rice. These receptors are sensors for salt and drought conditions by relaying signals to downstream pathways. One such receptor, the lectin RLK known as Salt Intolerance 1 (SIT1), is mainly expressed in root epidermal cells and enhances kinase activity to mediate salt stress responses. This kinase activity of SIT1 subsequently activates mitogen-activated protein kinases (MAPKs) 3 and 6 (Li et al. 2014). Another stress-gated cation channels function as hyperosmotic stress sensors are the OSCAs, calcium-permeable cation channels. The rice genome contains 11 OSCA genes, but only seven-OsOSCA1.1, OsOSCA1.2, OsOSCA2.1, OsOSCA2.4, OsOSCA2.5, OsOSCA3.1, and OsOSCA4.1-are upregulated in response to salt-induced osmotic stress and potentially act as osmosensors (Li et al. 2015a, b). It is hypothesized that rice possesses an unidentified cytosolic sensor capable of detecting intracellular Na^+ ions, as indicated by the elevated levels of free cytosolic Ca^{2+} ions in salt-stressed plants. Further research is needed to identify this cytosolic Na^+ sensor (Ponce et al. 2021).

15.3 Effects of Salinity Stress on Rice Plant

15.3.1 Effects of Salinity on Rice Biochemistry

The overall salt stress in the cell is a result of an imbalance in the cellular ion exchange pathway where high quantities of Na^+ act as a cellular toxin. There are three types of salt-induced stress; osmotic, ionic, and secondary stress. Osmotic stress reduces the water potential at the root surface, lowering plant water uptake and resulting in a water deficit in plants (Hasegawa et al. 2000). Ionic stress happens when high levels of Na^+ in the cytoplasm prevent other ions from entering plant cells (Fu and Luan 1998; Lazof and Bernstein 1999). Both osmotic and ionic stresses lead to secondary stresses in plants, including the build-up of toxic compounds, disturbances in nutrient balances, and the accumulation of ROS within plant cells (Hazman et al. 2015; Li et al. 2015a, b). The dynamic equilibrium between the creation and removal of ROS may be disrupted under salt stress conditions (Ashraf and Harris 2004; Munns and Tester 2008). At low concentrations, ROS plays a crucial role as intermediary signaling molecules, regulating various physiological processes such as gene expression and stomatal movement (Suzuki et al. 2012; Yi et al. 2014). However, under saline conditions, the increased production of ROS leads to cell death and damages cellular components including nucleic acids, enzymes, pigments, and membrane lipids (Ashraf and Foolad 2005; Liu et al. 2010; Goud and Kachole 2011). Lipid peroxidation, particularly, is a critical mechanism under salinity stress (Yang et al. 2015). This process results in the accumulation of malondialdehyde (MDA), a key marker of oxidative stress. As salt stress intensifies, MDA levels rise in many crops (Khan and Panda 2008). Salinity also disrupts a plant's hormonal balance, reducing levels of auxins, gibberellins, cytokinins, and salicylic acid, while increasing jasmonates and abscisic acid (Javid et al. 2011; Miransari and Smith 2014).

15.3.2 Salinity Effects on Rice Morphology

Salinity reduces the rate of leaf elongation, leaf length, and leaf width primarily due to changes in root water potential that lead to insufficient water supply to the leaves and decreased turgor pressure (Yeo et al. 1991; Roshandel and Flowers 2009; Wankhade et al. 2010). Salinity stress also diminishes shoot length and leaf area (Läuchli and Grattan 2007). Typically, salt exposure inhibits shoot growth, fresh and dry weight, NaCl uptake, disrupts ion homeostasis, and decreases leaf area, all contributing to lower photosynthesis and fewer photoassimilates. These effects are more severe in salt-sensitive genotypes (Rahman et al. 2016; Korres et al. 2022). Additionally, salinity stress hampers cell expansion and lateral bud development, negatively affecting shoot growth and development (Munns and Tester 2008). The various impacts of salinity, such as reductions in chlorophyll content, plant height, fresh and dry weight, and total carotenoids, also serve as key parameters for assessing salt tolerance (Daneshmand et al. 2010).

However, research has shown that shoots are more susceptible to salinity stress than roots (Munns 1993). Due to reduction in the fresh and dry weights of roots, as well as diminished root growth, rice plants experience decreased water and mineral uptake when exposed to salinity stress at various developmental stages (Liu et al. 2019a, b; Korres et al. 2022). To cope with salt stress, plants increase the number of lateral roots to compensate for the loss of root hydraulics (Krishnamurthy et al. 2011). The anatomical characteristics of the root system, such as root length, diameter, and the orientation of the xylem and phloem, influence the root's efficiency in absorbing water and nutrients under saline conditions (Farooq et al. 2019). Salinity stress induces several anatomical changes, including reduced root epidermal cell division, decreased osmotic effects, slower elongation rates, reduced primary growth, and the initiation of lateral root development (Jung and McCouch 2013). Additionally, salt stress reduces cortex thickness but increases xylem thickness (Dolatabadian et al. 2011). The suberization of the hypodermis and endodermis enhances the formation of the Casparian strip under salt stress (Shen et al. 2015).

15.3.3 Effects on Photosynthesis

The high salt content in a plant's rhizosphere can affect its physiological characteristics. Rice plants exposed to salt stress displayed a variety of physiological impacts, including a reduction in net photosynthesis, pigment degradation, stomatal conductance, relative water content (RWC), and transpiration rate (Hussain et al. 2017). Furthermore, salt stress causes PSII reaction centers to be disrupted, the oxygen-evolving complex to be interrupted, and the activity of the plastoquinone acceptor to be inhibited (Sobhanian et al. 2011). High salt stress leads to a reduction in total chlorophyll, chloroplasts and light-harvesting complexes degradation, photosynthetic quantum yield, and energy dissipation through non-photochemical pathways in sugar beet leaves (Hossain et al. 2017; Zhang et al. 2019; Zhu et al. 2020. A 3D model of whole rice mesophyll cells showed that under salt stress (100 mM for 4 days), swollen chloroplasts became spherical without an increase in volume (Oi et al. 2020).

Salinity also significantly affects the rice plant's water usage efficiency (WUE) (Ramezani et al. 2012). WUE has a positive relationship with both seed yield and biomass (Gholipoor et al. 2002). At the molecular and biochemical levels, each of these factors has detrimental pleiotropic effects on rice physiology and development. Stomatal closure serves as a protective measure to mitigate water loss, but it concurrently impedes the crucial uptake of CO_2 essential for photosynthesis. Moreover, high salinity stress disrupts the structure of chlorophyll, causing a reduction in chlorophyll content. This, in turn, compromises the plant's capacity to harness light energy, leading to a decline in photosynthetic efficiency and an overall impairment in plant growth when exposed to salt-stress conditions. Salt stress alters a plant's water relationship by reducing its leaf water potential (w), osmotic potential (o), and turgor pressure (p) (Wu et al. 2015; Arif et al. 2020).

15.3.4 Salt-induced Biochemical Changes in Seed

Research indicates that salinity stress affects both macro- and micronutrient absorption and uptake. Specifically, it impacts the intake of sodium (Na^+), potassium (K^+), calcium (Ca^{2+}), and overall cation content (Negrão et al. 2017). In rice, salinity stress influences biochemical properties such as proline, anthocyanins, peroxidase activity, calcium, sodium, potassium, chlorophyll, and H_2O_2 levels (Saeed 2018). Rice grains are rich in magnesium, calcium, phosphorus, and trace minerals like zinc, copper, iron, and manganese as well as vitamins including thiamine, riboflavin, and niacin, all of which are affected by salinity stress (Razzaq et al. 2020). Studies have shown that salinity increases sodium, iron, and zinc levels while reducing phosphorus and manganese. However, nitrogen, magnesium, copper, potassium, and calcium concentrations in tested genotypes remain unaffected (Verma et al. 1994). Additionally, salinity disrupts the structure of enzymes and other macromolecules, damages cell organelles and the plasma membrane, and hinders processes like respiration, photosynthesis, and protein synthesis (Parida and Das 2005; Panda and Khan 2009; Daszkowska-Golec 2011).

15.3.5 Effects on Seed Germination

Seed germination is affected by salinity due to osmotic stress, ion-specific effects, and oxidative damage (Flowers et al. 1977; Munns 2002). Numerous reports have shown that salinity has detrimental impacts on rice seedling establishment as well as germination percentage, germination index, germination rate, and mean germination time (Hakim et al. 2010; Korres et al. 2022). Na^+ and Cl^- affect embryo viability (Jahromi et al. 2008; Daszkowska-Golec 2011). Since water imbibition is a crucial step in the germination process, salts in the growing medium would inevitably lower its osmotic potential, preventing water from entering the seed and, as a result, the activation of the seed metabolism that would otherwise start the germination process (Shereen et al. 2011). Increased salinity has also been found to slow or prevent seed germination by inhibiting metabolic processes including protein degradation and mobilization to the developing embryo or starch degradation through amylase enzymes (Shereen et al. 2011; Liu et al. 2018).

15.3.6 Salinity Effect on Rice Yield

Rice development and productivity are inhibited by abiotic constraints such as salinity resulting in severe yield reductions known as the 'yield penalty' (Tripathi et al. 2012; Sahoo et al. 2013). Rice is one of the most salt-sensitive cereal crops, with plant development and yield markedly reduced at 30 mmol L^{-1} NaCl (3 dS m^{-1}) (Lutts and Guerrier 1995). Modern high-yielding rice cultivars exhibited a yield reduction of roughly 12% over a salt level of 3 dS m^{-1}. A saline level of 6 dS m^{-1} resulted in a 50% yield loss (Linh et al. 2012). Soil salinity has impacted >800 million hectares of agricultural land worldwide. When a soil's electric conductivity (EC) hits 4 dS m^{-1}, or 40 mM NaCl, it is deemed to be salinized. This results in an osmotic pressure of roughly 0.2 MPa and markedly lower crop yields for most crops (Munns and Tester 2008). Consequently, ion toxicity causes chlorosis and necrosis, primarily because of Na^+ accumulation (Munns 2002). The number of spikelets per panicle and the number of tillers per plant are highly responsive yield components to salinity. Tillering is influenced by stress in the early growth phase, while the number of spikelets per panicle is affected during the period from panicle formation to booting stage (Zeng et al. 2001). Furthermore, salt stress was found to reduce tiller numbers as soon as 3 weeks after germination, with more pronounced reductions observed in Japonica cultivars compared to Indica cultivars (Liu et al. 2019a, b). At either the vegetative or reproductive stages, salt stress has been shown to significantly reduce panicle dry weight, panicle length, and panicle initiation (Panda et al. 2015; Korres et al. 2022). Another yield factor, spikelet number per panicle, is significantly impacted by salt stress, especially when plants are exposed to salt stress between the 3-leaf and panicle onset (Zeng and Shannon 2000b). Panda et al. (2015) also showed that salinity exposure during the late booting stage to *Japonica* rice cultivar resulted in fewer spikelets per panicle. Grain weight reductions were seen in rice plants at the seedling stage (Korres et al. 2022).

Under conditions of salt stress, the issue of panicle sterility poses a significant challenge to rice grain production (Flowers and Yeo 1981). Various studies have confirmed that salt stress during fertilization can lead to panicle sterility, which diminishes grain setting either through reduced pollen viability, impaired stigmatic function, or both (Abdullah et al. 2001). Researchers such as Zeng and Shannon (2000a, b) have observed a negative correlation between increasing salinity levels and key yield components such as tiller number, spikelets per panicle, and the proportion of sterile florets (Zeng and Shannon 2000a; Hussain et al. 2017). Additionally, the inhibition of starch synthetase activity during grain development and decreased transfer of soluble sugars to both superior and inferior spikelets are primary factors contributing to reduced rice grain production under salt stress (Abdullah et al. 2001). In another study, an escalation in salt stress from 2 to 8 dS m-1 resulted in a decline in the number of filled grains per panicle (Aref and Rad 2012). Each of these factors influences overall grain yield, underscoring the importance of understanding how salt stress impacts rice production decline.

15.3.7 Response Mechanism of Salt in Plants

Plants employ various physiological and molecular mechanisms to manage stress conditions, such as osmotic, ionic, and tissue tolerance (Isayenkov and Maathuis 2019). When exposed to salt stress, plants exhibit two distinct phases of response. The initial phase, which is ion-independent, occurs rapidly within minutes to days, impacting shoot cell growth and inhibiting stomatal closure (Isayenkov and Maathuis 2019). Subsequently, the ion-dependent phase unfolds over days to weeks, intensifying ion cytotoxicity, which leads to metabolic suppression, premature leaf aging, and cell death (Isayenkov and Maathuis 2019).

To counteract the detrimental effects of ROS, plants have developed an antioxidant defense system. This system includes enzymes like superoxide dismutase (SOD), peroxidase, catalase (CAT), as well as non-enzymatic compounds (Gong et al. 2008; Zhu et al. 2011). SOD plays a crucial role by breaking down superoxide radicals into H_2O_2. CAT then converts H_2O_2 into water and oxygen, thus preventing the formation of hydroxyl radicals, which are involved in crucial processes like membrane restructuring during seed development (Goel and Sheoran 2003; Wang et al. 2009; Gill and Tuteja 2010).

15.4 microRNA Expression in Response to Salt Stress

Studies on plants have highlighted the importance of miRNAs in regulating responses to salt stress (Table 15.1). Previous research in *Arabidopsis* has confirmed that under saline conditions, miRNAs such as miR167, miR319, miR156, miR165, miR169, miR168, miR171, miR393, miR159, miR158, miR394, miR396, and miR397 were upregulated, whereas miR398 was downregulated (Liu et al. 2008). In salt-sensitive maize genotypes, increased levels of miR162, miR168, miR395, and miR474 were observed upon exposure to salt stress, whereas salt-tolerant maize line NC286 roots exhibited decreased expression of miR156, miR164, miR167, and miR396 (Ding et al. 2009). In *P. vulgaris*, the expression of miRS1 and miR159.2 was found to increase when subjected to salinity stress (Arenas-Huertero et al. 2009). Conversely, in *P. trichocarpa* under salinity conditions, miR482.2 and miR1450 were downregulated, while miR530a, miR1445, miR1446a-e, miR1447, and miR171 l-n were upregulated (Liu et al. 2008). Using next-generation sequencing, another study examined the differential expression of 50 miRNAs in soybeans exposed to salinity (Li et al. 2011). Furthermore, during salt stress, higher expression of miR159 and miR319 was observed in artichoke tissues (De Paola et al. 2012).

Our group has comprehensively profiled the changes in miRNAs in different rice varieties (Goswami et al. 2017). The mapping of the rice miRNome in naturally salt-tolerant pokkali and salt-tolerant pusa basmati roots provided new insights into understanding the molecular mechanisms operative in rice cultivars showing contrasting responses to salt stress (Goswami et al. 2020). The analysis revealed the modulation of many biological pathways, which are involved in salt-tolerance and play an important role in plant phenotype and physiology. Osa-miR396c was found to be downregulated under saline conditions in an ABA-dependent manner, and its overexpression exacerbated sensitivity to salt stress (Gao et al. 2011). In rice, miR408 expression decreased under salinity and drought stress, while its overexpression enhanced the plant's ability to tolerate salt stress (Mutum et al. 2013; Ma et al. 2015). Studies on wheat subjected to salinity revealed an upregulation of miR171 and miR393 (Gupta et al. 2014; Wang et al. 2014). Activation of the MYB transcription factor, which regulates miR171, was observed under salt stress in *Arabidopsis*, wheat, and barley (Liu et al. 2008; Wang et al. 2014; Deng et al. 2015). Furthermore, overexpression of Osa-miR393 increased salt tolerance, highlighting its regulatory role in enhancing salinity tolerance in *Arabidopsis* (Gao et al. 2011).

The Ath-miR172c, which regulates AP2-like transcription factors, is induced by salt stress (Li et al. 2016). Additionally, miR319 affects salt exclusion through HKT transporters (Zhou et al. 2013). In *T. salsuginea*, miR156a, miR156k, miR162a, miR164a, miR171c, and miR824 exhibit reduced expression during salt stress, whereas miR160a, miR16a, miR395a, and miR395b show increased expression in roots. Conversely, leaves under salt stress exhibit increased expression of miR160a, miR164a, miR168a, miR359a, and miR359b, and decreased expression of miR156k, miR162a, miR171c, and miR824 (Zhou et al. 2013). These observations collectively underscore the significant role of the miRNA family in responding to salt stress.

Table 15.1 List of representative miRNAs identified from different plant species having a role in salinity stress response and plant growth

miRNA family	Target genes	Plant	References
miR156	*SBP*	*Arabidopsis*	Liu et al. (2008)
	SBP-domain protein 6	*Zea mays*	Ding et al. (2009)
miR159	*MYB*	*Arabidopsis*	Liu et al. (2008)
	MADS box interactor	*Zea mays*	Ding et al. (2009)
miR160	*Two-component response regulator ARR11*		Ding et al. (2009)
miR162	*Endoribonuclease Dicer*		Ding et al. (2009)
miR164	*NAC1*		Ding et al. (2009)
miR166	*HD-ZIP rolled leaf1*		Ding et al. (2009)
miR167	*ARF8*	*Arabidopsis*	Liu et al. (2008)
	Auxin response factor 8	*Zea mays*	Ding et al. (2009)
miR168	*AGO1*	*Arabidopsis*	Liu et al. (2008)
	AGO1-1	*Zea mays*	Ding et al. (2009)
miR169	*NFY/MtHAP2-1*	*Arabidopsis*	Liu et al. (2008)
miR171	*SCL*		Liu et al. (2008)
	Spry-2		Ding et al. (2009)
miR319	*TCP*		Liu et al. (2008)
	Transcription factor GAMyb	*Zea mays*	Ding et al. (2009)
	NAC	*Bentgrass*	Zhou et al. (2013)
miR393	*TIR1/F-box AFB*	*Arabidopsis*	Liu et al. (2008)
miR395	*ATP sulfurylase*	*Zea mays*	Ding et al. (2009)
miR396	*GRF*	*Arabidopsis*	Liu et al. (2008)
	Cytochrome oxidase subunit I	*Zea mays*	Ding et al. (2009)
miR398	*CSD1 and CSD2*	*Oryza sativa*	Lu et al. (2010)
miR820	*DRM2*		Sharma et al. (2015)
miR172	*IDS1*		Cheng et al. (2021)
miR172a	*GMA*	*Glycine max*	Panda et al. (2015)
miR399	*PHOSPHATE 2 (PHO2)*	*Arabidopsis*	Baek et al. (2016)
MiR393	*TIR1/AFBs*	*Oryza sativa*	Liu et al. (2017)
miR395	*SULTR2;1 and (APS)*	*Arabidopsis*	Kawashima et al. (2009), Liang et al. (2010)
miR160	*ARFs*		Yang et al. (2019)
miR160b	*GhARF17/18*		Gao et al. (2022)
miR156	*SPL*		Cui et al. (2014)
miR398	*CSD1 and CSD2*		Dugas and Bartel (2008)
miR319	*PvPCF5*		Liu et al. (2019a, b)
miR394	*LCR*		Song et al. (2013)
miR408	*CSD1 and CSD2*		Ma et al. (2015)

15.5 miRNA Technology for Increasing Rice Grain Yield Under Saline Conditions

Since salinity adversely affects plant growth, development, and agricultural production on a worldwide scale, it is a substantial barrier to plant breeding (Wani et al. 2020). It is crucial to genetically modify rice to withstand environmental challenges to increase rice crop output potential (Sun et al. 2019). Fine-tuning the miRNA- target interaction is one of the potent biotechnology strategies to enhance the agronomic performance of crops (Khraiwesh et al. 2012; Basso et al. 2019).

15.5.1 Effects of the miRNA on Salt Stress Regulation

15.5.1.1 miR156

The initial discovery of miRNA156 in plants revealed its selective targeting of *SPL* transcription factors containing the SBP-box domain. Interaction between miR156 and *SPL* leads to diverse phenotypic alterations, including enhancements in plant height, leaf morphology, floral organ development, yield, and stress responses. Overexpression of miR156, and consequently modulation of the miR156-SPL-DFR pathway, has been associated with improved salt tolerance in transgenic *Arabidopsis* (Cui et al. 2014). The CaMV 35S promoter and the stress-responsive ZmRab17 promoter have both been utilized to overexpress Zma-miR156 in tobacco, resulting in increased salt tolerance in these transgenic plants (Kang et al. 2020).

In apple plants, overexpression of miR156a reduces salt tolerance, whereas overexpression of its target gene MdSPL13 enhances salt tolerance (150 mM NaCl). Additionally, the miR156-SPL module can upregulate MdWRKY100 to further regulate salt tolerance in apples, as evidenced by RLM-5′RACE analysis linking miR156 and MdSPL13 (Ma et al. 2021). Increased miR156 expression in alfalfa alters ion balance and appears to limit Na^+ uptake during severe salt stress (Arshad and Hannoufa 2022). The miR156-SPL module exerts control over salt stress tolerance by promoting MdWRKY100 expression, underscoring the potential of miRNA and its targets as indicators of salt tolerance capacity (Ma et al. 2021). In *Arabidopsis*, miR156 overexpression enhances salinity tolerance through regulation of the miR156-SPL-DFR pathway (Cui et al. 2014).

Furthermore, differences in salt tolerance between *Medicago truncatula* and *M. sativa* have been linked to distinct expression patterns of specific miRNAs such as miR166 and miR156 (Long et al. 2015).

15.5.1.2 miR397

In *Arabidopsis*, overexpression of SvMIR397 or AtMIR397 could suppress the expression of the target gene *LACCASE* (*LAC*). In particular it considerably decreased *LAC2*, *LAC4*, and *LAC17*, which resulted in lower lignin concentration and increased susceptibility to salt stress (Nguyen et al. 2020). Laccases are glycoproteins with a lot of copper atoms that are found in plants. High NaCl concentrations cause the expression of the laccase gene in the roots of tomato, maize, and

Arabidopsis (Cai et al. 2006). When salt stress is applied to artichoke roots, miR397a expression is low, which may cause *laccase* expression to be increased (De Paola et al. 2012). *Arabidopsis*'s constitutive expression of *Sv-miR397* resulted in down-regulation of the target gene *LAC2/4/17*. *SvMIR397* transformants displayed decreased rosette fresh weight and poor salt tolerance when subjected to 150 mM NaCl for seven days (Nguyen et al. 2020).

15.5.1.3 miR172

One of the downstream targets of miR172, *IDS1*, interacts with the promoters of genes involved in ROS scavenging, thereby inhibiting their transcription. This interaction underscores the regulatory role of miR172-IDS1 in maintaining ROS balance during salt stress in staple cereal crops such as rice and wheat (Cheng et al. 2021). In soybean, enhanced expression of miR172a suppresses the expression of its target gene *SSAC1* (*salt suppressed AP2 domain-containing*), leading to improved salt tolerance in transgenic plants (Pan et al. 2016). Conversely, reducing miR172c expression using short tandem target mimic (STTM) technology in soybeans may diminish their salt tolerance (Sahito et al. 2017). Suppression of miR172a/b through STTM and CRISPR/Cas9 technologies intensifies salt sensitivity in plants, highlighting the critical role of miR172a/b in salt stress response (Cheng et al. 2021).

15.5.1.4 miR399

In plants, miR399 represents one of the most ancient and highly conserved families of miRNAs. It targets the gene, *phosphate 2* (*PHO2*), which encodes an E2 ubiquitin-conjugating enzyme involved in ubiquitin-mediated protein degradation. The primary role of miR399 is to regulate the balance of inorganic phosphate (Pi) levels by reducing the expression of PHO2. The enzyme, PHO2, in turn, modulates the expression of phosphate starvation-induced genes (PSIs) and prevents excessive Pi accumulation. Overexpression of miR399f in Arabidopsis enhances tolerance to salinity (Baek et al. 2016).

15.5.1.5 miR393

miR393, a highly conserved miRNA in plants, targets the auxin receptor genes *transport inhibitor response1* (*TIR1*) and *auxin signaling F box protein* (*AFB*), which belong to a specific subgroup of F-box proteins. In addition to its involvement in rice tillering, miR393 plays roles in root development, overall growth regulation, and modulation of biotic and abiotic stress responses (Liu et al. 2017). Upon exposure to saline-alkali conditions, miR393a expression is reduced, whereas miR393b expression remains unaffected (Gao et al. 2011). miR393 is implicated in salt stress responses through its interaction with the scaffold protein RACK1A, which mediates ABA signaling pathways. The targeting of *TIR1/AFB2* by miR393 highlights RACK1 as a negative regulator of ABA phytohormone signaling. miR393 functions in salt acclimation by modulating the antagonistic interplay between ABA-mediated salt stress responses and auxin hormones (Denver and Ullah 2019).

Studies have identified 18 miRNAs exhibiting altered expression patterns in salt-tolerant maize lines and 25 in salt-sensitive lines, indicating differential regulation under stress conditions (Ding et al. 2009; Kumar et al. 2018). miR393 is conserved across various plant species, with rice possessing two members: osa-miR393a and osa-miR393b (Gao et al. 2011). Under salt and alkaline stress, over-expression of miR393a significantly alters its expression, contrasting with the stable expression of miR393b (Islam et al. 2022). Interestingly, transgenic plants overexpressing miR393 exhibit increased susceptibility to salt stress compared to non-transgenic plants (Gao et al. 2011). Multiple studies underscore how miR393 can negatively impact salt tolerance by modulating its expression levels in rice and *Arabidopsis* (Gao et al. 2011; Xia et al. 2012; Iglesias et al. 2014; Long et al. 2017).

15.5.1.6 miR395

The miR395 family is an old and highly conserved group of miRNAs, initially identified as being crucial for nutrient assimilation and absorption. The target genes of miR395 are *SULTR2;1* and *ATP-sulfurylases* (*APS*). miR395 participates in the transport and assimilation of sulphate (Kawashima et al. 2009; Liang et al. 2010). In *Arabidopsis*, Ath-miR395c overexpression inhibited seed germination in situations of excessive salt or dehydration, but Ath-miR395e overexpression promoted seed germination in saline environments (Kim et al. 2010). Target gene expression analysis indicates that miR395c downregulates the expression of *APS1, APS3, APS4*, and *SULTR2;1*, whereas miR395e affects the expression of *APS3* and *SULTR2;1* specifically (Kim et al. 2010).

15.5.1.7 miR396

Recent research has demonstrated that miR396 can effectively regulate salt tolerance. Under environmental cues such as exogenous ABA and salinity, miR399f tends to be up-regulated. In *Arabidopsis*, overexpression of miR399f results in increased tolerance to salt and ABA, although with heightened sensitivity to drought (Baek et al. 2016). When maize is subjected to salt stress, the Zma-miR396 family, which targets genes encoding *cytochrome oxidase subunit I*, was suppressed (Ding et al. 2009). In *P. cathayana* experiencing salt stress, miR396f shows decreased expression while its target genes encoding GRF are up-regulated. Conversely, salt treatment induces a significant increase in miR396 expression in *Arabidopsis* (Liu et al. 2008).

Moreover, Peu-miR396b exhibits upregulated expression levels in leaves, but not in roots, in response to salt stress (Liu et al. 2009). Overexpression of osa-miR396c in *Arabidopsis* and rice renders these plants sensitive to salinity (Gao et al. 2010). Conversely, overexpression of Osa-miR396c in creeping bentgrass enhances salinity tolerance, although it significantly decreases the expression of AsGRF3-6 in transgenic plants (Yuan et al. 2019). This down-regulation of AsGRF3-6 leads to notable alterations in growth patterns, including reduced shoot and root biomass, leaf area, internodes, leaf veins, and epidermal cells per unit area. Furthermore, these transgenic plants exhibit enhanced salt tolerance due to improved water retention, higher chlorophyll content, enhanced cell membrane integrity, and increased exclusion of Na^+ ions under high salinity conditions (Yuan et al. 2019).

15.5.1.8 miR417

miR417 targets *helicase domain protein* which has role in stress response. Research revealed that miR417 negatively affects *Arabidopsis* plant survival rates and germination in high saline environments. *Arabidopsis* plants overexpressing miR417 showed delayed seed germination under high salt concentrations or ABA (Jung and Kang 2007).

15.5.1.9 miR160

miR160 specifically targets genes encoding B3 DNA-binding proteins and ARFs. ARFs influence plant development, growth, and adaptation to environmental changes (Achard et al. 2006). Auxins in plants are critical for processes such as lateral root formation, tropisms, and apical dominance, regulating cell division, elongation, and differentiation (Leyser 2006). Studies have demonstrated that miR160 can modulate responses to salt stress (Zhang et al. 2021). In *Arabidopsis*, the miR160 family regulates the *auxin responsive factor* (*ARF*) genes (Yang et al. 2019). Expression of Vun-miR160a increases under salt stress conditions (Paul et al. 2011). Cotton plants with silenced Ghr-miR160b are more vulnerable to salt stress. Conversely, enhancing miR160b expression or suppressing the target genes GhARF17/18 enhances salt stress resistance, promoting plant growth and increasing root dry weight (Gao et al. 2022). miR160 plays a crucial role in root tip growth and gravity responses through its regulation of target genes *ARF10, ARF16*, and *ARF17* (Wang et al. 2005). miR167 targets *ARF6* and *ARF8*, which are positive regulators of adventitious root growth and lateral root development (Gutierrez et al. 2009).

15.5.1.10 miR398

The first miRNA to be specifically associated with plant resistance to environmental stress was miR398. Two genes, *CSD1* and *CSD2* in *Arabidopsis* that encode Copper/Zinc (Cu/Zn) and superoxide dismutases (SODs), are specifically targeted by miR398 (Dugas and Bartel 2008). It was found that transgenic lines were better able to withstand oxidative stress than wild-type plants when miR398 was downregulated (Sunkar et al. 2006).

15.5.1.11 miR169

miR-169 comprises a diverse miRNA family encompassing 17 known members, characterized by nine distinct mature sequences exhibiting minimal variation (Zhao et al. 2007). In their investigation of miRNA expression patterns in response to salinity stress in *Oryza sativa*, researchers identified two salt-responsive miRNAs, namely miR169n and miR169o. These miRNAs were clustered together and arose from a recent tandem duplication event. Zhao et al. (2007) demonstrated that members of the salt-induced miR169 family target the *NF-YA* transcription factor, leading to its cleavage following exposure to high salinity levels.

In another study involving *P. euphratica* subjected to salt stress, significant changes were observed in the expression levels of three miRNAs: miR164, miR166, and miR169 (Si et al. 2014). Specifically, miR169g and miR169n from the miR169 family exhibited altered expression under salt stress conditions (Luan et al. 2015). Furthermore, Zhao et al. (2009) identified an upstream cis-acting region responsive to ABA that regulates the expression of miR169n, suggesting a role for this stress-responsive hormone in modulating miR169n expression.

15.5.1.12 miR319

The first miRNA to be examined by forward genetics was miR319. The genes that miR319 targets belong to the *TCP* (*teosinte branched/cycloidea/proliferating cell factors*) TF family (Palatnik et al. 2003). Two studies have shown that miR319 is a positive regulator of salinity tolerance through transgenic experiments. Overexpression of the osa-miR319a, b genes in creeping bentgrass and switchgrass, respectively, enhanced salt tolerance of transgenic plants (Zhou et al. 2013; Liu et al. 2019a, b). miR319 expression knockdown using STTM technology can reduce transgenic plants' tolerance to salt (Liu et al. 2019a, b). In switch grass, miR319 transgenic plants suppress the expression of the target gene *PvPCF5*, causing elevated levels of ethylene and salt tolerance (Liu et al. 2019a, b). Overexpression of ath-miR393a in tobacco and rice improves salt tolerance in transgenic plants (Long et al. 2017; Zhao et al. 2019). The expression of miR159 and miR319 increased in the artichoke tissues upon exposure to saline solution (Finetti Sialer et al. 2012). A study by Zhou et al. (2013) also showed transgenic *A. stolonifera* overexpressing rice miR319, attained improved tolerance to salinity and drought.

15.5.1.13 miR394

Arabidopsis possesses only two variants within the miR394 subfamily, namely miR394a and miR394b. miR394 plays a crucial role in regulating *leaf curling responsiveness* (LCR), a member of the F-box protein family involved in modulating ABA signaling and the response to abiotic stresses. The F-box protein acts as a component of the SKP1-Cullin/CDC53-F-box (SCF) complex, a polyprotein ubiquitin ligase that targets specific substrates for degradation by the proteasome. Expression of miR394a/b is induced under conditions of salt stress and ABA exposure. *Arabidopsis* plants overexpressing miR394a/b or lacking functional LCR exhibit heightened susceptibility to salt stress (Song et al. 2013).

15.5.1.14 miR408

Under high salinity conditions, miR408 exhibited increased expression in *A. thaliana*, which resulted in the down-regulation of its target gene *cupredoxin*. Overexpression of miR408 in these plants extended both primary and lateral root lengths and boosted the expression of CSD1 and CSD2, thereby enhancing their tolerance to salinity-induced oxidative stress (Ma et al. 2015). Conversely, rice plants experiencing salinity and drought stresses demonstrated reduced levels of miR408 expression, but augmenting miR408 expression improved their tolerance to salinity stress (Ma et al. 2015).

In *S. miltiorrhiza* exposed to salt stress, miR408 expression also increased. Overexpressing miR408 facilitated seed germination and mitigated ROS production under salinity stress by enhancing the expression and activity of antioxidative enzymes such as NbSOD, NbPOD, and NbCAT (Guo et al. 2018). Similarly, in *T. aestivum*, TaCLP1, a target gene of miR408, contributed significantly to coping with severe salinity, as evidenced by enhanced yeast cell proliferation under high salt conditions upon TaCLP1 overexpression (Feng et al. 2013).

Under salt stress conditions, miR408 was found to be down-regulated in *O. sativa*, while its target genes, a helicase *DSHCT* and *plastocyanin-like* were up-regulated. The regulation of miR408 expression varies inconsistently among different plant species under salinity stress (Macovei et al. 2012). In *M. truncatula*, miR408 expression was up-regulated in both roots and shoots during drought stress, leading to downregulation of plastocyanin transcripts and establishing a direct link between drought response and copper homeostasis in this species (Trindade et al. 2010).

In *T. aestivum*, tae-miR408 expression patterns were induced under salt stress and phosphate deprivation, with expression gradually decreasing upon recovery to normal conditions (Bai et al. 2018). Overexpressing miR408 in *Arabidopsis* improved tolerance to salinity, cold, and oxidative stress but increased sensitivity to drought and osmotic stress. Similar improvements in stress resistance were observed in *N. benthamiana* when heterologously expressing miR408 from *Salvia miltiorrhiza* (Ma et al. 2015; Guo et al. 2018). Additionally, miRNA profiling revealed significant roles for miR408 in shoot and root tissues of *S. viridis* exposed to salinity (Pegler et al. 2020), and miR408 expression levels were altered in Kentucky bluegrass following exposure to salt stress (Luo et al. 2020).

15.6 miRNA-Mediated Regulatory Network Responsive to Salt Stress

miRNAs influence how plants respond to abiotic stress by regulating crucial components of a complicated gene network such as transcription factors (Fig. 15.1). Different plant miRNAs target distinctive mRNAs that might belong to the same gene family (Zhang and Wang 2015). Transcription factors are frequently the target of miRNAs, which are essential for completing plants' lifecycle. Several regulatory proteins, such as MADS box and zinc-finger proteins are involved in plant responses to salt stress (Noman et al. 2019).

miR393a has been shown to target *F-box* and *basic helix-loop-helix* proteins, exhibiting increased interaction under high salinity conditions (Jones-Rhoades and Bartel 2004). Beyond transcription factors, miRNA-mediated regulation extends to various physiological and metabolic pathways. For instance, genes encoding enzymes like cytochrome oxidase and NADP-dependent malic enzymes in plants respond to salt stress and are likely targets of miRNA (Jagannadham et al. 2019). Recently, several crucial physiological genes have been identified as targets of stress-responsive miRNAs, including ARFs, AA/IAA-ARF-dimerization, Cytochrome P450, Chlorophyll A-B binding protein, NADPH-cytochrome P450 reductase, Homeobox leucine-zipper proteins, NF-YA, and MYBs (Naik and Varadahalli 2020). miR160 specifically targets ARFs and B3 DNA-binding domain

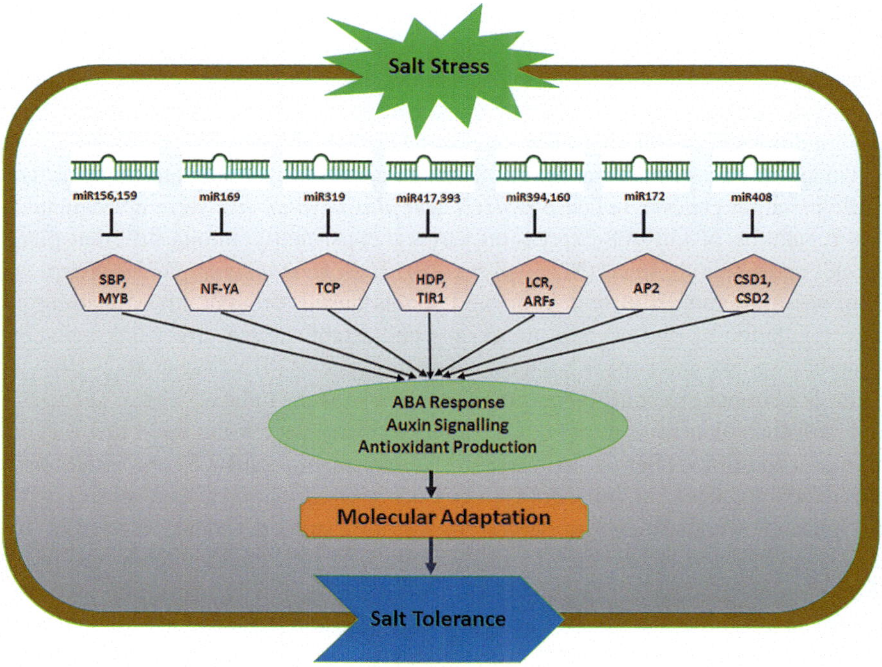

Fig. 15.1 Pictorial representation of miRNAs-mRNA gene interaction in response to salt stress

proteins, and its overexpression in transgenic plants has been linked to enhanced salt tolerance (Finet et al. 2013).

During salt stress, plants experience an accumulation of ROS due to oxidative stress. The primary cellular defense against oxidative stress involves molecular oxygen and H_2O_2, which are generated when superoxide radicals are dismutated by SOD. The activities of Cu/Zn-SOD and Laccases (LACs) are downregulated by microRNAs (miR397, miR398, miR408, and miR528), which modulate ROS levels and regulate the availability of copper ions (Yang and Guo 2018). Additionally, miR159, miR828, and miR858 are implicated in the regulation of numerous MYB transcription factors (Pang et al. 2009; Guan et al. 2014).

15.7 Conclusion

Studying the role of miRNAs is essential for enhancing crop breeding to achieve superior quality, increased yield, and enhanced stress resilience. miRNAs serve as crucial regulators of gene expression post-transcriptionally, exerting control over diverse physiological processes, including plant growth and development in response to environmental challenges. These regulatory molecules control specific target genes, thereby influencing various cellular functions.

miRNAs such as miR156, miR159, miR164, miR167, miR397, miR393, miR398, miR396, and miR408 are implicated as mediators facilitating communication between salinity stress adaptation and enhanced crop productivity. Additionally, miRNAs can serve as biomarkers for identifying and monitoring environmental stresses broadly or specifically. For instance, miR408, a highly conserved miRNA across terrestrial plants, has been proposed as a biomarker for both biotic and abiotic stresses in major crop species due to its regulatory roles in drought, salinity, temperature responses, and management of biotic stresses.

Therefore a comprehensive understanding of miRNA-directed gene targets in model agricultural plants would pave the way for novel strategies to improve crop yields effectively.

References

Abdullah Z, Khan MA, Flowers T (2001) Causes of sterility in seed set of rice under salinity stress. J Agron Crop Sci 187(1):25–32

Abobatta WF (2020) Plant responses and tolerance to extreme salinity: learning from halophyte tolerance to extreme salinity. In: Salt and drought stress tolerance in plants: signaling networks and adaptive mechanisms. Springer, Cham, pp 177–210

Achard P, Cheng H, De Grauwe L, Decat J, Schoutteten H, Moritz T, Van Der Straeten D, Peng J, Harberd NP (2006) Integration of plant responses to environmentally activated phytohormonal signals. Science 311(5757):91–94

Aref F, Rad HE (2012) Physiological characterization of rice under salinity stress during vegetative and reproductive stages. Indian J Sci Technol 5(4):2578–2586

Arenas-Huertero C, Pérez B, Rabanal F, Blanco-Melo D, De la Rosa C, Estrada-Navarrete G, Sanchez F, Covarrubias AA, Reyes JL (2009) Conserved and novel miRNAs in the legume Phaseolus vulgaris in response to stress. Plant Mol Biol 70:385–401

Arif Y, Singh P, Siddiqui H, Bajguz A, Hayat S (2020) Salinity induced physiological and biochemical changes in plants: an omic approach towards salt stress tolerance. Plant Physiol Biochem 156:64–77

Arshad M, Hannoufa A (2022) Alfalfa transcriptome profiling provides insight into mir156-mediated molecular mechanisms of heat stress tolerance. Genome 65(6):315–330

Ashraf M, Foolad M (2005) Pre-sowing seed treatment-A shotgun approach to improve germination, plant growth, and crop yield under saline and non-saline conditions. Adv Agron 88:223–271

Ashraf M, Harris P (2004) Potential biochemical indicators of salinity tolerance in plants. Plant Sci 166(1):3–16

Baek D, Chun HJ, Kang S, Shin G, Park SJ, Hong H, Kim C, Kim DH, Lee SY, Kim MC (2016) A role for *Arabidopsis* miR399f in salt, drought, and ABA signaling. Mol Cells 39(2):111–118

Bai Q, Wang X, Chen X, Shi G, Liu Z, Guo C, Xiao K (2018) Wheat miRNA TaemiR408 acts as an essential mediator in plant tolerance to Pi deprivation and salt stress via modulating stress-associated physiological processes. Front Plant Sci 9:499

Basso MF, Ferreira PCG, Kobayashi AK, Harmon FG, Nepomuceno AL, Molinari HBC, Grossi-de-Sa MF (2019) Micro RNA s and new biotechnological tools for its modulation and improving stress tolerance in plants. Plant Biotechnol J 17(8):1482–1500

Begum Y (2022) Regulatory role of microRNAs (miRNAs) in the recent development of abiotic stress tolerance of plants. Gene 821:146283

Cai X, Davis EJ, Ballif J, Liang M, Bushman E, Haroldsen V, Torabinejad J, Wu Y (2006) Mutant identification and characterization of the laccase gene family in *Arabidopsis*. J Exp Bot 57(11):2563–2569

Cheng X, He Q, Tang S, Wang H, Zhang X, Lv M, Liu H, Gao Q, Zhou Y, Wang Q (2021) The miR172/IDS1 signaling module confers salt tolerance through maintaining ROS homeostasis in cereal crops. New Phytol 230(3):1017–1033

Cui LG, Shan JX, Shi M, Gao JP, Lin HX (2014) The miR156-SPL 9-DFR pathway coordinates the relationship between development and abiotic stress tolerance in plants. Plant J 80(6):1108–1117

Daneshmand F, Arvin MJ, Kalantari KM (2010) Physiological responses to NaCl stress in three wild species of potato i*n vitro*. Acta Physiol 32:91–101

Daszkowska-Golec A (2011) *Arabidopsis* seed germination under abiotic stress as a concert of action of phytohormones. OMICS 15(11):763–774

De Paola D, Cattonaro F, Pignone D, Sonnante G (2012) The miRNAome of globe artichoke: conserved and novel micro RNAs and target analysis. BMC Genomics 13(1):1–14

Deng P, Wang L, Cui L, Feng K, Liu F, Du X, Tong W, Nie X, Ji W, Weining S (2015) Global identification of microRNAs and their targets in barley under salinity stress. PLoS One 10(9):e0137990

Denver JB, Ullah H (2019) miR393s regulate salt stress response pathway in *Arabidopsis* thaliana through scaffold protein RACK1A mediated ABA signaling pathways. Plant Signal Behav 14(6):1600394

Dietz K-J, Tavakoli N, Kluge C, Mimura T, Sharma S, Harris G, Chardonnens A, Golldack D (2001) Significance of the V-type ATPase for the adaptation to stressful growth conditions and its regulation on the molecular and biochemical level. J Exp Bot 52(363):1969–1980

Ding D, Zhang L, Wang H, Liu Z, Zhang Z, Zheng Y (2009) Differential expression of miRNAs in response to salt stress in maize roots. Ann Bot 103(1):29–38

Djaman K, Mel V, Boye A, Diop L, Manneh B, El-Namaky R, Koudahe K, Futakuchi K (2020) Rice genotype and fertilizer management for improving rice productivity under saline soil conditions. Paddy Water Environ 18:43–57

Dolatabadian A, Sanavy SAMM, Ghanati F (2011) Effect of salinity on growth, xylem structure and anatomical characteristics of soybean. Not Sci Biol 3(1):41–45

Dugas DV, Bartel B (2008) Sucrose induction of *Arabidopsis* miR398 represses two Cu/Zn superoxide dismutases. Plant Mol Biol 67(4):403–417

Farooq M, Hussain M, Ul-Allah S, Siddique KH (2019) Physiological and agronomic approaches for improving water-use efficiency in crop plants. Agric Water Manag 219:95–108

Feng H, Zhang Q, Wang Q, Wang X, Liu J, Li M, Huang L, Kang Z (2013) Target of tae-miR408, a chemocyanin-like protein gene (TaCLP1), plays positive roles in wheat response to high-salinity, heavy cupric stress and stripe rust. Plant Mol Biol 83(4-5):433–443

Finet C, Berne-Dedieu A, Scutt CP, Marlétaz F (2013) Evolution of the ARF gene family in land plants: old domains, new tricks. Mol Biol Evol 30(1):45–56

Finetti Sialer M, Catalano D, Fracchiolla A, Morgese A, Pignone D, Sonnante G, De Paola D, Cattonaro F (2012) MicroRNAs in the globe artichoke: detection and analysis. VIII International Symposium on Artichoke, Cardoon and their Wild Relatives 983

Flowers T, Yeo A (1981) Variability in the resistance of sodium chloride salinity within rice (Oryza sativa L.) varieties. New Phytol 88(2):363–373

Flowers T, Troke P, Yeo A (1977) The mechanism of salt tolerance in halophytes. Annu Rev Plant Physiol 28(1):89–121

Fu H-H, Luan S (1998) AtKUP1: a dual-affinity K+ transporter from *Arabidopsis*. Plant Cell 10(1):63–73

Gao P, Bai X, Yang L, Lv D, Li Y, Cai H, Ji W, Guo D, Zhu Y (2010) Over-expression of osa-MIR396c decreases salt and alkali stress tolerance. Planta 231(5):991–1001

Gao P, Bai X, Yang L, Lv D, Pan X, Li Y, Cai H, Ji W, Chen Q, Zhu Y (2011) Osa-MIR393: a salinity-and alkaline stress-related microRNA gene. Mol Biol Rep 38(1):237–242

Gao Z, Ma C, Zheng C, Yao Y, Du Y (2022) Advances in the regulation of plant salt-stress tolerance by miRNA. Mol Biol Rep 49(6):5041–5055

Gholipoor M, Soltani A, Shekari F, Shekari F (2002) Effects of salinity on water use efficiency and its components in chickpea (*Cicer arietinum* L.). Acta Agron 50(2):127–134

Gill SS, Tuteja N (2010) Reactive oxygen species and antioxidant machinery in abiotic stress tolerance in crop plants. Plant Physiol Biochem 48(12):909–930

Goel A, Sheoran I (2003) Lipid peroxidation and peroxide-scavenging enzymes in cotton seeds under natural ageing. Biol Plant 46:429–434

Gong H, Chen K, Zhao Z, Chen G, Zhou W (2008) Effects of silicon on defense of wheat against oxidative stress under drought at different developmental stages. Biol Plant 52:592–596

Goswami K, Tripathi A, Sanan-Mishra N (2017) "Comparative miRomics of Salt-Tolerant and Salt-Sensitive Rice" *Journal of Integrative*. Bioinformatics 14(1):20170002

Goswami K, Mittal D, Gautam B, Sopory SK, Sanan-Mishra N (2020) Mapping the Salt Stress-Induced Changes in the Root miRNome in Pokkali Rice. *Biomolecules 10*:498

Goud PB, Kachole M (2011) Effect of hydrogen peroxide on protein content and catalase activity in *Cajanus cajan* (L.) Millsp. leaves. Int J Curr Res 3:171–268

Guan X, Pang M, Nah G, Shi X, Ye W, Stelly DM, Chen ZJ (2014) miR828 and miR858 regulate homoeologous MYB2 gene functions in *Arabidopsis* trichome and cotton fibre development. Nat Commun 5(1):3050

Guo X, Niu J, Cao X (2018) Heterologous expression of Salvia miltiorrhiza microRNA408 enhances tolerance to salt stress in *Nicotiana benthamiana*. Int J Mol Sci 19(12):3985

Gupta OP, Meena NL, Sharma I, Sharma P (2014) Differential regulation of microRNAs in response to osmotic, salt and cold stresses in wheat. Mol Biol Rep 41(7):4623–4629

Gutierrez L, Bussell JD, Pacurar DI, Schwambach J, Pacurar M, Bellini C (2009) Phenotypic plasticity of adventitious rooting in *Arabidopsis* is controlled by complex regulation of auxin response factor transcripts and microRNA abundance. Plant Cell 21(10):3119–3132

Hakim M, Juraimi A, Begum M, Hanafi M, Ismail MR, Selamat A (2010) Effect of salt stress on germination and early seedling growth of rice (Oryza sativa L.). Afr J Biotechnol 9(13):1911–1918

Hasegawa PM, Bressan RA, Zhu J-K, Bohnert HJ (2000) Plant cellular and molecular responses to high salinity. Annu Rev 51(1):463–499

Hazman M, Hause B, Eiche E, Nick P, Riemann M (2015) Increased tolerance to salt stress in OPDA-deficient rice allene oxide cyclase mutants is linked to an increased ROS-scavenging activity. J Exp Bot 66(11):3339–3352

Hopmans JW, Qureshi A, Kisekka I, Munns R, Grattan S, Rengasamy P, Ben-Gal A, Assouline S, Javaux M, Minhas P (2021) Critical knowledge gaps and research priorities in global soil salinity. Adv Agron 169:1–191

Hossain MS, Persicke M, ElSayed AI, Kalinowski J, Dietz K-J (2017) Metabolite profiling at the cellular and subcellular level reveals metabolites associated with salinity tolerance in sugar beet. J Exp Bot 68(21-22):5961–5976

Hussain S, Zhang J-H, Zhong C, Zhu L-F, Cao X-C, Yu S-M, Bohr JA, Hu J-J, Jin Q-Y (2017) Effects of salt stress on rice growth, development characteristics, and the regulating ways: a review. J Integr Agric 16(11):2357–2374

Iglesias MJ, Terrile MC, Windels D, Lombardo MC, Bartoli CG, Vazquez F, Estelle M, Casalongué CA (2014) MiR393 regulation of auxin signaling and redox-related components during acclimation to salinity in *Arabidopsis*. PLoS One 9(9):e107678

Isayenkov SV, Maathuis FJ (2019) Plant salinity stress: many unanswered questions remain. Front Plant Sci 10:80

Islam W, Waheed A, Naveed H, Zeng F (2022) MicroRNAs mediated plant responses to salt stress. Cells 11(18):2806

Jagannadham PTK, Muthusamy SK, Chidambaranathan P (2019) Micromics: a novel approach to understand the molecular mechanisms in plant stress tolerance. In: Recent approaches in omics for plant resilience to climate change. Springer, Cham, pp 93–108

Jahromi F, Aroca R, Porcel R, Ruiz-Lozano JM (2008) Influence of salinity on the in vitro development of Glomus intraradices and on the in vivo physiological and molecular responses of mycorrhizal lettuce plants. Microb Ecol 55:45–53

Javid MG, Sorooshzadeh A, Moradi F, Modarres Sanavy SAM, Allahdadi I (2011) The role of phytohormones in alleviating salt stress in crop plants. Aust J Crop Sci 5(6):726–734

Jones-Rhoades MW, Bartel DP (2004) Computational identification of plant microRNAs and their targets, including a stress-induced miRNA. Mol Cell 14(6):787–799

Jung HJ, Kang H (2007) Expression and functional analyses of microRNA417 in *Arabidopsis* thaliana under stress conditions. Plant Physiol Biochem 45(10-11):805–811

Jung JK, McCouch S (2013) Getting to the roots of it: genetic and hormonal control of root architecture. Front Plant Sci 4:186

Kang T, Yu C-Y, Liu Y, Song W-M, Bao Y, Guo X-T, Li B, Zhang H-X (2020) Subtly manipulated expression of ZmmiR156 in tobacco improves drought and salt tolerance without changing the architecture of transgenic plants. Front Plant Sci 10:1664

Kawashima CG, Yoshimoto N, Maruyama-Nakashita A, Tsuchiya YN, Saito K, Takahashi H, Dalmay T (2009) Sulphur starvation induces the expression of microRNA-395 and one of its target genes but in different cell types. Plant J 57(2):313–321

Khan M, Panda S (2008) Alterations in root lipid peroxidation and antioxidative responses in two rice cultivars under NaCl-salinity stress. Acta Physiol 30:81–89

Khraiwesh B, Zhu J-K, Zhu J (2012) Role of miRNAs and siRNAs in biotic and abiotic stress responses of plants. Biochim Biophys Acta 1819(2):137–148

Kim JY, Lee HJ, Jung HJ, Maruyama K, Suzuki N, Kang H (2010) Overexpression of microRNA395c or 395e affects differently the seed germination of *Arabidopsis* thaliana under stress conditions. Planta 232(6):1447–1454

Korres NE, Loka DA, Gitsopoulos TK, Varanasi VK, Chachalis D, Price A, Slaton NA (2022) Salinity effects on rice, rice weeds, and strategies to secure crop productivity and effective weed control. A review. Agron Sustain Dev 42(4):58

Krishnamurthy P, Ranathunge K, Nayak S, Schreiber L, Mathew M (2011) Root apoplastic barriers block Na^+ transport to shoots in rice (*Oryza sativa* L.). J Exp Bot 62(12):4215–4228

Kumar V, Khare T, Shriram V, Wani SH (2018) Plant small RNAs: the essential epigenetic regulators of gene expression for salt-stress responses and tolerance. Plant Cell Rep 37(1):61–75

Läuchli A, Grattan S (2007) Plant growth and development under salinity stress. In: Advances in molecular breeding toward drought and salt tolerant crops. Springer, Cham, pp 1–32

Lazof DB, Bernstein N (1999) Effects of salinization on nutrient transport to lettuce leaves: consideration of leaf developmental stage. New Phytol 144(1):85–94

Leyser O (2006) Dynamic integration of auxin transport and signalling. Curr Biol 16(11):R424–R433

Li T, Li H, Zhang Y-X, Liu J-Y (2011) Identification and analysis of seven H_2O_2-responsive miRNAs and 32 new miRNAs in the seedlings of rice (Oryza sativa L. ssp. indica). Nucleic Acids Res 39(7):2821–2833

Li C-H, Wang G, Zhao J-L, Zhang L-Q, Ai L-F, Han Y-F, Sun D-Y, Zhang S-W, Sun Y (2014) The receptor-like kinase SIT1 mediates salt sensitivity by activating MAPK3/6 and regulating ethylene homeostasis in rice. Plant Cell 26(6):2538–2553

Li Y, Yuan F, Wen Z, Li Y, Wang F, Zhu T, Zhuo W, Jin X, Wang Y, Zhao H (2015a) Genome-wide survey and expression analysis of the OSCA gene family in rice. BMC Plant Biol 15(1):1–13

Li H, Zhu Y, Hu Y, Han W, Gong H (2015b) Beneficial effects of silicon in alleviating salinity stress of tomato seedlings grown under sand culture. Acta Physiol 37:1–9

Li W, Wang T, Zhang Y, Li Y (2016) Overexpression of soybean miR172c confers tolerance to water deficit and salt stress, but increases ABA sensitivity in transgenic *Arabidopsis thaliana*. J Exp Bot 67(1):175–194

Liang G, Yang F, Yu D (2010) MicroRNA395 mediates regulation of sulfate accumulation and allocation in *Arabidopsis thaliana*. Plant J 62(6):1046–1057

Linh LH, Linh TH, Xuan TD, Ham LH, Ismail AM, Khanh TD (2012) Molecular breeding to improve salt tolerance of rice (*Oryza sativa* L.) in the red river delta of Vietnam. Int J Plant Genomics 2012:949038

Liu H-H, Tian X, Li Y-J, Wu C-A, Zheng C-C (2008) Microarray-based analysis of stress-regulated microRNAs in *Arabidopsis thaliana*. RNA 14(5):836–843

Liu Q, Zhang Y-C, Wang C-Y, Luo Y-C, Huang Q-J, Chen S-Y, Zhou H, Qu L-H, Chen Y-Q (2009) Expression analysis of phytohormone-regulated microRNAs in rice, implying their regulation roles in plant hormone signaling. FEBS Lett 583(4):723–728

Liu Y, Ye N, Liu R, Chen M, Zhang J (2010) H_2O_2 mediates the regulation of ABA catabolism and GA biosynthesis in *Arabidopsis* seed dormancy and germination. J Exp Bot 61(11):2979–2990

Liu Y, Wang K, Li D, Yan J, Zhang W (2017) Enhanced cold tolerance and tillering in switchgrass (*Panicum virgatum* L.) by heterologous expression of Osa-miR393a. Plant Cell Physiol 58(12):2226–2240

Liu L, Xia W, Li H, Zeng H, Wei B, Han S, Yin C (2018) Salinity inhibits rice seed germination by reducing α-amylase activity via decreased bioactive gibberellin content. Front Plant Sci 9:275

Liu Y, Li D, Yan J, Wang K, Luo H, Zhang W (2019a) MiR319 mediated salt tolerance by ethylene. Plant Biotechnol J 17(12):2370–2383

Liu J, Shabala S, Shabala L, Zhou M, Meinke H, Venkataraman G, Chen Z, Zeng F, Zhao Q (2019b) Tissue-specific regulation of Na+ and K+ transporters explains genotypic differences in salinity stress tolerance in rice. Front Plant Sci 10:1361

Long RC, Li MN, Kang JM, Zhang TJ, Sun Y, Yang QC (2015) Small RNA deep sequencing identifies novel and salt-stress-regulated microRNAs from roots of Medicago sativa and *Medicago truncatula*. Physiol Plant 154(1):13–27

Long R, Li M, Li X, Gao Y, Zhang T, Sun Y, Kang J, Wang T, Cong L, Yang Q (2017) A novel miRNA sponge form efficiently inhibits the activity of miR393 and enhances the salt tolerance and ABA insensitivity in *Arabidopsis thaliana*. Plant Mol Biol Report 35(4):409–415

Lu Y, Feng Z, Bian L, Xie H, Liang J (2010) miR398 regulation in rice of the responses to abiotic and biotic stresses depends on CSD1 and CSD2 expression. Funct Plant Biol 38(1):44–53

Luan M, Xu M, Lu Y, Zhang L, Fan Y, Wang L (2015) Expression of zma-miR169 miRNAs and their target ZmNF-YA genes in response to abiotic stress in maize leaves. Gene 555(2):178–185

Luo H, Zhou Z, Song G, Yao H, Han L (2020) Antioxidant enzyme activity and microRNA are associated with growth of Poa pratensis callus under salt stress. Plant Biotechnol Rep 14(4):429–438

Lutts S, Guerrier G (1995) Peroxidase activities of two rice cultivars differing in salinity tolerance as affected by proline and NaCl. Biol Plant 37:577–586

Ma C, Burd S, Lers A (2015) miR408 is involved in abiotic stress responses in *Arabidopsis*. Plant J 84(1):169–187

Ma Y, Xue H, Zhang F, Jiang Q, Yang S, Yue P, Wang F, Zhang Y, Li L, He P (2021) The miR156/SPL module regulates apple salt stress tolerance by activating MdWRKY100 expression. Plant Biotechnol J 19(2):311–323

Macovei A, Gill SS, Tuteja N (2012) microRNAs as promising tools for improving stress tolerance in rice. Plant Signal Behav 7(10):1296–1301

Miransari M, Smith D (2014) Plant hormones and seed germination. Environ Exp Bot 99:110–121

Mumtaz MZ, Saqib M, Abbas G, Akhtar J, Qamar Z-U (2018) Genotypic variation in rice for grain yield and quality as affected by salt-affected field conditions. J Plant Nutr 41(2):233–242

Munns R (1993) Physiological processes limiting plant growth in saline soils: some dogmas and hypotheses. Plant Cell Environ 16(1):15–24

Munns R (2002) Comparative physiology of salt and water stress. Plant Cell Environ 25(2):239–250

Munns R, Tester M (2008) Mechanisms of salinity tolerance. Annu Rev Plant Biol 59:651–681

Mutum RD, Balyan SC, Kansal S, Agarwal P, Kumar S, Kumar M, Raghuvanshi S (2013) Evolution of variety-specific regulatory schema for expression of osa-miR408 in indica rice varieties under drought stress. FEBS J 280(7):1717–1730

Naik HK, Varadahalli RD (2020) Genomic identification of salt induced microRNAs in niger (*Guizotia abyssinica* Cass.). Plant Gene 23:100242

Negrão S, Schmöckel S, Tester M (2017) Evaluating physiological responses of plants to salinity stress. Ann Bot 119(1):1–11

Nguyen DQ, Brown CW, Pegler JL, Eamens AL, Grof CP (2020) Molecular manipulation of microRNA397 abundance influences the development and salt stress response of *Arabidopsis thaliana*. Int J Mol Sci 21(21):7879

Noman A, Aqeel M, Khalid N, Islam W, Sanaullah T, Anwar M, Khan S, Ye W, Lou Y (2019) Zinc finger protein transcription factors: integrated line of action for plant antimicrobial activity. Microb Pathog 132:141–149

Oi T, Enomoto S, Nakao T, Arai S, Yamane K, Taniguchi M (2020) Three-dimensional ultrastructural change of chloroplasts in rice mesophyll cells responding to salt stress. Ann Bot 125(5):833–840

Palatnik JF, Allen E, Wu X, Schommer C, Schwab R, Carrington JC, Weigel D (2003) Control of leaf morphogenesis by microRNAs. Nature 425(6955):257–263

Pan W-J, Tao J-J, Cheng T, Bian X-H, Wei W, Zhang W-K, Ma B, Chen S-Y, Zhang J-S (2016) Soybean miR172a improves salt tolerance and can function as a long-distance signal. Mol Plant 9(9):1337–1340

Panda S, Khan M (2009) Growth, oxidative damage and antioxidant responses in green gram (*Vigna radiata* L.) under short-term salinity stress and its recovery. J Agron Crop Sci 195(6):442–454

Panda BB, Badoghar AK, Sekhar S, Kariali E, Mohapatra PK, Shaw BP (2015) Biochemical and molecular characterisation of salt-induced poor grain filling in a rice cultivar. Funct Plant Biol 43(3):266–277

Pang M, Woodward AW, Agarwal V, Guan X, Ha M, Ramachandran V, Chen X, Triplett BA, Stelly DM, Chen ZJ (2009) Genome-wide analysis reveals rapid and dynamic changes in miRNA and siRNA sequence and expression during ovule and fiber development in allotetraploid cotton (*Gossypium hirsutum* L.). Genome Biol 10(11):1–21

Parida AK, Das AB (2005) Salt tolerance and salinity effects on plants: a review. Ecotoxicol Environ Saf 60(3):324–349

Paul S, Kundu A, Pal A (2011) Identification and validation of conserved microRNAs along with their differential expression in roots of *Vigna unguiculata* grown under salt stress. Plant Cell Tissue Organ Cult 105(2):233–242

Pegler JL, Nguyen DQ, Grof CP, Eamens AL (2020) Profiling of the salt stress responsive microRNA landscape of C4 genetic model species *Setaria viridis* (L.) Beauv. Agronomy 10(6):837

Ponce KS, Guo L, Leng Y, Meng L, Ye G (2021) Advances in sensing, response and regulation mechanism of salt tolerance in rice. Int J Mol Sci 22(5):2254

Qadir M, Quillérou E, Nangia V, Murtaza G, Singh M, Thomas RJ, Drechsel P, Noble AD (2014) Economics of salt-induced land degradation and restoration. Nat Res Forum 38:282–295

Qin Y, Duan Z, Xia X, Yin W (2011) Expression profiles of precursor and mature microRNAs under dehydration and high salinity shock in Populus euphratica. Plant Cell Rep 30:1893–1907

Rahman MA, Thomson MJ, Shah-E-Alam M, de Ocampo M, Egdane J, Ismail AM (2016) Exploring novel genetic sources of salinity tolerance in rice through molecular and physiological characterization. Ann Bot 117(6):1083–1097

Ramezani M, Seghatoleslami M, Mousavi G, Sayyari-Zahan M (2012) Effect of salinity and foliar application of iron and zinc on yield and water use efficiency of Ajowan (*Carum copticum*). Int J Agric Crop Sci 4(7):421–426

Razzaq A, Ali A, Safdar LB, Zafar MM, Rui Y, Shakeel A, Shaukat A, Ashraf M, Gong W, Yuan Y (2020) Salt stress induces physiochemical alterations in rice grain composition and quality. J Food Sci 85(1):14–20

Roshandel P, Flowers T (2009) The ionic effects of NaCl on physiology and gene expression in rice genotypes differing in salt tolerance. Plant Soil 315:135–147

Saeed M (2018) Abiotic stress tolerance in Rice (Oryza sativa L.): a genomics perspective of salinity tolerance. In: Rice crop-current developments. IntechOpen, London, pp 181–192

Sahito ZA, Wang L, Sun Z, Yan Q, Zhang X, Jiang Q, Ullah I, Tong Y, Li X (2017) The miR172c-NNC1 module modulates root plastic development in response to salt in soybean. BMC Plant Biol 17(1):1–12

Sahoo KK, Tripathi AK, Pareek A, Singla-Pareek SL (2013) Taming drought stress in rice through genetic engineering of transcription factors and protein kinases. Plant Stress 7(1):60–72

Sharma N, Tripathi A, Sanan-Mishra N (2015) Profiling the expression domains of a rice-specific microRNA under stress. Front Plant Sci 6:139552

Shen J, Xu G, Zheng HQ (2015) Apoplastic barrier development and water transport in Zea mays seedling roots under salt and osmotic stresses. Protoplasma 252:173–180

Shereen A, Ansari R, Raza S, Mumtaz S, Khan M, Khan MA (2011) Salinity induced metabolic changes in rice (*Oryza sativa* L.) seeds during germination. Pak J Bot 43(3):1659–1661

Shrivastava P, Kumar R (2015) Soil salinity: a serious environmental issue and plant growth promoting bacteria as one of the tools for its alleviation. Saudi J Biol Sci 22(2):123–131

Si J, Zhou T, Bo W, Xu F, Wu R (2014) Genome-wide analysis of salt-responsive and novel microRNAs in Populus euphratica by deep sequencing. BMC Genet 15:S6

Sobhanian H, Aghaei K, Komatsu S (2011) Changes in the plant proteome resulting from salt stress: toward the creation of salt-tolerant crops? J Proteome 74(8):1323–1337

Song JB, Gao S, Sun D, Li H, Shu XX, Yang ZM (2013) miR394 and LCR are involved in *Arabidopsis* salt and drought stress responses in an abscisic acid-dependent manner. BMC Plant Biol 13:1–16

Sun X, Lin L, Sui N (2019) Regulation mechanism of microRNA in plant response to abiotic stress and breeding. Mol Biol Rep 46(1):1447–1457

Sunkar R, Kapoor A, Zhu J-K (2006) Posttranscriptional induction of two Cu/Zn superoxide dismutase genes in *Arabidopsis* is mediated by downregulation of miR398 and important for oxidative stress tolerance. Plant Cell 18(8):2051–2065

Suzuki N, Koussevitzky S, Mittler R, Miller G (2012) ROS and redox signalling in the response of plants to abiotic stress. Plant Cell Environ 35(2):259–270

Trindade I, Capitão C, Dalmay T, Fevereiro MP, Santos DM, d. (2010) miR398 and miR408 are up-regulated in response to water deficit in *Medicago truncatula*. Planta 231(3):705–716

Tripathi AK, Pareek A, Sopory SK, Singla-Pareek SL (2012) Narrowing down the targets for yield improvement in rice under normal and abiotic stress conditions via expression profiling of yield-related genes. Rice 5(1):1–12

Verma K, Saxena R, Barthwal A, Deshmukh S (1994) Remote sensing technique for mapping salt affected soils. Int J Remote Sens 15(9):1901–1914

Wang J-W, Wang L-J, Mao Y-B, Cai W-J, Xue H-W, Chen X-Y (2005) Control of root cap formation by microRNA-targeted auxin response factors in *Arabidopsis*. Plant Cell 17(8):2204–2216

Wang Y-N, Chen J, Li X, Wang B, Cai X, Huang L (2009) Predicting rate constants of hydroxyl radical reactions with organic pollutants: algorithm, validation, applicability domain, and mechanistic interpretation. Atmos Environ 43(5):1131–1135

Wang B, Sun Y-F, Song N, Wei J-P, Wang X-J, Feng H, Yin Z-Y, Kang Z-S (2014) MicroRNAs involving in cold, wounding and salt stresses in *Triticum aestivum* L. Plant Physiol Biochem 80:90–96

Wani SH, Kumar V, Khare T, Guddimalli R, Parveda M, Solymosi K, Suprasanna P, Kavi Kishor P (2020) Engineering salinity tolerance in plants: progress and prospects. Planta 251(4):1–29

Wankhade SD, Bahaji A, Mateu-Andrés I, Cornejo M-J (2010) Phenotypic indicators of NaCl tolerance levels in rice seedlings: variations in development and leaf anatomy. Acta Physiol 32:1161–1169

Wu H, Zhu M, Shabala L, Zhou M, Shabala S (2015) K+ retention in leaf mesophyll, an overlooked component of salinity tolerance mechanism: a case study for barley. J Integr Plant Biol 57(2):171–185

Xia K, Wang R, Ou X, Fang Z, Tian C, Duan J, Wang Y, Zhang M (2012) OsTIR1 and OsAFB2 downregulation via OsmiR393 overexpression leads to more tillers, early flowering and less tolerance to salt and drought in rice. PLoS One 7(1):e30039

Yang Y, Guo Y (2018) Unravelling salt stress signaling in plants. J Integr Plant Biol 60(9):796–804

Yang Y, Yao Y, Zhang X (2015) Comparison of growth and physiological responses to severe drought between two altitudinal Hippophae rhamnoides populations. Growth 44(603):614

Yang T, Wang Y, Teotia S, Wang Z, Shi C, Sun H, Gu Y, Zhang Z, Tang G (2019) The interaction between miR160 and miR165/166 in the control of leaf development and drought tolerance in *Arabidopsis*. Sci Rep 9(1):2832

Yeo A, Lee A-S, Izard P, Boursier P, Flowers T (1991) Short-and long-term effects of salinity on leaf growth in rice (*Oryza sativa* L.). J Exp Bot 42(7):881–889

Yi H, Liu X, Yi M, Chen G (2014) Dual role of hydrogen peroxide in *Arabidopsis* guard cells in response to sulfur dioxide. Adv Toxicol 2014:6

Yuan S, Zhao J, Li Z, Hu Q, Yuan N, Zhou M, Xia X, Noorai R, Saski C, Li S (2019) MicroRNA396-mediated alteration in plant development and salinity stress response in creeping bentgrass. Hortic Res 6:48

Zeng L, Shannon MC (2000a) Effects of salinity on grain yield and yield components of rice at different seeding densities. Agron J 92(3):418–423

Zeng L, Shannon MC (2000b) Salinity effects on seedling growth and yield components of rice. Crop Sci 40(4):996–1003

Zeng L, Shannon MC, Lesch SM (2001) Timing of salinity stress affects rice growth and yield components. Agric Water Manag 48(3):191–206

Zhang B, Wang Q (2015) MicroRNA-based biotechnology for plant improvement. J Cell Physiol 230(1):1–15

Zhang K, Shi Y, Cui X, Yue P, Li K, Liu X, Tripathi BM, Chu H (2019) Salinity is a key determinant for soil microbial communities in a desert ecosystem. mSystems 4:18. https://doi.org/10.1128/msystems.00225-18

Zhang X, Shen J, Xu Q, Dong J, Song L, Wang W, Shen F (2021) Long noncoding RNA lncRNA354 functions as a competing endogenous RNA of miR160b to regulate ARF genes in response to salt stress in upland cotton. Plant Cell Environ 44(10):3302–3321

Zhao B, Liang R, Ge L, Li W, Xiao H, Lin H, Ruan K, Jin Y (2007) Identification of drought-induced microRNAs in rice. Biochem Biophys Res Commun 354(2):585–590

Zhao B, Ge L, Liang R, Li W, Ruan K, Lin H, Jin Y (2009) Members of miR-169 family are induced by high salinity and transiently inhibit the NF-YA transcription factor. BMC Mol Biol 10(1):1–10

Zhao J, Yuan S, Zhou M, Yuan N, Li Z, Hu Q, Bethea FG Jr, Liu H, Li S, Luo H (2019) Transgenic creeping bentgrass overexpressing Osa-miR393a exhibits altered plant development and improved multiple stress tolerance. Plant Biotechnol J 17(1):233–251

Zheng C, Liu C, Liu L, Tan Y, Sheng X, Yu D, Sun Z, Sun X, Chen J, Yuan D (2023) Effect of salinity stress on rice yield and grain quality: a meta-analysis. Eur J Agron 144:126765

Zhou M, Li D, Li Z, Hu Q, Yang C, Zhu L, Luo H (2013) Constitutive expression of a miR319 gene alters plant development and enhances salt and drought tolerance in transgenic creeping bentgrass. Plant Physiol 161(3):1375–1391

Zhu C, Ding Y, Liu H (2011) MiR398 and plant stress responses. Physiol Plant 143(1):1–9

Zhu Y, Jiang X, Zhang J, He Y, Zhu X, Zhou X, Gong H, Yin J, Liu Y (2020) Silicon confers cucumber resistance to salinity stress through regulation of proline and cytokinins. Plant Physiol Biochem 156:209–220

Glossary

3' polyadenylation Addition of a poly(A) tail to the 3' end of an RNA transcript.

5' capping Addition of a modified nucleotide to the 5' end of an RNA transcript.

ABA (Abscisic Acid) A plant hormone that plays a crucial role in stress responses, particularly in regulating stomatal closure.

Abiotic stress Environmental stresses, such as drought, heat, cold, salinity, or nutrient deficiency that can negatively impact plant growth and development.

Abiotic stress tolerance The ability of crops to withstand environmental stresses such as drought, salinity, cold, and heat, often through the regulation of stress-responsive genes.

Adaptation The process by which plants adjust to environmental cues to enhance their tolerance and survival under stressful conditions.

Adapter sequence Short DNA sequences used in DNA sequencing library preparation to attach DNA fragments to the sequencing platform.

Ago (Argonaute) A key component of the RNA-induced silencing complex (RISC) that binds to siRNAs or miRNAs and mediates their interaction with target mRNAs, leading to mRNA cleavage or translational repression.

Agri-biotechnology The applications of biotechnology techniques in agriculture to improve crop yields, enhance food security, and address agricultural challenges.

Agricultural production The process of cultivating crops and raising livestock for human consumption or other uses.

Alternative splicing A process by which different combinations of exons are joined together to generate multiple mRNA isoforms from a single gene, leading to proteomic diversity.

amiRNAs (Artificial microRNAs) microRNAs designed to target and inhibit the expression of specific miRNA genes.

AMOs (anti-miRNA oligonucleotides) Oligonucleotides designed to bind to and inhibit the activity of specific mature miRNAs.

Antioxidant defense system Enzymatic and non-enzymatic processes in plants that detoxify reactive oxygen species and protect cells from oxidative damage.

Antisense oligonucleotides Synthetic single-stranded nucleic acid molecules that bind to target mRNAs with high specificity, either inhibiting their translation or

K. Goswami et al. (eds.), *Non-coding RNAs for Crop Improvement*,
https://doi.org/10.1007/978-981-97-7213-1

inducing their degradation, thus offering a potential alternative to RNAi-based approaches for gene silencing.

Antisense RNA RNA molecules that are complementary to specific mRNA sequences and can inhibit their translation.

Antiviral defense mechanisms Mechanisms plants use to defend against viral infections.

Antiviral immunity The body's defense mechanisms specifically targeted against viral infections, which may be influenced by RNAi-based products.

Antiviral transgene A transgene encoding a small RNA molecule, such as a short hairpin RNA (shRNA) or artificial microRNA (amiRNA), designed to specifically target and silence viral gene expression. Antiviral transgenes are incorporated into host plant genomes to confer resistance against viral infections and reduce disease severity.

AP2-like TF (APETALA2-like transcription factor) Transcription factors that regulate gene expression in response to environmental stresses.

Apoplastic pathway The movement of ions and water through the non-living spaces (apoplast) outside the plasma membrane of plant cells.

Arbuscular mycorrhizal fungi Fungi that form symbiotic relationships with plant roots, enhancing nutrient uptake and stress tolerance.

ARF (Auxin response factor) Transcription factors involved in mediating auxin responses.

Assembly-based methods Techniques for reconstructing genomic or transcriptomic sequences by assembling short sequencing reads into longer contiguous sequences.

Auxin Plant hormone involved in various aspects of plant growth and development.

Basic Local Alignment Search Tool (BLAST) A bioinformatics tool used to search nucleotide or protein databases for sequences similar to a query sequence, often used in selecting target sequences for RNAi experiments.

Biogenesis The process by which ncRNAs are synthesized or produced within a cell.

Bioinformatics The application of computational techniques and statistical methods to analyze and interpret biological data, including genomic and transcriptomic sequences.

Bioinformatics tools Computational tools and algorithms used for the analysis and interpretation of biological data, including sequence analysis, structure prediction, and functional annotation.

Biological processes The series of events and interactions within a cell or organism that lead to specific outcomes, such as development, disease, and gene regulation.

Biomass regulation The control of plant biomass production, which is important for crop yield and resource allocation.

Biopesticide A type of pesticide derived from natural materials such as plants, animals, bacteria, or minerals, used to control pests in agriculture while minimizing environmental impact.

BLAST analysis A bioinformatics tool used to compare nucleotide or protein sequences against a database to identify similarities and assess functional or evolutionary relationships.

Broad-spectrum Pesticides Pesticides that are effective against a wide range of pests but may also harm non-target organisms.

CaM pathway Calcium/calmodulin pathway, another calcium-mediated signaling pathway.

Canonical pathway The conventional process of RNA synthesis and processing.

cDNA (Complementary DNA) DNA synthesized from a single-stranded RNA template in a reaction catalyzed by the enzyme reverse transcriptase. It represents the expressed genes in a cell.

CDPK pathway Calcium-dependent protein kinase pathway, involved in stress response.

Cellular homeostasis The maintenance of stable internal conditions within a cell or organism despite external changes or challenges.

Central Dogma of Molecular Biology The concept proposed by Francis Crick in 1958, stating that genetic information flows from DNA to RNA to protein, with proteins being the primary functional molecules in cells. The discovery of non-coding RNAs challenged this idea by revealing the regulatory roles of RNA molecules beyond protein synthesis.

ceRNAs (Competing endogenous RNAs) RNAs, including lncRNAs, that compete for binding to microRNAs, thereby regulating the availability of microRNAs for target mRNAs.

Chemical pesticides Synthetic substances used to control or eliminate pests in agriculture, often with harmful effects on human health and the environment.

ChIP-Seq (Chromatin immunoprecipitation sequencing) Technique used to identify DNA regions bound by specific proteins, such as histones or transcription factors.

Chromatin modification Alterations to the structure of chromatin, impacting gene expression by regulating access to DNA.

Chromatin modification signatures Epigenetic marks, such as methylation patterns, on DNA or histone proteins, that regulate gene expression.

circRNAs (Circular RNAs) RNA molecules with a covalently closed loop structure, which have been found to play regulatory roles in gene expression, including alternative splicing.

Cis-priming Priming where the stimulus and the stress are of the same kind.

Cis-regulation Regulation of gene expression by nearby regulatory elements on the same chromosome.

CLIP-Seq (Cross-linking immunoprecipitation sequencing) Technique used to identify RNA molecules bound by specific RNA-binding proteins.

Cloning and sequencing of small RNAs Experimental techniques used to isolate and identify small RNA molecules, such as miRNAs, in plant cells.

COG (Clusters of Orthologous Groups) Groups of orthologous genes that likely share a common ancestor.

Complementary base pairing The specific binding between nucleotide sequences of RNA molecules, wherein nucleotides with complementary bases (A with U, G with C) form hydrogen bonds, enabling the formation of RNA-RNA duplexes and RNA-protein interactions.

Computational approaches Techniques and methods that utilize computer algorithms and software tools to analyze biological data, including genomic and transcriptomic sequences, with the aim of understanding biological processes and predicting molecular interactions.

Computational tools Software programs and algorithms designed to analyze, interpret, and visualize large-scale genomic and transcriptomic data, including those generated by NGS technologies.

Contigs Overlapping DNA segments produced by assembling sequenced fragments.

Copper/Zinc SOD (Superoxide dismutase) Enzyme involved in detoxifying superoxide radicals.

COPS (Convergent overlapping pairs) Genomic regions where sense and antisense transcripts overlap, potentially forming double-stranded RNA molecules that regulate gene expression.

Co-suppression A phenomenon observed in plants where the overexpression of a gene leads to the suppression of both the endogenous and exogenous copies of that gene.

CPB (Cocoa pod borer) A major insect pest, specifically the moth species *Conopomorpha cramerella*, that infests cocoa plantations in South-East Asia, causing significant damage to cocoa pods and reducing crop yield.

CRISPR Clustered regularly interspaced short palindromic repeats, a genome editing tool that can be used to induce mutations in miRNA sequences.

CRISPR/Cas9 A genome editing tool used to study the functional roles of noncoding RNAs by introducing targeted mutations or deletions in the genomic loci encoding the RNA molecules.

CRISPRi (CRISPR interference) A genome engineering technique that uses catalytically inactive Cas9 (dCas9) fused to transcriptional repressors to target specific DNA sequences and inhibit gene expression at the transcriptional level, offering a complementary approach to RNAi for gene silencing.

Crop improvement The enhancement of crop traits such as yield, productivity, stress tolerance, and nutritional quality through various genetic and biotechnological approaches, including RNAi.

Crop plants Plants cultivated for agricultural purposes, including food crops, fiber crops, and ornamental crops.

Crop protection Measures taken to safeguard agricultural crops from damage caused by pests, diseases, and environmental stressors, ensuring optimal yield and quality.

Cytochrome oxidase Enzyme involved in electron transport in plant mitochondria.

Databases Structured collections of biological data, often accessible and searchable online, used for research and analysis.

D-bodies Nuclear dicing bodies where miRNA processing occurs.

DCL (Dicer-like) ribonucleases Enzymes involved in the processing of dsRNA molecules into small RNAs, particularly miRNAs and siRNAs.

De novo assembly A bioinformatics method used to reconstruct full-length RNA transcripts from short sequencing reads without the need for a reference genome. It is articularly useful for identifying novel non-coding RNA transcripts that may not be annotated in existing genome databases.

Deep learning models Machine learning algorithms used to analyze complex biological data, including RNA sequencing data.

DEG analysis (Differential Expressed Gene Analysis) Identification and comparison of genes expressed differentially between different experimental conditions.

DEGs (Differential expression genes) Genes that show differences in expression levels between different samples or conditions, indicating potential regulatory roles.

DEPC-treated water Water treated with diethyl pyrocarbonate (DEPC) to inactivate RNases and DNases, ensuring RNA integrity.

Deregulated miRNAs MicroRNAs whose expression is altered in response to stress.

Dicer An endoribonuclease enzyme that cleaves long double-stranded RNA molecules into short interfering RNAs (siRNAs) or precursor microRNAs (pre-miRNAs) into mature miRNAs, which are essential for RNAi-mediated gene silencing.

Differential expression analysis A bioinformatics approach used to identify genes or RNA transcripts that are differentially expressed between experimental conditions or biological samples allowing for the identification of ncRNAs that are dynamically regulated under specific conditions or in response to external stimuli.

Disease management Strategies and practices aimed at preventing, mitigating, or controlling the spread of plant diseases, including viral infections. RNAi-based approaches represent promising tools for disease management in crop plants by targeting viral genes and reducing virus transmission within agricultural systems.

DNA methylation The addition of methyl groups to DNA molecules, often associated with gene silencing and regulation of gene expression.

DNase I An enzyme used to degrade or remove DNA that may be contaminating RNA samples.

DRBs (dsRNA binding proteins) Proteins that bind to dsRNA and facilitate RNAi.

Drought avoidance and tolerance Plant mechanisms for coping with drought stress, including reducing transpiration, increasing water uptake, and maintaining water potential.

Drought escape A plant strategy to avoid drought stress by completing its life cycle before the onset of dry conditions.

Drought stress A significant environmental stress that affects plant growth, development, and yield by reducing water availability.

dsRNA (Double-stranded RNA) RNA molecules consisting of two complementary strands of RNA nucleotides. dsRNA molecules can trigger RNAi pathways in cells, leading to the degradation of targeted mRNA sequences and gene silencing.

easiRNAs Epigenetically activated siRNAs, enriched in pollen vegetative cells.

EC (Electrical conductivity) A measure of the ability of a solution to conduct an electric current, often used as an indicator of soil salinity.

Environmental impact The effects of human activities on the natural environment, including factors such as pollution, habitat destruction, and biodiversity loss, which can be mitigated through the use of sustainable pest control methods like RNAi-based biopesticides.

Epialleles Variants of genes influenced by epigenetic modifications, capable of controlling gene expression and phenotype.

Epigenetic changes Exogenous RNAi techniques may cause unanticipated epigenetic changes in plants, necessitating thorough evaluation of potential epigenetic alterations.

Epigenetic memory The ability of epigenetic modifications to persist through cell divisions and influence gene expression over time.

Epigenetic regulation Regulation of gene expression through modifications to chromatin structure, such as DNA methylation, histone modifications, and chromatin remodeling, influenced by the interactions between small and lncRNA-sand chromatin-associated proteins.

Epigenetics The study of reversible changes in gene expression that are heritable during cell division, without alterations to the DNA sequence itself.

Epigenome A collective term for biochemical alterations to nuclear DNA, post-translational modifications to histone proteins, and variations in noncoding RNA biogenesis, all of which can affect gene expression without altering the underlying DNA sequence.

Expression profiling The measurement of the activity (expression) of genes simultaneously to create a global picture of cellular function.

Food security The state in which all people, at all times, have physical, social, and economic access to sufficient, safe, and nutritious food to meet their dietary needs and food preferences for an active and healthy life.

Functional analysis Experimental approaches aimed at elucidating the biological roles and mechanisms of action of non-coding RNAs, often involving the manipulation of RNA expression levels and studying their effects on cellular processes.

Functional annotation The process of assigning biological functions or regulatory roles to RNA transcripts based on sequence similarity, structural features, and experimental evidence.

Functional characterization The process of determining the biological roles and activities of molecules within a biological system, such as non-coding RNAs, by studying their effects on cellular processes, gene expression, and phenotype.

Functional genomics The study of gene function at the genome-wide level, including the identification of genes involved in specific biological processes, pathways, or diseases, often through the integration of genomic, transcriptomic, proteomic, and metabolomic data.

Gain-of-function Experimental manipulation resulting in the enhancement or acquisition of a specific molecular function or phenotype, often used to study the effects of overexpression or activation of non-coding RNAs.

Gene expression The process by which information from a gene is used to synthesize a functional gene product, typically a protein or RNA molecule, which contributes to the phenotype of an organism.

Gene regulation The control of gene expression, including transcription, post-transcriptional modifications, and translation, which allows organisms to respond to internal and external cues and maintain cellular homeostasis.

Gene silencing The process of inhibiting the expression of a gene at the transcriptional or translational level, typically achieved through RNAi mechanisms.

Genetic Engineering The manipulation of an organism's genetic material to introduce desired traits or modify existing ones.

Genome databases Repositories of genetic information, including DNA sequences, annotations, and other related data.

Genomic "dark matter" Regions of the genome that were previously poorly understood or thought to have no functional significance, including non-coding regions and non-coding RNAs.

Genomics The study of entire genomes, including gene sequences, variations, and structures, which facilitates the identification of target genes and prediction of off-target effects in RNAi studies.

GMO (Genetically modified organisms) Organisms whose genetic material has been altered using genetic engineering techniques.

GO Analysis (Gene Ontology Analysis) A method used to classify genes into categories based on their associated biological processes, molecular functions, and cellular components.

hc-siRNAs (Heterochromatic small interfering RNAs) siRNAs derived from transposons in heterochromatic regions.

High-confidence candidates Genomic elements, such as ncRNAs, that have been identified with a high level of certainty using bioinformatics approaches, indicating a high likelihood of biological significance.

High-Throughput Sequencing Technologies that enable the sequencing of a large volume of DNA or RNA molecules simultaneously, also known as next-generation sequencing.

High-throughput transcriptome sequencing Large-scale sequencing of RNA transcripts to study gene expression patterns.

HIGS (Host-induced gene silencing) A strategy in RNAi technology where plants are engineered to express dsRNA targeting specific genes of invading pests, leading to their gene silencing upon ingestion by the pests.

Histone acetylation Addition or removal of acetyl groups to histone proteins, influencing chromatin structure and gene expression.

Histone deacetylation Removal of acetyl groups from histone proteins, typically associated with gene repression.

Histone modifications Chemical alterations to histone proteins, such as methylation, acetylation, phosphorylation, etc., which influence chromatin structure and gene expression.

Homology Similarity in DNA or protein sequences between different organisms, suggesting a common evolutionary origin.

Homology-based methods Bioinformatics approaches that rely on sequence or structural similarity to known sequences or families for identifying and classifying novel sequences.

Homology-dependent process A process in which nucleotide sequence similarity between two RNA molecules determines their interaction and subsequent degradation.

hpRNA (Hairpin RNA) RNA molecules with self-complementary regions that form a hairpin structure, often used to trigger RNAi.

HSPs (Heat Shock Proteins) Proteins that assist in managing cellular stress by aiding in protein folding and preventing protein aggregation under HTS conditions.

HTS (High temperature stress) Environmental conditions above a certain threshold that induce stress on plants.

In silico Referring to experiments or analyses performed using computer-based simulations or algorithms rather than in a laboratory setting.

In vitro Referring to experiments conducted outside a living organism, typically in a test tube or culture dish.

Insect Pest Infestation The presence and proliferation of insects that damage crops or other agricultural products.

In-situ Hybridization A method used to visualize the localization of specific RNA molecules within cells or tissues by hybridizing them with complementary labeled probes.

Interaction The process by which small and lncRNAs interact with each other, as well as with other RNA molecules, DNA, and proteins, to regulate gene expression and cellular processes.

Inter-generational priming Priming effects seen in the current generation.

Ion toxicity The harmful effects of excessive accumulation of specific ions, such as sodium (Na^+), chloride (Cl^-), or potassium (K^+), inside plant cells, disrupting cellular functions and metabolism.

Ionic stress Stress resulting from the accumulation of specific ions, such as sodium (Na^+) or chloride (Cl^-), inside plant cells, disrupting ion homeostasis and metabolic processes.

IPM (Integrated Pest Management) A sustainable approach to pest management that combines biological, cultural, physical, and chemical methods to control pests effectively while minimizing economic, health, and environmental risks.

Isoforms Different forms of a gene or protein produced by alternative splicing or post-translational modifications.

Knockdown A method used to reduce the expression of a specific gene or RNA molecule, typically achieved by introducing short interfering RNAs (siRNAs) or antisense oligonucleotides that target the transcript for degradation.

Knockout A genetic technique used to completely eliminate the expression of a gene or RNA molecule, often achieved through the deletion or disruption of the gene sequence.

LAC (Laccase) Enzyme involved in lignin biosynthesis and stress responses.

Lepidopteran An order of insects that includes butterflies and moths, to which the cocoa pod borer belongs.

Lignin Complex polymer that provides structural support in plant cell walls.

Linux A popular open-source operating system used for software development and bioinformatics analysis.

lncRNAs (long non coding RNAs) RNA transcripts longer than 200 nucleotides that do not encode proteins. They regulate gene expression at various levels, including epigenetic, transcriptional, post-transcriptional, or translational levels.

Luciferase and GFP Reporter Assays Techniques used to study miRNA-mediated gene regulation by fusing target sequences with reporter genes like luciferase or GFP and measuring their expression levels.

Machine learning A branch of artificial intelligence that involves training algorithms to recognize patterns or make predictions based on data.

Malondialdehyde (MDA) A marker of lipid peroxidation and oxidative stress, formed by the degradation of polyunsaturated fatty acids in cell membranes.

MAP Kinase Pathways Signaling pathways involved in cellular responses to various stimuli, including stress.

Marker genes Genes with known traits or characteristics that are used to identify or select for specific genotypes.

Markers Genetic sequences used to track the presence of transgenes.

Meiotic stress memory Priming effects passed down to subsequent generations.

Membrane structure The arrangement of lipids and proteins that form the cell membrane.

Meta-analysis A statistical method that integrates and analyzes data from multiple independent studies to generate a common result, often used to identify common patterns or trends across studies.

Metabolomics The study of small molecules involved in metabolic processes, providing insights into the metabolic consequences of gene silencing and potential biomarkers for RNAi effects.

Microarray A high-throughput technique for analyzing gene expression by hybridizing labeled RNA samples to a surface containing thousands of DNA probes.

Microbiome The collective genomes of microorganisms living in and on an organism.

Micro-bombardment A method for delivering RNA molecules into plant cells by bombarding them with particles coated with RNA.

Microinjection of dsRNA A method of delivering dsRNA directly into the gut of insect pests to trigger RNAi-mediated gene silencing.

MicroRNA (miRNA or miR) Small non-coding RNA molecules involved in post-transcriptional regulation of gene expression by base-pairing with target mRNAs.

miR genes Genes that encode for microRNAs.

miRBase A database that provides comprehensive information on microRNAs, including sequences, annotations, and predicted targets.

miRNA-mRNA Pull-down Assay An experimental method used to identify miRNA binding mRNAs by capturing the miRNA-mRNA complex using biotinylated miRNAs as bait.

Modern breeding approaches Contemporary techniques for crop improvement, including genetic engineering and marker-assisted selection.

Molecular biology The field of biology that focuses on the structure and function of biological molecules, particularly nucleic acids and proteins, and their roles in genetics and cell function.

Molecular crop breeding The use of molecular techniques to improve crop varieties for desirable traits such as yield, stress tolerance, and nutritional content.

Molecular mechanism The biochemical and biophysical processes underlying biological phenomena, including gene regulation, signaling pathways, and metabolic reactions, which determine the behavior and functions of biological systems.

mRNA degradation The process by which mRNAs are enzymatically degraded, leading to a decrease in their abundance and ultimately reducing protein expression levels.

mRNA (Messenger RNA) RNA molecules that carry genetic information from DNA to the ribosomes, where they serve as templates for protein synthesis.

N (Nitrogen) An essential nutrient for plant growth, efficient uptake of which is critical for maximizing crop yield and minimizing environmental damage.

Nanodrop A spectrophotometer used for the quantification and purity assessment of nucleic acid samples.

Nanoparticle-mediated dsRNA delivery A delivery method where dsRNA molecules are encapsulated in nanoparticles to increase stability and efficacy in triggering RNAi in pests.

Nanotechnology The manipulation of materials at the nanoscale to produce desired properties or effects.

natsiRNAs (Naturally Antisense Transcript siRNAs) siRNAs generated from endogenous self-complementary transcripts.

NCBI Database The National Center for Biotechnology Information, which provides access to various biological databases, including gene sequences.

ncRNA (noncoding RNA(ncRNA) RNA molecules transcribed from DNA that do not encode proteins but play important regulatory roles in gene expression and various cellular processes.

Network dynamics The dynamic nature of interactions between small and lncRNAswithin regulatory networks, which can be influenced by cellular conditions, developmental stages, and environmental stimuli.

NGS (Next-Generation Sequencing) High-throughput sequencing technologies used to sequence millions of DNA fragments simultaneouslythat enable rapid generation of large amounts of DNA or RNA sequence data.

Nitrogen deprivation Condition in which plants experience a shortage of nitrogen, an essential nutrient for growth and development.

NMD (Nonsense-mediated mRNA decay) A cellular surveillance mechanism that degrades mRNAs containing premature termination codons, preventing the translation of potentially harmful truncated proteins.

Non-canonical Pathway Unconventional or irregular routes of RNA synthesis and processing.

Non-transformative RNAi An RNAi approach that does not involve genetic modification of host organisms, often using SIGS to deliver dsRNA to pests without altering the plant's genome.

Northern Blot A classical molecular biology technique used to detect RNA molecules. It involves separating RNA samples by gel electrophoresis, transferring them to a membrane, hybridizing them with specific probes, and detecting the signals.

NR (Nitrate Reductase) Enzymes involved in the conversion of nitrate to nitrite during nitrogen assimilation.

NRT (Nitrate Transporters) Proteins involved in the uptake and transport of nitrate ions in plants.

NUE (Nutrient Use Efficiency) Refers to an organism's ability to effectively acquire, assimilate, and utilize nutrients, particularly crucial in agriculture for optimizing crop yield while minimizing environmental impact.

Nutrient Uptake Efficiency The ability of plants to absorb and utilize nutrients efficiently from the soil.

Off-target effects Unintended silencing of genes due to the imperfect complementarity between siRNAs or miRNAs and non-target mRNAs, which can lead to unintended consequences in RNAi experiments.

OMICS Studies Studies that employ high-throughput technologies, such as genomics, transcriptomics, proteomics, and metabolomics, to comprehensively analyze biological molecules and their interactions within a biological system.

ORF (Open reading frame) A sequence of DNA or RNA that potentially encodes a protein, bounded by start and stop codons.

Osmolyte Accumulation The accumulation of organic molecules (osmolytes) that help maintain cell structure and function under osmotic stress.

Osmoregulation The control of solute concentrations and water balance within cells.

Osmotic potential The pressure required to prevent the flow of water across a semipermeable membrane due to differences in solute concentration, influencing water uptake by plant roots.

Osmotic stress Stress caused by the presence of solutes in the soil solution, reducing water availability to plant roots and affecting cellular hydration and turgor pressure.

Overexpression The ectopic expression of a gene or RNA molecule at higher-than-normal levels, typically achieved by introducing additional copies of the gene or using strong promoters to drive transcription.

Oxidative stress Imbalance between the production of reactive oxygen species (ROS) and the ability of cells to detoxify or repair the resulting damage, leading to oxidative damage to lipids, proteins, and DNA.

Panicle The branched cluster of flowers on certain types of grasses, including rice, where the seeds develop and mature.

Panicle sterility The condition where the flowers in a rice panicle fail to produce viable pollen or receive viable pollen for fertilization, resulting in reduced grain set and yield.

PARE (Parallel Analysis of RNA Ends) A method used to identify the cleavage sites of small RNAs, such as miRNAs and siRNAs, within target mRNAs.

PCR (Polymerase Chain Reaction) A laboratory technique used to make millions to billions of copies of a specific DNA sample, allowing researchers to amplify and study small amounts of DNA.

PDR (Pathogen-Derived Resistance) Plant resistance to viral infections conferred by the expression of small RNA molecules derived from viral sequences.

Pest control The management of pest populations to minimize their impact on crops and agricultural productivity, often achieved through the application of pesticides, including biopesticides such as RNAi-based formulations.

PGPR-Plant Interaction The symbiotic relationship between plants and beneficial rhizobacteria, which can enhance plant growth and stress tolerance.

phasiRNAs Phased secondary siRNAs, derived from long RNA precursors with a precise head-to-tail arrangement.

Phenotype The observable characteristics of an organism, determined by its genotype and environmental factors.

Phenotypic Modifications Observable changes in plant traits or characteristics.

Photosynthesis The process by which green plants, algae, and some bacteria convert light energy into chemical energy in the form of glucose, using carbon dioxide (CO_2) and water (H_2O) in the presence of chlorophyll and other pigments.

piRNA (piwi-interacting RNA) A class of small RNA molecules, typically 24-32 nucleotides in length, that interact with piwi proteins to suppress the activity of transposable elements and maintain genome stability in germ cells.

Plant innate immunity The pre-existing defense mechanisms plants have against pathogens.

Plant stress response Complex physiological, biochemical, and molecular changes that occur in plants in response to environmental stresses, such as drought, to mitigate the adverse effects and enhance survival.

Pol IV and Pol V RNA polymerase IV and RNA polymerase V are plant-specific DNA-dependent RNA polymerase.

Polyadenylation The addition of a polyadenylate tail to the 3' end of mRNA molecules, which stabilizes the mRNA and is involved in nuclear export and translation initiation.

Post-transcriptional regulation Regulation of gene expression that occurs after transcription, including processes such as mRNA splicing, editing, transport, localization, and degradation, mediated by interactions between small and long non-coding RNAs.

Pre-miRNAs Precursor forms of microRNAs.

pri-miRNAs Primary transcripts of miR genes.

Promoter A region of DNA located upstream of a gene that serves as a binding site for RNA polymerase and transcription factors, initiating the transcription of the gene into RNA.

Protein homeostasis The maintenance of proper protein folding, localization, and degradation within cells, which is disrupted under HTS conditions leading to the activation of stress response pathways.

Proteomics The study of proteins and their functions, which helps elucidate downstream effects of gene silencing at the protein level and uncover alterations in signaling pathways.

PTGS (Posttranscriptional gene silencing) A mechanism of RNAi where gene expression is suppressed after transcription, typically through mRNA degradation.

Pull-down assays A technique used to identify RNA-protein interactions by pulling down RNA molecules bound to specific proteins using immobilized antibodies or RNA-binding proteins.

qRT-PCR (Quantitative Reverse Transcription Polymerase Chain Reaction) A sensitive technique used to quantify the amount of specific RNA in a sample, providing insights into gene expression levels.

RACE (Rapid Amplification of cDNA Ends) A technique used to obtain the full-length sequence of an RNA transcript by amplifying its ends.

RBP (RNA-binding proteins) Proteins that bind to RNA molecules and regulate their processing, transport, stability, and translation. They play crucial roles in post-transcriptional gene regulation by interacting with both coding and non-coding RNAs.

RdDM (RNA-directed DNA methylation) A process in plants involving small RNAs guiding DNA methylation.

RDR2 RNA-Dependent RNA Polymerase 2, involved in generating the second RNA strand for hc-siRNA biogenesis.

Regulatory network Interconnected system of genes, transcription factors, and regulatory elements that coordinate gene expression in response to internal and external stimuli, facilitating adaptation to changing environmental conditions.

Regulatory non-coding RNAs ncRNAs that play a role in controlling gene expression, often by interacting with other nucleic acids or proteins.

Regulatory role The function of miRNAs in modulating gene expression by binding to complementary sequences in target mRNAs, leading to their degradation or translational repression.

Reporter Assays Experimental techniques used to assess the activity or function of non-coding RNAs by measuring the expression or activity of reporter genes that are regulated by the RNA of interest.

Resistance mechanisms Biological strategies employed by plants to defend against viral infections, including physical barriers, immune responses, and gene silencing mechanisms such as RNAi.

Ribozymes RNA molecules with catalytic activity that can cleave other RNA molecules.

RIP (RNA Immunoprecipitation) A method used to identify RNA molecules that interact with specific proteins by immunoprecipitating the protein-RNA complexes and analyzing the associated RNA molecules.

RISC (RNA-induced silencing complex) Multiprotein complexes containing small RNAs such as siRNAs or miRNAs, which mediate gene silencing through mRNA degradation or translational repression.

Risk Assessment The process of identifying, evaluating, and mitigating potential risks associated with the use of technologies or products, such as RNAi-based biopesticides.

RNA (Ribonucleic Acid) A molecule that is present in the majority of living organisms and is made up of nucleotides, which are ribose sugars attached to nitrogenous bases and phosphate groups.

RNA:DNA R-loop A structure formed by the interaction between RNA and DNA molecules, which can affect gene expression, including alternative splicing.

RNAi (RNA interference) A biological mechanism in which small non coding RNA molecules interfere with gene expression and subsequent translation, potentially leading to gene silencing by targeting mRNA for degradation or translational silencing.

RNAi-based biopesticide A pesticide derived from RNAi technology, where specific RNA molecules are used to target and silence genes essential for the survival or reproduction of pests, providing a sustainable and environmentally friendly approach to crop protection.

RNA-seq (RNA sequencing) A high-throughput sequencing method used to analyze the transcriptome and identify novel RNA transcripts.

RNA silencing Process by which small RNAs regulate gene expression by targeting specific mRNAs for degradation or translational repression.

RNA silencing suppressors Proteins encoded by some viruses that counteract the host RNAi machinery, allowing the virus to evade silencing.

ROS (Reactive oxygen Species) Highly reactive molecules produced in cells under stress conditions, which can cause oxidative damage to cellular components.

ROS scavenging enzymes Enzymes that neutralize reactive oxygen species.

rRNA (Ribosomal RNA) RNA molecules that are components of ribosomes, essential for protein synthesis in cells.

Salinity stress Environmental stress caused by high levels of salt in the soil or irrigation water, which can adversely affect plant growth and productivity by disrupting ion balance, nutrient uptake, and water absorption.

Secondary stress Additional stresses, such as oxidative stress or nutrient imbalances, triggered by primary stresses like osmotic or ionic stress.

Seed priming A technique to enhance seed germination and stress tolerance by pre-treating seeds with various substances.

Sequence-specific inhibition Inhibition of gene expression based on sequence complementarity.

Sequence-specific mechanism Refers to the ability of molecules or processes to recognize and interact with specific sequences of nucleotides in DNA or RNA.

shRNAs (Short Hairpin RNAs) RNA molecules that can induce RNAi by producing small interfering RNAs (siRNAs) within cells, often engineered to target specific viral sequences and are expressed in host plants to confer resistance against viral infections.

SIGS (Spray-induced gene silencing) A method for delivering RNAi triggers to plants via foliar spray, aiming to control viral infections.

siRNA (Short interfering RNAs) Small ncRNAs typically around 20-25 nucleotides long, involved inRNAi pathways and defense against viruses and transposons.

sncRNAs (Small non-coding RNAs) RNA molecules typically less than 20-30 nucleotides in length, including miRNAs and siRNAs, which do not code for proteins and play significant roles in regulating developmental processes and responses to biological stress.

snRNAs (Small nuclear RNAs) Small RNA molecules found within the nucleus that are involved in RNA processing and splicing.

SPL (SQUAMOSA PROMOTER BINDING PROTEIN-LIKE) A family of transcription factors involved in the regulation of miRNA transcription.

Stem-loop PCR A PCR technique specifically designed for amplifying short RNA sequences, such as miRNAs, by utilizing stem-loop primers.

Stomatal conductance The rate at which stomata, small pores on plant leaves and stems, open and close to regulate gas exchange, particularly the uptake of carbon dioxide (CO2) for photosynthesis and the release of water vapor.

Stress Signaling Network A network of signaling pathways and molecules involved in transmitting stress signals and initiating adaptive responses in plants.

Stress signals Molecular cues triggered by environmental stressors.

Stress Tolerance The ability of plants to withstand and adapt to adverse environmental conditions, such as biotic and abiotic stresses through physiological, biochemical, and molecular mechanisms.

STTM (Short tandem target mimic) A technique used to create stable knockdown lines in plants by mimicking the target of a specific miRNA, thereby reducing its activity.

Sustainable agriculture Agricultural practices that aim to maximize productivity while minimizing environmental impact, promoting long-term ecological balance and resource conservation.

Symplastic pathway The movement of ions and water through the symplast, the interconnected network of plant cells via plasmodesmata.

TALENs (Transcription activator-like effector nucleases) Engineered nucleases used for genome editing.

Target mimicry Mechanism by which non-coding RNAs mimic the sequence of target mRNAs, thereby sequestering regulatory molecules and preventing them from exerting their effects.

ta-siRNA (Trans-acting siRNAs) Endogenous ncRNAs generated through mRNA cleavage in trans, belonging to the family of phasiRNAs. Involved in plant-virus interactions and systemic acquired resistance (SAR).

TF (Transcription factor) A protein that binds to specific DNA sequences, thereby regulating the transcription of nearby genes.

TGS (Transcriptional gene silencing) A mechanism of RNAi where gene expression is suppressed at the transcriptional level.

Thermo-priming A process where plants are pre-exposed to shorter durations of high temperature to improve their adaptation to subsequent HTS conditions.

Thermotolerance Ability of an organism to withstand high temperatures.

Tools Software programs or algorithms designed to perform specific tasks or analyses in bioinformatics or genomics.

Traditional breeding approaches Conventional methods used to improve crop traits, such as selective breeding.

Transcriptional Gene Regulation Processes that control the rate of transcription of genetic information from DNA to RNA.

Transcriptome The complete set of RNA transcripts produced by the genome under specific circumstances or in a specific cell using high-throughput sequencing techniques.

Transcriptomics The study of gene expression at the transcript level, providing insights into mRNA abundance, alternative splicing, and non-coding RNA expression.

Transgene-induced RNAi RNAi induced by the introduction of transgenic sequences into an organism.

Transgenerational Inheritance The passing down of epigenetic traits from one generation to the next, influencing phenotype without alterations to the DNA sequence.

Transgenic plants Plants containing foreign genetic material introduced through genetic engineering.

Transient Silencing RNAi-mediated gene silencing is often transient in plants, requiring repeated applications or other strategies to maintain the desired phenotype.

Translational repression The inhibition of protein synthesis from mRNAs, typically mediated by factors such as miRNAs or RNA-binding proteins that interfere with ribosome function or mRNA stability.

Trans-priming Priming where the stimulus is different from the stress.

Trans-regulation Regulatory mechanisms in which lncRNAs interact with sncRNAs to modulate their activity or stability, often leading to changes in gene expression.

Triazol method A method for isolating RNA using Triazol reagent, which disrupts cells and denatures proteins, allowing the extraction of RNA.

tRNA (Transfer RNA) RNA molecules that transport amino acids to the ribosome during protein synthesis.

Vector A DNA molecule used as a vehicle to transfer foreign genetic material into another cell.

VIGS (Virus-induced gene silencing) A technique that uses modified viruses to trigger RNAi-mediated gene silencing in plants.

Viral multiplication The process by which viruses replicate and spread within host organisms.

Viral pathogens Organisms that cause plant diseases, with significant economic impact on agriculture.

vsiRNA (Virus-Derived Small RNAs) Small RNA molecules produced by the host plant in response to viral infection. Are generated through the processing of viral double-stranded RNA intermediates by Dicer-like enzymes and play a role in RNAi-mediated antiviral defense mechanisms.

WUE (Water use efficiency) The ratio of biomass or yield produced per unit of water consumed, indicating the plant's ability to convert water into biomass or harvestable products.

Yield The number of grainsproduced per unit area of land over a specified period, typically measured in kilograms per hectare (kg/ha) or metric tons per hectare (t/ha).

Yield The amount of agricultural product (e.g., grain, fruit, or vegetable) harvested per unit of land area, often measured in kilograms per hectare (kg/ha) or tons per hectare (t/ha).

ZFN (Zinc finger nuclease) Engineered nucleases used for genome editing.

The manufacturer's authorised representative in the EU is Springer Nature Customer Service Centre GmbH, Europaplatz 3, 69115 Heidelberg, Germany. If you have any concerns regarding our products, please contact ProductSafety@springernature.com

Printed and bound by CPI Group (UK) Ltd, Croydon, CR0 4YY
30/09/2026
02225253-0011